STUDY GUIDE

to accompany

BIOLOGY
The Science of Life

Study Guide to accompany Biology: The Science of Life, Third Edition

92 93 9 8 7 6 5 4 3

ISBN 0-673-38045-9

STUDY GUIDE

to accompany

THIRD EDITION

W. J. LEVERICH

St. Louis University

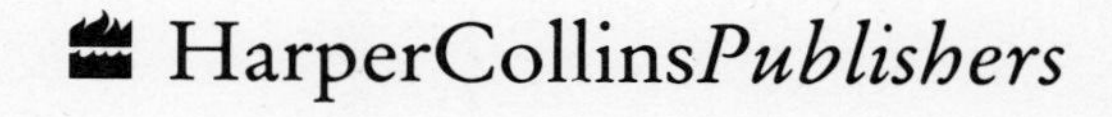

TABLE OF CONTENTS

TO THE STUDENT

Welcome to the study of Biology! It is a subject which covers many broad areas. Your text and your instructor have divided Biology into manageable units which cover much more defined sub-areas. This Study Guide follows the text, chapter by chapter, to help you organize your efforts at mastering the important concepts of Biology.

How to Study Biology

There are very many suitable approaches of studying biology, and you should try to find the techniques that will work best for you. One key element is organization: if you try to master complex topics in a scattered approach, it will be more difficult. Here are a few hints you may wish to consider.

- Lecture: Be sure to attend class faithfully. Listening to a working biologist discuss concepts new to you is sure to help. If you miss class, you miss a first-hand explanation. Be prepared before you go to class so you can spend more time listening rather than trying to frantically record everything in your notes.

- Reading: Read each assignment *before* you go to class. This will allow you to listen more effectively in class. It will not be the first time you have heard the topics being covered. You will be amazed at how much more you get from class if you can anticipate the material being discussed.

- Vocabulary: As you read, make sure you master the meaning of each new term you encounter. When you are in class, the new terms will be familiar to you already. You may wish to keep a personal glossary for terms you might confuse. Remember, if you don't understand the words, you won't understand the concepts.

- Time Budget: You must allow adequate time to study. A few hours a day or two before an exam is not adequate. Set aside now, on your daily calendar, enough time to read the text, review your notes, and organize your study work. A specific time period on specific days is an ideal way to ensure adequate time. Keep to this schedule. Don't allow your friends to tempt you to cheat on your commitment to study. (Just be sure to also set aside time to do other things!)

- Study With Others? For some people, occasional review sessions with other students are useful. You can compare notes to see if you are achieving the same understanding. A friend may catch some important point in class that you missed. But don't rely on social studying entirely - you must master the material yourself at some point.

- Homework: If homework is assigned, do it. The reason for homework is to allow you to practice thinking about the new concepts you are learning. The more you practice thinking critically, the easier it becomes.

Special Problems: If some special circumstance interferes with your studying, address the problem sooner, not later. If you don't understand some concept, if you are ill, if you miss a class, if you have a social disaster that keeps you from studying, a problem may arise. If it does, get in touch with your instructor or your advisor right away. Most problems can be solved easily. Our goal is to make it easy for you to learn Biology, not to make it harder. Any problem is more easily solved the sooner you deal with it.

How to Use This Study Guide

Each chapter in this Study Guide has several parts which, when used regularly, will help you master the material and gauge your progress. Here are some suggestion about each part.

- Learning Objectives: These statements describe in general terms tasks you should be able to accomplish after you have mastered the material covered in the chapter. You might think of these as essay questions. Can you perform each of the tasks listed?

- Chapter Outline: This will give you an overview of the material in the chapter and how it is organized. You may use this to help divide the material into smaller units to study. You may also use this outline as a "checklist" of topics, to keep track of your progress.

- Key Words: In this section are listed relevant terms, roughly in the order they appear in the text. You can use this list in several ways. First, do you know the *meaning* each of the listed terms? If not, then find out the meaning. Secondly, do you know *why* the term is listed here? What is the context of the word or phrase, and how does it relate to other terms? Many students find it useful to use a list of key words as a checklist after they have studied. Run down the list and ask youself for each term "Do I know the meaning? Do I know its significance?" If you answer (to yourself, of course) "no," then you have identified an area that you need to work on a little more.

- Exercises: In the exercises you will be asked to organize some material you have learned and present it in some summary form, or to practice some skill you should have gained. In many chapters, there are blank diagrams for you to label. Sometimes you must sketch a diagram yourself. Other times you may be asked to complete a summary chart. Many of these exercises reflect figures or tables in the text. You will gain the most from these exercises if you work to accomplish them *without* consulting the text until you are ready to check your work. These exercises should suggest other exercises you might do: work at finding ways to summarize a lot of information.

- Self-Exam: The self exams are collections of questions much like those you might see on an hour exam or a final exam. Most of the questions address some particular point covered in the text. These questions are drawn from material throughout each chapter, so you can use the self-exam to jusge how well you have studied each of the text chapters. Mark your responses to the the questions before you check the answer section. If you do miss a question, re-study the text on the point in question.

- Questions for Discussion: These questions are generally more thought-provoking than the short questions of the self-exam. Very often, the response is open-ended. There may be no one "correct" answer, but your response should be better organized and more insightful if you have carefully studied the chapter. Use these questions to practice organizing a discussion of topics covered in the chapter.

How will you know when you have studied enough? That's hard to say. If you use this Study Guide to help check yourself, you should have a pretty good idea if you need more work. If you can cruise the Key Words list easily, if you can finish the Exercises without peeking in the text, if you have no trouble with the Self-Test questions (or others just like them), then you have learned quite a bit of Biology. But if you can still see room for improvement, don't cheat yourself; go over it one more time.

Study hard and good luck!

CHAPTER 1 Mr. Darwin and the Meaning of Life

Learning Objectives

After mastering the material covered in chapter 1, you should be able to confidently do the following tasks:

- Explain how Charles Darwin's observations during his trip on the *Beagle* led to his ideas of evolution and descent with modification.
- Identify the important ideas in the work of Lyell and Malthus that influenced Darwin's ideas.
- Distinguish between inductive reasoning and deductive reasoning.
- Explain how the scientific method of acquiring knowledge is applied.
- Distinguish hypotheses, theories, laws, and the standards applied to each.
- Describe the substance of Darwin's conception of natural selection.
- Identify the characteristics that are normally associated with life.

Chapter Outline

I. Where Are the Rabbits? Darwin's Voyage on the *Beagle*.

II. The Workings of Science.

- A. Inductive reasoning.
- B. Deductive reasoning.
- C. Scientific method.
- D. Hypothesis, theory, and law.

III. Evolution: The Theory Develops.

- A. The theory of natural selection.
- B. Darwin's *On the Origin of Species*.
- C. Testing evolutionary hypotheses.
- D. The impact of Darwin.
- E. Reducing and synthesizing.

IV. Characteristics of Life.

V. Biology, the Study of Life.

Key Words

H.M.S. *Beagle*
Charles Darwin
James Fitzroy
armadillo
sloth
Charles Lyell
species
evolution
descent with modification
observation
inductive reasoning
deductive reasoning
coral atoll
scientific method
hypothesis
theory
law
prediction
control
controlled experiment
variable
placebo
uncontrolled variable
conclusion
mark-and-recapture
Galapagos Islands
natural selection
artificial selection
Thomas Malthus
On the Origin of Species
Joseph Hooker
Alfred Russel Wallace
Thomas Huxley
historicity
reductionism
synthesists
homeostasis

Self-Exam

You should be able to easily answer the following questions after learning the material in chapter 1. If you have difficulty with any question, study the appropriate section in the text and try again.

A. Multiple Choice Questions

Circle one alternative that best completes the statement or answers the question.

1. As a youth Charles Darwin showed interest in
   a. medicine.
   b. insect and rock collecting.
   c. law.
   d. evolution.

2. What observation in the South American grasslands piqued Darwin's interest?
   a. Rabbits were nowhere to be observed.
   b. Rabbits in South America could not swim.
   c. In South America, rabbits lived only in the mountains.
   d. The rabbits there were different from British rabbits.

3. Which of the following did Darwin actually find in South America?
   a. living rabbits.
   b. living hippopatamus-sized armadillos.
   c. giant ground sloths.
   d. tree-dwelling sloths.

4. In Charles Lyell's view of physical history of the earth,
   a. present-day landforms were the result of creation.
   b. earth-changing forces in prehistoric times were different from contemporary times.
   c. forces that had changed the earth in the past were still acting in the present.
   d. cataclysms like floods and earthquakes gave form to the the earth only in prehistoric times.

5. The Galapagos Islands were quite important in shaping Darwin's ideas. Where are they?
   a. In the Atlantic Ocean, off the coast of Africa.
   b. In the Atlantic Ocean, near Argentina.
   c. In the South Pacific, near southern Chile.
   d. In the equatorial Pacific Ocean, somewhat west of Ecuador.

6. Where did Darwin think the ancestors of species he observed on the Galapagos Islands had come from?
   a. Europe.

b. The African mainland.
c. The Cape Verde Islands.
d. The South American mainland.

7. The process of reaching a conclusion on the basis of a number of observations, moving from the specific to the general, is best termed
   a. inductive reasoning.
   b. deductive reasoning.
   c. the scientific method.
   d. a controlled experiment.

8. In an experiment, the control group might be given a substitute for the variable called
   a. the experimental treatment.
   b. a placebo.
   c. an uncontrolled variable.
   d. a dependent variable.

9. What did Darwin observe regarding the finches he collected in the Galapagos?
   a. They were all very similar to one another; they differed from one another very little.
   b. They were very unlike any of the finches in South America.
   c. They were all similar to species in South America, but differed from one another in important ways.
   d. There were only two distinct species.

10. In Darwin's concept of natural selection, what was the *selective agent*?
    a. A supernatural factor.
    b. Random chance.
    c. The environment.
    d. The organisms themselves.

11. Another biologist working in the tropics independently arrived at essentially the same conclusion as Darwin concerning natural selection. This biologist was
    a. Joseph Hooker.
    b. Alfred Russel Wallace.
    c. Thomas Huxley.
    d. Thomas Malthus.

12. Scientists that seek the big picture, who are more interested in understanding how disparate observations can be related, are best termed
    a. reductionists.
    b. synthesists.
    c. naturalists.
    d. apologists.

13. Living things have a number of common traits; the tendency to maintain a "steady state" is called
    a. organization.
    b. reproduction.
    c. adaptation.
    d. homeostasis.

B. True or False Questions

Mark the following statements either T (True) or F (False).

_____ 14. James Fitzroy, the captain of the *Beagle*, was an ideal companion for Darwin, since he too believed strongly in evolution.

_____ 15. Both inductive reasoning and deductive reasoning have a basis in observations.

_____ 16. Although speculation is important in generating new ideas, facts plus speculation do not make science.

_____ 17. A good experiment is set up in such a way that there will be two alternative explanations for the observations.

_____ 18. Many plant and animal breeders before Darwin's time knew that "like begets like," and they were able to practice artificial selection.

_____ 19. In good scientific practice, a hypothesis cannot be tested with observations made prior to formation of the hypothesis.

_____ 20. Most scientific progress during this century has been through conceptual breakthroughs equivalent to Darwin's insights of the role of natural selection in evolution. Such breakthroughs are common.

_____ 21. Living things can adapt only through evolutionary change; individuals cannot adapt by themselves.

_____ 22. In a scientific paper, the "results" section normally follows the "material and methods" section.

C. Fill in the Blanks

Answer the question or complete the statement by filling in the blanks with the correct word or words.

23. The scientific term given to the millions of unique kinds of organisms, each of which differs from the others in some important ways, is ________________.

24. The idea that populations can change through time, with new species eventually arising, is termed ________________.

25. The process which draws specific conclusions from some larger assumption, leading from the general to the specific, is ________________ reasoning.

26. After some observation has been made, a ________________ may be proposed to explain that observation.

27. If a hypothesis is supported by a considerable body of evidence, it may be given the status of a ________________.

28. An explanation of nature which is supported by so much evidence as to be virtually irrefutable is termed a scientific.________________.

29. A properly designed experiment should have an experimental treatment and a ________________ treatment.

30. On the Galapagos Islands, Darwin saw a number of different species of ________________, birds which had apparently evolved in various directions from some South American mainland ancestor.

31. The geologist whose work strongly influenced Darwin's thinking was ________________.

32. The scientific ________________ would seek to understand biological systems in terms of the laws of physics and chemistry.

33. In a scientific paper, an author would interpret the results and might indulge in speculation in the ________________ section.

34. Details of how an experiment was conducted would be place in the ________________ section of a paper.

Questions for Discussion

1. Did biological science exist before Darwin's time? What is it about Darwin's contribution that warrants beginning the study of biology with a review of his work? Did Darwin arrive at his conclusions about natural selection via inductive reasoning or deductive reasoning?

2. Imagine that you walk into your home one day, and you make an observation with your nose (it can happen). Your observation is "I smell bread baking." Formulate and state a hypothesis that will explain this observation. How would you test this hypothesis? Can you make predictions of future observations? What kind of experiment would you do to confirm or refute your hypothesis?

3. List the characteristic that we normally associate with living things. Do we normally have any difficulty classifying things as living or non-living? Is an apple living or non-living? An apple-seed? A potato? A fresh beefsteak? Bovine cells in tissue culture? A deep-frozen embryo? The human immunodeficiency virus (HIV) which causes AIDS? If any of these fails to meet the criteria for living, which characteristics does it lack?

Answers to Self-Exam

Multiple Choice Questions

1. b
2. a
3. d
4. c
5. d
6. d
7. a
8. b
9. c
10. c
11. b
12. b
13. b

True or False Questions

14. F
15. T
16. F
17. F
18. T
19. T
20. F
21. F
22. T

Fill in the Blank Questions

23. species
24. evolution
25. deductive
26. hypothesis
27. theory
28. law
29. control
30. finches
31. Charles Lyell
32. reductionist
33. discussion
34. methods and materials

CHAPTER 2 Small Molecules

Learning Objectives

After mastering the material covered in chapter 2, you should be able to confidently do the following tasks:

- Describe the basic structure of matter, including living matter, in terms of atoms and molecules of the naturally occurring chemical elements.

- Explain how the electron configurations of atoms affect the chemical characteristics of the different elements, and explain the roles of electron configurations in formation of chemical bonds.

- Identify the biologically important molecules and ions in which we find carbon, hydrogen, oxygen, nitrogen, phosphorus, and sulfur.

- Identify the key functional groups that occur in biologically important molecules.

- Outline the chemical nature of water, and explain the physical and chemical properties that make it such an important substance for living organisms.

- Distinguish the properties of acidic, neutral, and basic solutions, and show how the pH scale is used to measure the relative acidity or alkalinity of a solution.

Chapter Outline

I. Elements, Atoms, and Molecules

A. Atomic structure.
B. Neutrons, isotopes, and biology.

II. Electrons and the Chemical Characteristics of Elements

A. Electron energy levels and orbitals.
   1. Electron shells and orbitals.
B. Energetic tendencies.
C. Chemical reactions: Filling the outer electron shell.
   1. Ions and the ionic bond.
   2. The covalent bond: Sharing electron pairs.

III. Chemical Bonds and the SPONCH Elements

A. Carbon and the covalent bond.
   1. Methane.
   2. Carbon backbones.

B. Nitrogen.
C. Phosphorus.
D. Sulfur.

IV. The Water Molecule and Hydrogen Bonding

A. The polar nature of water.
B. Water and the hydrogen bond.
C. Characteristics of water.
D. pH, acids, and bases.

Key Words

atom
molecule
compound
subatomic particles
neutrons
protons
electrons
nucleus
strong nuclear forces
atomic mass
atomic weight
dalton
molecular mass
atomic number
isotopes
radioactive isotopes
radioisotopes
alpha particles
beta particles
photon
half-life
tracers
electron energy levels
orbitals
energy shells
1s orbital
2s orbital
2p orbitals
3s orbital
3p orbitals
4s orbital
octet rule
energetic tendencies
noble elements
chemical reactions
reactants
products
electronegativity
ions
polar molecule
nonpolar molecule
ionic bond
covalent bond
hydrogen bond
van der Waals forces
electron donor
electron acceptor
dissolve
solution
nonpolar covalent bond
polar covalent bond
organic molecules
methane
marsh gas
chemical formula
structural formula
bond angle
tetrahedron
double bond
triple bond
macromolecule
carbon backbone
hydrocarbon
butane
saturated fat
unsaturated fat
fatty acid
N_2
molecular nitrogen
NH_3
ammonia
NH_4^+
ammonium ion
NO_3^-
nitrate ion
nitrogen fixation
legume
nitrogen-fixing bacteria
phosphate
phosphate group
HPO_4^{-2}
$H_2PO_4^-$
phosphate ions
P_i
inorganic phosphate
phospholipids
adenosine triphosphate
ATP
adenosine diphosphate
ADP
amino acids
functional groups
sulfhydryl group
carboxyl group
amino group
methyl group
aldehyde
ketone
hydrologic cycle
hydration shells
hydrophobic
hydrophilic
lecithin
adhesion
cohesion
capillary action
meniscus
imbibition
adsorption
surface tension
specific heat
heat of vaporization
density
hydrogen ion
H^+
hydroxide ion
OH^-
molar concentration
mole
scientific notation
hydronium ion
H_3O^+
acid
hydrochloric acid
HCl
pH scale
H^+ concentration
base
alkali
acidic solution
neutral solution
basic solution

carbon dioxide
CO_2
carbonic acid
H_2CO_3

bicarbonate ion
HCO_3^-
sodium bicarbonate

$NaHCO_3$

Exercises

1. Identify each of the following functional groups, using terms from the following list: hydroxyl group, amino group, carboxyl group, methyl group, aldehyde group, sulfhydryl group, ketone group, phosphate group. For each, also give the chemical formula (e.g., NH_2 for amino group). When you are finished.check your answers with table 2.2 in the text.

a. ______________________ (—N(H)—H)

b. ______________________ (>C=O)

c. ______________________ (—C(=O)—H)

d. ______________________ (—C(=O)—OH)

e. ______________________ (—O—H)

f. ______________________ (—S—H)

g. ______________________ (—O—P(=O)(—O—H)—O—H)

h. ______________________ (—C(H)(H)—H)

2. Complete the diagrams below to illustrate two fatty acids, each with 12 carbon atoms in their carbon backbones. Complete the diagrams so that one fatty acid has the formula $(C_{11}H_{23})COOH$ and the other has the formula $(C_{11}H_{17})COOH$. Which of these is saturated, and which is unsaturated? Why?

HO(O=)C—C(H)(H)—

HO(O=)C—C(H)(H)—

3. The line below represents a scale of pH values but it is calibrated to show the molar concentration of H^+ ions. Place pH values on the scale, and then arrange the items in the following list to show their relative acidities or alkalinities. To check your work, see table 2.3 in the text.

1 molar nitric acid
1 molar sodium hydroxide
ammonia solution
baking soda
black coffee
gastric juices
lemon juice
milk
oven cleaner
pure water
seawater
soap solution
tomato juice
vinegar
washing soda

H^+ concentration	pH	Examples
10^{-14}		________
10^{-13}		________
10^{-12}		________
10^{-11}		________
10^{-10}		________
10^{-9}		________
10^{-8}		________
10^{-7}		________
10^{-6}		________
10^{-5}		________
10^{-4}		________
10^{-3}		________
10^{-2}		________
10^{-1}		________
10^{0}		________

Self-Exam

You should be able to easily answer the following questions after learning the material in chapter 2. If you have difficulty with any question, study the appropriate section in the text and try again.

A. Multiple Choice Questions

Circle one alternative that best completes the statement or answers the question.

1. A substance that cannot be separated into simpler substances simply by using chemical means is
   a. a molecule.
   b. a compound.
   c. an element.
   d. a nucleus.

2. A typical atomic nucleus may contain
   a. protons and electrons.

b. neutrons and electrons.
c. protons and neutrons.
d. protons, neutrons, and electrons.

3. Single atoms of an element always have
   a. equal numbers of protons and electrons.
   b. equal numbers of neutrons and electrons.
   c. equal numbers of protons and neutrons.
   d. equal numbers of protons, neutrons, and electrons.

4. The atomic mass of an element is determined by
   a. the number of neutrons plus protons, and it is measured in grams.
   b. the number of neutrons plus protons, and it is measured in daltons.
   c. the number of protons plus electrons, and it is measured in grams.
   d. the number of protons plus electrons, and it is measured in daltons.

5. Isotopes of an element differ in
   a. the number of electrons in different isotopes.
   b. the number of protons in different isotopes.
   c. the number of neutrons in different isotopes.
   d. their atomic numbers.

6. Electrons with the least energy will occupy which of the following orbitals?
   a. 1s.
   b. 2s.
   c. 2p.
   d. 3p.

7. What is the greatest number of electrons that may occupy a 2p orbital?
   a. one.
   b. two.
   c. six.
   d. eight.

8. Which of the following statements is *not* one of the energetic tendencies observed in atomic structure?
   a. Orbiting electrons tend to form pairs.
   b. Positive and negative charges tend to balance.
   c. Atoms tend to form full outer shells.
   d. All matter tends to move toward higher energy states.

9. Ions are defined as atoms
   a. with neutral electronegativity.
   b. with differing atomic numbers.
   c. that have gained or lost one or more electrons.
   d. that are chemically nonreactive.

10. Which of the following items is *not* one of the means by which the different atoms of molecules are held together with one another?
   a. ionic bonds.
   b. covalent bonds.
   c. van der Waals forces.
   d. strong nuclear forces.

11. When sodium and chlorine react to form sodium chloride, the sodium atom
   a. gives up one electron, and takes on a net negative charge.
   b. gives up one electron, and takes on a net positive charge.
   c. accepts one electron, and takes on a net negative charge.
   d. accepts one electron, and takes on a net positive charge.

12. Which of the following statements is correct, regarding ionic bonding?
    a. In sodium chloride crystals, unpaired electrons remain.
    b. Electrostatic charges are unbalanced in the sodium chloride crystal.
    c. The outer shells of both sodium and chlorine become filled to capacity.
    d. Ionically bonded sodium chloride will not dissolve in water.

13. Covalent bonds can be polar or nonpolar; which of the following is correct?
    a. In molecular hydrogen, atoms are joined by nonpolar covalent bonds, while the bonds in water are polar.
    b. In molecular hydrogen, atoms are joined by polar covalent bonds, while the bonds in water are nonpolar.
    c. In both water and molecular hydrogen, the atoms are joined by nonpolar covalent bonds.
    d. In both water and molecular hydrogen, the atoms are joined by polar covalent bonds.

14. In methane , the hydrogen atoms are joined to the carbon atoms by
    a. ionic bonds.
    b. covalent bonds.
    c. hydrogen bonds.
    d. atomic bonds.

15. Which of the following is the correct structural formula for methane?
    a. CH_4.
    b. H_4C.
    c. H-H-C-H-H.
    d.
```
  H
  |
H-C-H
  |
  H
```

16. The molecule
```
  H H H H
  | | | |
H-C-C-C-C-H
  | | | |
  H H H H
```
is termed

    a. methane, a hydrocarbon.
    b. propane, a hydrocarbon.
    c. butane, a hydrocarbon.
    d. acetylene, a hydrocarbon.

17. Nitrogen in the atmosphere is in the form of
    a. molecular nitrogen, N_2.
    b. ammonia, NH_3.
    c. ammonium ions, NH_4^+.
    d. nitrate ions, NO_3^-.

18. In living organisms, phosphorus typically occurs in
    a. protein.
    b. hydrocarbons.
    c. ATP.
    d. carbohydrate.

19. In living organisms, sulfur is found in
    a. certain amino acids of protein.
    b. hydrocarbons like methane.

c. the phosphate groups of ATP.
d. only in carbohydrate.

20. Water molecules are polar because
    a. when they get very cold, they can turn to ice.
    b. the hydrogen atoms are small compared to the oxygen atom.
    c. there is an uneven distribution of positive and negative charge, and the oxygen side is negatively charged.
    d. there is an uneven distribution of positive and negative charge, and the oxygen side is positively charged.

21. In the liquid phase, different water molecules tend to be held together by
    a. ionic bonds.
    b. covalent bonds.
    c. hydrogen bonds.
    d. van der Waals forces.

22. When sodium chloride dissolves in water, which portion of the water molecules tend to be oriented toward the chloride ion?
    a. The electronegative hydrogen atoms.
    b. The electropositive hydrogen atoms.
    c. The electronegative oxygen atoms.
    d. The electropositive oxygen atoms.

23. When a nonpolar substance is added to water, it will typically
    a. dissolve quickly.
    b. form crystals.
    c. break down the hydrogen bonds between water molecules.
    d. form clumps that do not mix with water.

24. Molecules of lecithin are polar on one end and nonpolar on the other end. Which of the following statements accurately describes the properties of a lecithin molecule?
    a. Its polar hydrophilic end will mix with vinegar.
    b. Its nonpolar hydrophilic end will mix with oil.
    c. Its polar hydrophobic end will mix with oil.
    d. Its nonpolar hydrophobic end will mix with vinegar.

25. Capillary action in a small glass tube is the result of which properties of water?
    a. absorption by the glass tube.
    b. the vaporization of water and absorption by glass.
    c. adhesion of water to the glass surface and cohesion of water molecules to one another.
    d. the decreasing density of water as it cools while evaporating.

26. The surface tension that allows small insects to walk on the surface of water is due to
    a. the density of water being less than the density of the insect.
    b. the cohesion of water molecules to one another.
    c. the latent heat of vaporization that is characteristic of water.
    d. the molar concentration of solutes.

27. Which of the following substances has a greater specific heat than water?
    a. lead.
    b. table sugar.
    c. ethyl alcohol.
    d. ammonia.

28. If a particular solution is determined to have a pH of 10,
    a. the molar concentration of H^+ ions is 10^{-10} and the molar concentration of OH^- ions is 10^{-4}.
    b. the molar concentration of H^+ ions is 10^{-4} and the molar concentration of OH^- ions is 10^{-10}.
    c. the molar concentration of H^+ ions is 10^{+10} and the molar concentration of OH^- ions is 10^{-10}.
    d. the molar concentration of H^+ ions is 10^{-10} and the molar concentration of OH^- ions is 10^{+4}.

29. Which of the following lists is correctly arranged from most acidic to most basic?
   a. ammonia solution ⇒ seawater ⇒ vinegar ⇒ gastric juices.
   b. gastric juices ⇒ vinegar ⇒ seawater ⇒ ammonia solution.
   c. vinegar ⇒ gastric juices ⇒ seawater ⇒ ammonia solution.
   d. seawater ⇒ gastric juices ⇒ vinegar ⇒ ammonia solution.

30. The bubbles that rise to the surface of beer contain
   a. oxygen.
   b. carbon dioxide.
   c. carbonic acid.
   d. bicarbonate ion.

B. True or False Questions

Mark the following statements either T (True) or F (False).

_____ 31. A molecule is a substance that cannot be separated into simpler substances by purely chemical means.

_____ 32. The most numerous atoms in humans are hydrogen atoms.

_____ 33 Although magnesium is commonly required in small amounts by animals, it is not needed by plants.

_____ 34. Protons and electron are held together in atoms by strong nuclear forces.

_____ 35. A molecule's molecular mass is the sum of the atomic masses of its constituent atoms.

_____ 36. An atom with 6 electrons, 6 protons, and 6 neutrons has an atomic number of 12.

_____ 37. In a single oxygen atom, the 2s orbital is filled with electrons, but not all 2p orbitals are filled.

_____ 38. Atoms tend to have completely filled outer electron shells, even if some lower-energy orbitals must remain unfilled.

_____ 39. Elements with atomic numbers of 2, 10, and 18 have completely filled outer electron shells.

_____ 40. Ions have an electrostatic charge because they are joined by covalent bonds.

_____ 41. In table salt, NaCl, chlorine acts as an electron acceptor and sodium acts as an electron donor.

_____ 42. In crystalline sodium chloride, charges are balanced, but some electrons remain unpaired.

_____ 43. A carbon atom can typically form four covalent bonds to other atoms.

_____ 44. DNA is an example of a macromolecule.

_____ 45. Most plants can easily utilize atmospheric nitrogen, N_2, as a direct source of nitrogen.

_____ 46. A sulfhydryl group is characteristic of all amino acids.

_____ 47. Water molecules are polar; the oxygen end is positively charged, and the hydrogen end is negative.

_____ 48. Nonpolar molecules are generally not soluble in water.

_____ 49. Water reaches its highest density at 0°C.

_____ 50. If the pH of a soap solution is 10, the molar concentration of H^+ ions is 10^{-10}.

C. Fill in the Blanks

Answer the question or complete the statement by filling in the blanks with the correct word or words.

51. A unit formed by two or more atoms joined together is termed a ________________.

52. The most common elements in living matter are carbon, hydrogen, and ________________.

53. The stable subatomic particles that make up the nuclei of atoms are ________________.

54. The number of protons and neutrons in an atom's nucleus determines its ________________.

55. U-235 and U-238 differ in the number of neutrons in their nuclei; they are ________________ of uranium.

56. The electrons of a helium atom (atomic number = 2) occupy the ________________.

57. The starting substances for a chemical reaction are the ________________, and the substances produced by the reaction are the ________________.

58. A bond formed by the electrostatic attraction after an electron is transferred from a donor to an acceptor is a(an) ________________ bond.

59. When a pair of electrons is shared unequally, a covalent bond is ________________.

60. When a hydrocarbon has no double bonds between carbon atoms, and the number of hydrogen atoms bonded to the carbon atoms is maximum, the hydrocarbon is said to be ________________.

61. Plants that harbor nitrogen-fixing bacteria in their roots are common in the ________________ family.

62. Membranes of cells are rich in phosphate-containing fatty substances called ________________.

63. Carboxyls, aldehydes, and ketones all contain a double-bonded ________________ atom.

64. Different molecules of water tend to be held together by ________________

65. The material in egg yolk that allows the mixing of vinegar and oil in mayonnaise is ________________.

66. Water rises in tubes with small internal diameters via ________________.

67. Large bodies of water tend to vary relatively little in temperature because water has a relatively high ________________.

68. Acids tend to release ________________ ions, and bases tend to release ________________ ions.

69. Pure water at pH 7 has ________________ moles of H^+ per liter.

70. Most biological processes in animals take place at pH values between ________________.

Questions for Discussion

1. What are the three energetic tendencies followed by atoms? Explain how ionic bonding and covalent bonding are consistent with these tendencies. Why are the "noble elements" so unreactive?

2. Why do ionically bonded molecules, such as sodium chloride, dissolve in water so readily? Why are covalently bonded hydrocarbons relatively insoluble in water? Water is sometimes considered to be a "universal solvent" for living systems. Explain what is meant by this.

3. Volumes have been written praising water as uniquely suited to be the fluid of life. Outline the properties of water that are so important to living things on earth. It has been suggested that ammonia might have an analogous role as a basis for some other form of life on some other planet. What properties of ammonia make it a candidate for such a role?

4. Human blood has a slightly alkaline pH, and the cells of many other organisms have similar pH values. Not surprisingly, this is the optimal pH for many of the chemical reactions of life. It has been observed that this pH is similar to that of seawater. How might this observation bear on our understanding of the origin of living systems and where the chemical processes of life evolved?

Answers to Self Exam

Multiple Choice Questions

1. c
2. c
3. a
4. b
5. c
6. a
7. b
8. d
9. c
10. d
11. b
12. c
13. a
14. b
15. d
16. c
17. a
18. c
19. a
20. c
21. c
22. b
23. d
24. a
25. c
26. b
27. d
28. a
29. b
30. b

True or False Questions

31. F
32. T
33. F
34. F
35. T
36. F
37. T
38. F
39. T
40. F
41. T
42. F
43. T
44. T
45. F
46. F
47. F
48. T
49. F
50. T

Fill in the Blank Questions

51. molecule
52. oxygen
53. protons and neutrons
54. atomic mass
55. isotopes
56. 1s orbital
57. reactants, products
58. ionic
59. polar
60. saturated
61. legume
62 phospholipids
63. oxygen
64. hydrogen bonds
65. lecithin
66. capillary action
67. heat of vaporization
68. H^+, OH^-
69. 10^{-7}
70. 6 and 8

CHAPTER 3 The Molecules of Life

Learning Objectives

After mastering the material covered in chapter 3, you should be able to confidently do the following tasks:

- Identify the molecular building blocks of the major groups of compounds important in living organisms: the carbohydrates, the lipids, the polypeptides and proteins, and the nucleic acids,
- Specify the structural characteristics of the most common types of carbohydrates, and give examples of each type.
- Show how fatty acids, glycerol, and other components are combined to form lipids.
- Explain how peptide bonds are formed between amino acids to form polypeptides.
- Describe the primary, secondary, tertiary, and quaternary structure of proteins, and give examples of the functions that are typical of different kinds proteins.
- Give the basic structure of a nucleotide, and show how nucleotides are bonded to one another to form the macromolecules DNA and RNA.

Chapter Outline

I. The Carbohydrates

A. Monosaccharides and disaccharides.
B. Polysaccharides.
  1. Plant and animal starches.
    a. amylose.
    b. amylopectin.
    c. glycogen.
    d. primary and secondary structure of starches.
  2. Structural polysaccharides.
    a. cellulose.
    b. chitin.
  3. Mucopolysaccharides.

II. The Lipids

A. Triglycerides.
B. Saturated, unsaturated, and polyunsaturated fats.
C. Phospholipids.
D. Waxes and steroids.

III. The Proteins

A. Amino acids.
B. Polypeptides.
   1. Levels of protein structure.
C. Conjugated proteins and prosthetic groups.
D. Binding proteins.
E. Structural proteins.

IV. The Nucleic Acids.

A. Nucleic acid function.
B. Nucleic acid structure.

Key Words

carbohydrate
$(CH_2O)_n$
monosaccharide
glucose
disaccharide
polysaccharide
aldehyde group
ketone group
blood sugar
corn sugar
grape sugar
dextrose
sucrose
table sugar
fructose
dehydration linkage
dehydration synthesis
enzyme
hydrolytic cleavage
glycosidic linkage
lactose
milk sugar
lactose intolerance
plant starch
amylose
amylopectin
potato starch
maltose
glycogen
cellulose
alpha glucose
beta glucose
beta-glycosidic linkage
cellulase
fiber
microfibrils
tensile strength
lignin
hemicellulose
pectin
algin
agar
carageenan
chitin
exoskeleton
N-acetyl glucosamine
proteoglycan
mucopolysaccharide
heparin
lipid
triglyceride
fat
oil
glycerol
fatty acid
alcohol
ester linkage
saturated fatty acids
unsaturated fatty acids
polyunsaturated fatty acids
hydrogenation
oleic acid
linoleic acid
linolenic acid
phospholipid
phosphate group
detergent
wetting agent
lecithin
plasma membrane
glycolipid
wax
steroid
sterols
lanolin
cholesterol
vitamin D
steroid hormones
protein
enzyme
polypeptide
amino acid
peptide bond
dipeptide
alpha carbon
amino group
carboxylic acid group
sulfhydryl group
disulfide linkage
N-terminal end
amino end
C-terminal end
carboxyl end
primary level organization
amino acid sequence
secondary level organization
alpha helix
beta sheet
tertiary level organization
hydrogen bonding
ionic attractions
hydrophobic interactions
covalent bonding
keratin
monomeric proteins
quaternary level organization
hemoglobin
alpha chain
beta chain
heme group
denaturation
urea
simple protein
conjugated protein
prosthetic group
chromoprotein
glycoprotein
lipoprotein
cofactors
binding protein

transferrin
haptoglobin
avidin
biotin
structural proteins
collagen
elastin
nucleic acid
DNA
deoxyribonucleic acid
RNA
ribonucleic acid
genes
double helix
nucleotide
phosphate group
5-carbon sugar
nitrogen base
adenine
thymine
guanine
cytosine
uracil
transcription

Exercises

1. Identify the following molecular diagrams, using the terms: adenine nucleotide, alanine, cholesterol, fatty acid, glucose, glycerol, N-acetyl glucosamine, sucrose.

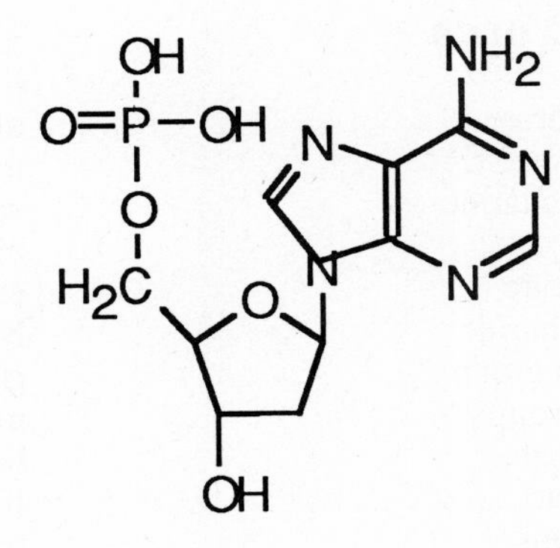

a. ____________________ b. ____________________ c. ____________________

d. ____________________ e. ____________________ f. ____________________

g. ____________________ h. ____________________

2. In the space below, join the building blocks that are given to form the required molecule. Be sure to show where the bond is formed to connect the building blocks, and state the type of linkage formed.

a. Join glucose and fructose subunits to form sucrose:

b. Join three alpha glucose subunits with alpha 1-4 linkages:

c. Now, join three **beta** glucose subunits with 1-4 linkages.

d. Combine glycerol with three fatty acids to form a triglyceride.

e. Combine the amino acids glycine and alanine to form a dipeptide.

Self-Exam

You should be able to easily answer the following questions after learning the material in chapter 3. If you have difficulty with any question, study the appropriate section in the text and try again.

A. Multiple Choice Questions

Circle one alternative that best completes the statement or answers the question.

1. The sugar lactose is
   a. a monosaccharide.
   b. a disaccharide made up of two glucose subunits.
   c. a disaccharide made up of a glucose subunit and a fructose subunit.

d. a disaccharide made up of a glucose subunit and a galactose subunit.

2. The two simple subunits of sucrose are joined by
   a. a peptide linkage.
   b. a dehydration linkage.
   c. a hydrolytic linkage.
   d. a hydrogen bond.

3. The storage form of glucose in animals is most often
   a. amylose and amylopectin.
   b. glycogen.
   c. cellulose.
   d. chitin.

4. The polysaccharide amylopectin is made up from glucose subunits joined to one another with
   a. only 1-4 linkages.
   b. only 1-6 linkages.
   c. 1-4 linkages and 1-6 linkages.
   d. 1-2 linkages and 1-4 linkages.

5. The polysaccharides characterized by alpha 1-4 glycosidic linkages and beta 1-4 glycosidic linkages are, respectively,
   a. amylose and cellulose.
   b. amylose and glycogen.
   c. cellulose and amylose.
   d. cellulose and glycogen.

6. The cellulose-digesting enzyme cellulase is produced by
   a. termites and fungi.
   b. garden snails and fungi.
   c. cows and termites.
   d. humans and garden snails.

7. The exoskeleton of insects is best characterized as
   a. a structural polysaccharide.
   b. a structural protein.
   c. a structural phospholipid.
   d. structural cellulose.

8. A common characteristic of the lipids is
   a. all contain fatty acids.
   b. all contain glycerol.
   c. all are organic molecules that tend to be fat-soluble.
   d. all are organic molecules that tend to be water-soluble.

9. Triglycerides are formed by ester linkages
   a. involving the hydroxyl groups of glycerol and the carboxyl groups of fatty acids.
   b. involving the carboxyl groups of glycerol and the hydroxyl groups of fatty acids.
   c. involving the hydroxyl groups of glycerol and the hydrogen atoms of fatty acids.
   d. involving the hydrogen atoms of glycerol and the carboxyl groups of fatty acids.

10. Although fats are generally not soluble in water, soap is. Soaps include
   a. calcium salts of fatty acids.
   b. sodium and potassium salts of fatty acids.
   c. hard water.
   d. water-insoluble salts of fats.

11. If two fatty acids have the same number of carbon atoms, the one with more double bonds between carbon atoms is
    a. more saturated.
    b. more unsaturated.
    c. more hydrogenated.
    d. found in harder fats.

12. Linoleic acid is an example of
    a. a saturated fatty acid that is essential in human diet.
    b. an unsaturated fatty acid that is essential in human diet.
    c. a saturated fatty acid that should be avoided in human diet.
    d. an unsaturated fatty acid that should be avoided in human diet.

13. Which of the following accurately describes phospholipids?
    a. The phosphate group is hydrophilic and water soluble, but the fatty acid tails are hydrophobic.
    b. The phosphate group is hydrophobic and soluble in nonpolar substances, unlike the fatty acid tails.
    c. The fatty acid tails are hydrophilic and water soluble, but the phosphate group is hydrophobic.
    d. The fatty acid tails are soluble in polar substances, but the phosphate group is not.

14. The plasma membrane of cells is made up of
    a. phospholipids and cellulose.
    b. polymers of N-acetyl glucosamine and phospholipids.
    c. phospholipids with glycolipids and glycoproteins.
    d. cellulose and amylopectin.

15. Waxes can be described as
    a. large triglycerides.
    b. esters of fatty acids with glycerol.
    c. esters of fatty acids with long straight-chain alcohols.
    d. phospholipids esters with alcohols.

16. The skin-softening substance lanolin is chemically
    a. an oil.
    b. a fat.
    c. a sterol.
    d. a phospholipid.

17. Proteins are typically made up of
    a. amino acids connected with disulfide linkages.
    b. amino acids connected with peptide bonds.
    c. amino acids connected with ester linkages.
    d. amino acids that are ionically bonded to one another.

18. Which of the following statements is generally accurate for all amino acids?
    a. Amino acids have an amino group and a carboxyl group.
    b. Amino acids are non-polar, uncharged molecules.
    c. Amino acids are polar, uncharged molecules.
    d. Amino acids near positive or negative charges when dissolved in water.

19. The beta pleated sheet is an example of protein
    a. primary level of organization.
    b. secondary level of organization.
    c. tertiary level of organization.
    d. quaternary level of organization.

20. Disulfide linkages are important in determining a protein's
    a. primary level of organization.
    b. secondary level of organization.

c. tertiary level of organization.
d. quaternary level of organization.

21. Bends or folds in a proteins structure often occur where there is no alpha helix, such as the site of
a. cysteine.
b. proline.
c. glycine.
d. alanine

22. The hemoglobin molecule is an example of
a. a protein with only primary and secondary levels of organization.
b. a protein that requires a cofactor.
c. a simple protein that is a chromoprotein.
d. a conjugated protein that is a chromoprotein.

23. Collagen, elastin, and keratin are examples of
a. conjugated proteins.
b. chromoproteins.
c. binding proteins.
d. structural proteins.

24. Nucleic acids like DNA are polymers of nucleotides that are composed of
a. phosphates, sugars, and nitrogen-containing bases.
b. phosphates and sugars, but no nitrogen-containing parts.
c. 5-carbon sugars and nitrogen-containing bases, but no parts with phosphorus.
d. nitrogen-containing bases and phosphates groups, but no sugars.

25. The genetic information in nucleic acids is encoded in
a. the sequence of amino acids in DNA and RNA.
b. the sequence of the saccharides that make up the molecule.
c. the particular peptide sequence in DNA or RNA.
d. the particular nucleotide sequence in DNA or RNA.

B. True or False Questions

Mark the following statements either T (True) or F (False).

_____ 26. Glucose and fructose both have the formula $C_6H_{12}O_6$.

_____ 27. Glucose is one of the simplest disaccharides.

_____ 28. An oxygen double-bonded to a carbon can form either an aldehyde group or a ketone group.

_____ 29. During digestion, sucrose is broken down by hydrolytic cleavage into two glucose molecules.

_____ 30. Sucrose, lactose, and maltose are all disaccharides.

_____ 31. Amylopectin is a molecule of 1-4 glucose chains cross-linked with occasional 1-6 linkages.

_____ 32. Amylose and amylopectin are the main storage carbohydrates for both plants and animals.

_____ 33. Enzymes in human saliva can easily break beta-glycosidic linkages.

_____ 34. The "fiber" in whole-bran cereal is actually cellulose.

_____ 35. Although cellulose microfibrils are very strong, they have relatively little tensile strength.

_____ 36. The exoskeleton of a lobster is rich in polysaccharide.

_____ 37. The fluid that lubricates skeletal joints in humans includes mucopolysaccharides.

_____ 38. Three molecules of water are formed for each triglyceride produced by dehydration synthesis.

_____ 39. Fatty acids are actually alcohols.

_____ 40. Unsaturated fatty acids have more carbon-carbon double bonds than do saturated fatty acids.

_____ 41. Phospholipids differ from triglycerides in having a phosphate group in place of glycerol.

_____ 42. Waxes are generally harder and more hydrophobic than fats.

_____ 43. Cholesterol can serve as a precursor for vitamin D and steroid hormones.

_____ 44. All of the naturally occurring amino acids are hydrophobic, and tend to cluster together.

_____ 45. In many polypeptides, the amino acids coil and form an alpha helix.

C. Fill in the Blanks

Answer the question or complete the statement by filling in the blanks with the correct word or words.

46. Most carbohydrate have the empirical formula _________________.

47. Dextrose, corn sugar, and blood sugar are all names for _________________.

48. Maltose is built by a _________________ between two glucose molecules.

49. Potato starch is about 20% _________________ and about 80% _________________.

50. A linear polymer of beta glucose subunits joined with 1-4 linkages is termed _________________.

51. Glucose can occur in an alpha ring form, a beta ring form, and in a _________________ form.

52. Agar is a structural polysaccharide found in some _________________.

53. The polymer of N-acetyl glucosamine is called _________________.

54. The jellylike substance that fills the human eye is rich in _________________.

55. Lipids tend to be _________________-soluble, and not _________________-soluble.

56. Triglycerides are formed by ester linkages between the alcohol _________________ and three _________________.

57. After it is hydrogenated, a vegetable oil will be more _________________ at room temperature.

58. The phosphate group of a phospholipid is _________________ and will mix with water.

59. A substance that will reduce the surface tension of water is termed a _________________.

60. Lecithin is a storage phospholipid found in _________________.

61. The outer surfaces of many insects and plants are covered with _________________ to limit water loss.

62. Steroids all have in common a core of ________________ interlocking rings.

63. Protein catalysts are termed ________________.

64. A peptide bond connects the carboxyl group carbon of one amino acid with the ________________ atom of another amino acid.

65. All polypeptides have an N-terminal ________________ end and a C-terminal ________________ end.

Questions for Discussion

1. We eat many plants, yet we cannot digest cellulose. Explain the molecular basis for this. How are cattle able to gain energy from the cellulose of plants?

2. The three-dimensional shape of enzymes is very important for their proper functioning. Explain the levels of organization in enzyme structure that give enzymes the specific shapes they have. How does heating or cooking affect the activity of enzymes?

3. Many different molecules are used for energy storage by different organisms. Outline the most common storage molecules. Fats have about twice as many calories per gram as do carbohydrates. Does this have anything to do with the role of fats as storage molecule? Do plants (other than couch potatoes) store fats or oils?

Answers to Self Exam

Multiple Choice Questions

1. d
2. b
3. b
4. c
5. a
6. b
7. a
8. c
9. a
10. b
11. b
12. b
13. a
14. c
15. c
16. c
17. b
18. a
19. b
20. c
21. b
22. d
23. d
24. a
25. d

True or False Questions

26. T
27. F
28. T
29. F
30. T
31. T
32. F
33. F
34. T
35. F
36. T
37. T
38. T
39. F
40. T
41. F
42. T
43. T
44. F
45. T

Fill in the Blank Questions

46. $(CH_2O)_n$
47. glucose
48. dehydration synthesis
49. amylose, amylopectin
50. cellulose
51. straight chain
52. seaweed
53. chitin
54. proteoglycans
55. fat-, water-
56. glycerol, fatty acids
57 solid
58. hydrophilic
59. wetting agent
60. egg yolks
61. wax
62. four
63. enzymes
64. nitrogen
65. amino, carboxyl

CHAPTER 4 Cell Structure

Learning Objectives

After mastering the material covered in chapter 4, you should be able to confidently do the following tasks:

- Explain our current understanding of cell theory - that cells are the fundamental unit of life, and that they always arise from pre-existing cells.
- Outline the features that distinguish prokaryotic cells from eukaryotic cells.
- Describe the organelles that have been described and studied in eukaryotic cells.
- Sketch the basic structural arrangement of a typical animal cell and a typical plant cell.
- Distinguish plant cells and animal cells in those features where they differ.
- Evaluate the serial endosymbiosis hypothesis proposed to explain the origins of mitochondria and chloroplasts in eukaryotic cells.

Chapter Outline

I. Cell Theory.

A. What is a cell?
B. Cell size and surface-volume relations.

II. Prokaryotic Cells.

III. Eukaryotic Cells.

A. Surface structures.
1. cell wall.
2. plasma membrane.
B. Internal support: the cytoskeleton.
1. microfilaments.
2. intermediate fibers.
3. microtubules.
C. Control and cell reproduction: the nucleus.
D. Organelles of synthesis, storage, and cytoplasmic transport.
1. ribosomes and protein synthesis.
2. endoplasmic reticulum.
3. Golgi complex.
4. lysosomes.

5. microbodies.
6. vacuoles.

E. Energy-generating organelles.
1. chloroplasts.
2. plastids.
3. mitochondria.

F. Organelles of movement.
1. centrioles and basal bodies.
2. cilia and flagella.

IV. Essays.

A. Looking at cells.
1. light microscopy.
2. electron microscopy.

B. Evolution of the eukaryotic cell.
1. autogenous hypothesis.
2. endosymbiosis hypothesis.

Key Words

cells
cell theory
Schleiden
Schwann
Virchow
neuron
micrometer (μm)
vacuoles
cytoplasm
surface-volume hypothesis
plasma membrane
cytoplasmic streaming
brush borders
prokaryote
bacteria
eukaryote
heterotrophs
autotrophs
pathogen
decomposer
photoautotrophic
photosynthesis
chemoautotrophs
thylakoid
cell wall
sheath
capsule
nucleoid
chromosome
plasmid
mesosome
fission
asexual reproduction
sexual reproduction
conjugation
sex pilus (-i)
flagellum (-a)
organelle
nucleus
Monera
Protista
Fungi
Plantae
Animalia
plant cell wall
middle lamella
cellulose
lignin
pectin
suberin
cutin
phospholipid bilayer
cytoskeleton
microfilaments
actin
microvilli
amoeboid movement
pseudopods
intermediate filaments
epithelial cells
keratin
microtubules
tubulin
nucleolus
ribosomal RNA
mitosis
nuclear envelope
nuclear pores
endoplasmic reticulum (ER)
ER lumen
rough endoplasmic reticulum
ribosomes
antibiotics
tetracycline
streptomycin
polyribosomes
polysomes
smooth endoplasmic reticulum
Golgi complex
cisterna (-ae)
dictyosomes
vesicles
cis face
trans face
lysosome
autophagy
heterophagy
digestion vacuole
lysosomal storage disease
Tay-Sachs disease
N-acetyl hexosaminidase
ganglioside
microbodies
peroxisomes
catalase
glyoxysome
vacuoles
cell sap
anthocyanins
endomembranal system
chloroplast
chlorophyll
stroma
grana
granum thylakoids
stroma thylakoids
CF_0CF_1 complex
ATP synthetase

plastids
proplastids
leucoplasts
chromoplasts
mitochondrion (-ia)
endosymbiont
serial endosymbiosis hypothesis
autogenous hypothesis
inner compartment
matrix
crista (-ae)
outer compartment
F_0F_1 complex
centriole
basal body
cilium (-ia)
flagellum (-a)
Paramecium
9 + 2 structure
dynein arms
nexin
radial spokes
inner sheath
axoneme
light microscope
compound microscope
objective lens
eyepiece
ocular
condenser
iris diaphragm
stain
histology
section
resolving power
nanometer (nm)
electron microscope (EM)
transmission EM
freeze-fracturing
scanning EM
high voltage EM
ultracentrifugation

Exercises

1. Although all organisms are composed of cells, some key features distinguish cells of prokaryotes from eukaryotic cells. In addition, there are some characteristic differences between typical plant and animal cells. Complete the following table to distinguish these groups by indicating if a feature is present or absent, or by describing differences in features. Check your work in Table 4.1 of the text.

	Prokaryotic Cells	Plant Cells	Animal Cells
Cell Wall		Present, cellulose	
Centrioles	Absent		
Chromosomes			
Cytoskeleton			
Flagella and cilia			
Membranes			
Nucleus			
Organelles			
Ribosomes			
Vacuoles			

2. These are diagrams of representative plant and animal cells. Indicate which is a plant cell and which is an animal cell, and provide labels for the various structure indicated. Use terms from the following list. Check your work with figure 4.1 in the text.

basal body
cell wall
central vacuole
centriole
chloroplast
cytoskeleton
diffuse chromosomes
endoplasmic reticulum
flagellum
Golgi complex
grana
lysosome
microtubules
microvilli
mitochondrion
nuclear envelope
nucleolus
nucleus
peroxisome
plasma membrane
plastid
pseudopod
ribosomes
vacuole

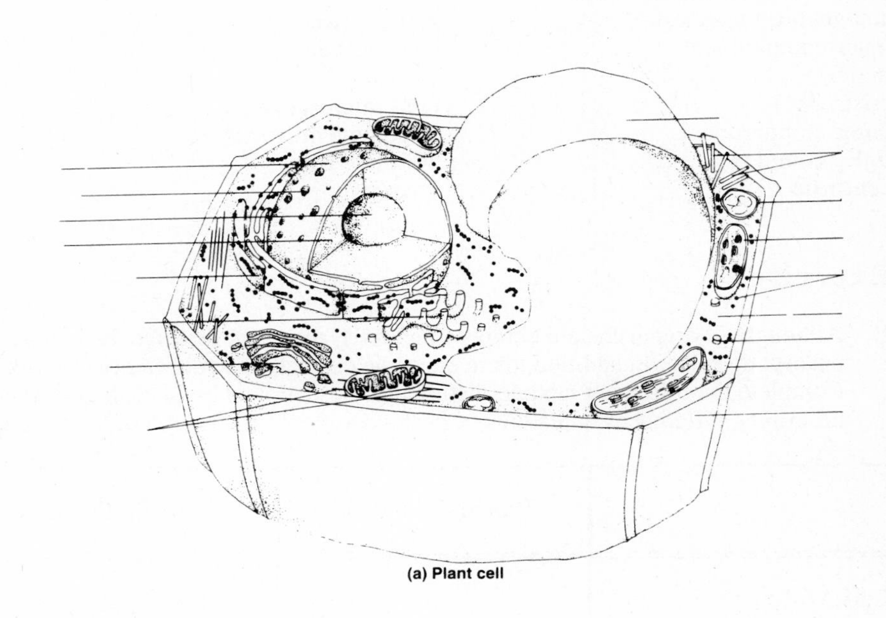

(a) Plant cell

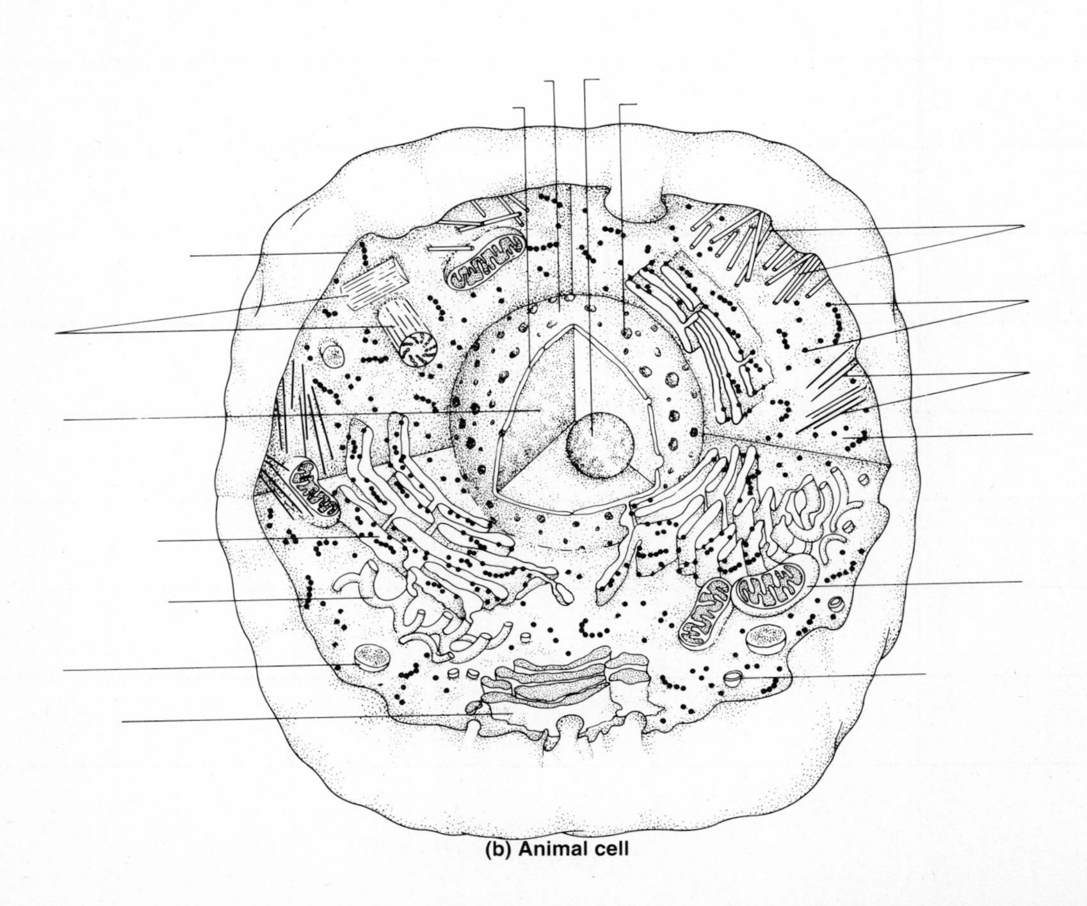

(b) Animal cell

Self-Exam

You should be able to easily answer the following questions after learning the material in chapter 4. If you have difficulty with any question, study the appropriate section in the text and try again.

A. Multiple Choice Questions

Circle one alternative that best completes the statement or answers the question.

1. Which of the following is *not* characteristic of cells today?
   a. Cells can extract useful energy from raw materials they have taken in.
   b. Cells can synthesize many organic molecules from inorganic materials.
   c. Cells can arise independently from nonliving matter in the environment.
   d. Cells can respond to stimuli from their surroundings.

2. Typical cells range in size
   a. from 10 µm to 100 µm.
   b. from 10 mm to 100 mm
   c. from 100 µm to 1000 µm.
   d. from 100 mm to 1000 mm.

3. If a cuboidal cell is approximately a cube, as the length of each edge increases from 10µm to 20µm, its volume will
   a. increase from 100 μm^2 to 400 μm^2.
   b. increase from 1000 μm^3 to 8000 μm^3.
   c. increase from 1000 μm^3 to 10,000 μm^3.
   d. increase from 1000 μm^3 to 2000 μm^3.

4. Examples of cells with relatively large surface areas compared to cell volumes might be
   a. cells involved in storage.
   b. plant cells with large central vacuoles.
   c. cells that function in absorption and transport of materials.
   d. large cells like bird eggs.

5. Bacteria that dwell in the deep ocean and extract energy from chemical reactions involving inorganic substances are considered
   a. heterotrophic prokaryotes.
   b. photoautotrophic eukaryotes.
   c. photoautotrophic prokaryotes.
   d. chemoautotrophic prokaryotes.

6. Bacterial cells are surrounded by a plasma membrane,
   a. and many have a cell wall of cellulose or chitin.
   b. and many have cells walls of peptidoglycan or protein.
   c. but only a few have cell walls.
   d. but they lack cell walls.

7. The bacteria are placed in the kingdom
   a. Monera.
   b. Protista
   c. Fungi.
   d. Plantae.

8. The feature that is *not* present in eukaryotic cells is
   a. membrane-bounded organelles.
   b. nuclear envelopes.
   c. endoplasmic reticulum.

   d. solid, rotating flagella.

9. The waxy substance that is secreted to the upper surface of plant leaves for waterproofing is
   a. cellulose.
   b. lignin.
   c. suberin.
   d. cutin.

10. The cytoskeleton of eukaryotic cells is made of
    a. the plasma membrane and flagella.
    b. microfilaments, intermediate filaments, and microtubules.
    c. the cell wall.
    d. the endoplasmic reticulum and the nucleus.

11. Microfilaments are made up of units of the protein
    a. actin.
    b. keratin.
    c. tubulin.
    d. lignin.

12. The longitudinal elements with a 9 + 2 arrangement in cilia are
    a. microfilaments.
    b. microfibrils.
    c. microtubules.
    d. microsomes.

13. The nucleolus is a structure in the nucleus of eukaryotic cells that is particularly rich in
    a. phospholipid bilayer membrane.
    b. ribosomal RNA.
    c. ribosomes.
    d. microtubules.

14. Which of the following statements best describes the eukaryotic nucleus?
    a. The nucleus is bounded by a single continuous membrane.
    b. The nucleus is bounded by two continuous layers of membrane.
    c. The nucleus is bounded by two layers of membrane with pores or openings to the cytoplasm.
    d. The eukaryotic nucleus is not separated from the cytoplasm by any membranes.

15. Where are many ribosomes located in eukaryotic cells?
    a. on the outer surface of mitochondria.
    b. in the nucleus.
    c. on the outer surface of endoplasmic reticulum.
    d. in the Golgi apparatus.

16. The Golgi apparatus in plant cells is sometimes termed
    a. a dictyosome.
    b. a thylakoid.
    c. a plastid.
    d. a glyoxysome.

17. Vesicles formed from the *trans* face of the Golgi complex
    a. typically come from the rough endoplasmic reticulum.
    b. are usually added to the next cisterna in line,
    c. leave the Golgi complex altogether, moving to various destinations.
    d. usually rejoin the *cis* face of the complex.

18. The devastating Tay-Sachs disease is a genetic disorder among those termed
    a. chromosomal storage diseases.

b. lysosomal storage diseases.
c. infectious diseases.
d. benign diseases.

19. A principal enzyme in liver peroxisomes is
    a. catalase.
    b. amylase.
    c. ATPase.
    d. N-acetyl hexosaminidase.

20. The organelles involved in keeping the stems and leaves of plants like celery firm and erect are the
    a. nuclei.
    b. microsomes.
    c. plastids.
    d. water-filled vacuoles.

21. Which organelles have their own DNA, in addition to nuclear DNA?
    a. cilia and flagella.
    b. ribosomes and mitochondria.
    c. chloroplasts and mitochondria.
    d. centrioles and chloroplasts.

22. ATP synthetase is located
    a. in peroxisomes.
    b. in CF_1 complexes and F_1 complexes.
    c. in lysosomes.
    d. in ribosomes.

23. In many plants, the brightly colored carotenoid pigments are found in
    a. chromosomes.
    b. chromoplasts.
    c. leucoplasts.
    d. chloroplasts.

24. Which of the following cell types might be predicted to have the *fewest* mitochondria?
    a. liver cell.
    b. muscle cell.
    c. fat cell.
    d. sperm cell.

25. Centriole structure is characterized by
    a. a ring of nine pairs of microtubules, surrounding two central pairs of microtubules.
    b. a ring of nine pairs of microtubules, surrounding two solitary central microtubules.
    c. a ring of nine triplets of microtubules, surrounding two solitary central microtubules.
    d. a ring of nine triplets of microtubules, with no central microtubules.

B. True or False Questions

Mark the following statements either T (True) or F (False).

_____ 26. A single cell shows all the properties that we normally associate with living things.

_____ 27. There are no cells larger than about 100 μm diameter.

_____ 28. It is generally believed that the size of cells is limited by the decreasing ratio of surface area to volume as the diameter of a cell increases.

_____ 29. All bacteria are prokaryotes, as well as some of the singled-celled Protista.

_____ 30. Bacteria can be heterotrophic, photoautotrophic, or chemoautotrophic.

_____ 31. Bacterial cell walls are peptidoglycan or protein rather than cellulose or chitin.

_____ 32. One of the similarities between prokaryotes and eukaryotes is the similar structure of their flagella.

_____ 33. Eukaryotes are easily characterized by their membrane-surrounded nucleus.

_____ 34. Cell walls of woody plants are impregnated with lignin.

_____ 35. The cytoskeletal elements have several roles: they provide support for organelles, they aid in cell movement, and they anchor the plasma membrane.

_____ 36. Microfilaments are composed of molecules of the protein tubulin.

_____ 37. Microtubules are important in centrioles, basal bodies, cilia, flagella, and the mitotic spindle.

_____ 38. The lumen of the endoplasmic reticulum is continuous with the interior of the nucleus.

_____ 39. We could locate the intracellular site of protein synthesis on rough ER.

_____ 40. The Golgi complex is another name for smooth endoplasmic reticulum.

_____ 41. Golgi cisternae originate from portions of nearby rough ER.

_____ 42. In mature plant cells, chloroplasts always arise from pre-existing chloroplasts.

_____ 43. In chloroplasts, the grana are actually stacks of thylakoids.

_____ 44. Centrioles are a prominent feature in most higher plant cells as mitosis begins.

_____ 45. The cells lining respiratory passages of nonsmoking humans have many active cilia.

C. Fill in the Blanks

Answer the question or complete the statement by filling in the blanks with the correct word or words.

46. Early cell theory is often ascribed to the 1839 writings of the German scientists __________________.

47. Organisms that use nonliving organic matter for an energy source are __________________.

48. In some bacteria, fission is accompanied by formation of the __________________, an inward extension of the plasma membrane.

49. Most single-celled eukaryotes are place in the kingdom __________________.

50. The layer between the cell walls of adjacent plant cells is the __________________.

51. Plasma membrane is essentially two layers of __________________.

52. In amoeboid movement, parts of a cell can become __________________, tube-like extensions through which more fluid cytoplasm flows.

53. Bead-like strings of ribosomes are sometimes called __________________.

54. The flattened bag-like sacs that compose the Golgi complex are ________________.

55. ________________ have been described as "suicide bags."

56. The breakdown of hydrogen peroxide into water and oxygen is catalyzed by ________________.

57. In plants, water soluble anthocyanin pigments are sometimes found in ________________.

58. The clear, watery interior of a chloroplast is its ________________.

59. The enzyme that catalyzes the production of ATP is ________________.

60. Chloroplasts, leucoplasts, and chromoplasts arise from ________________.

61. Many scientists believe that mitochondria are descended from prokaryotes that became ________________ in ancestral eukaryotic cells.

62. The ________________ is the inner compartment of a mitochondrion.

63. ________________ give rise to basal bodies, which underlay cilia and flagella.

64. The active core of a flagellum is its ________________.

65. A measure of a microscope's ability to distinguish two close objects is its ________________.

Questions for Discussion

1. Cells can carry out many functions, from assimilation and digestion of raw materials, to synthesis and transport of new molecules, and even to diverse functions such as reproduction, response, and movement. Some single celled organisms carry out all these functions, while in complex multicellular organisms like humans different cells are involved in different functions. Outline the essential functions of cells, and give examples of human cells that are specialized for these functions.

2. In spite of our tendency to emphasize the ways various groups of organisms *differ*, many scientists marvel at the unity of many features for so many living things. What are some important *similarities* between prokaryotes and eukaryote? What are some significant ways that plants and animals are *alike*? Do you think such fundamental similarities provide information concerning the ancient origin of life?

3. Viruses are biologically quite important. In simplest terms, they consist of a nucleic acid core surrounded by a protein coat of some sort. They multiply only by infecting living cells. Are they living organisms? How do they relate to cell theory? Can you place them in one of the five Kingdoms?

4. Cite the evidence that supports the endosymbiosis hypothesis proposed to explain the origins of intracellular organelles like chloroplasts and mitochondria. Do you think the evidence supports the hypothesis well enough to accept the hypothesis? Do you think such a symbiosis might occur today?

5. The surface-volume hypothesis bears on the observation that there is generally a narrow range of size for cells (with a few exceptions). Cells really never get too large. There are some really large organisms, but they achieve great size through multicellularity. How does the surface-volume hypothesis explain cell size limitation? Can you think of advantages to multicellularity in addition to allowing increased overall size?

Answers to Self Exam

Multiple Choice Questions

1. c
2. a
3. b
4. c
5. d
6. b
7. a
8. d
9. d
10. b
11. a
12. c
13. b
14. c
15. c.
16. a
17. c
18. b
19. a
20. d
21. c
22. b
23. b
24. c
25. d

True or False Questions

26. T
27. F
28. T
29. F
30. T
31. T
32. F
33. T
34. T
35. T
36. F
37. T
38. F
39. T
40. F
41. T
42. T
43. T
44. F
45. T

Fill in the Blank Questions

46. Schleiden and Schwann
47. decomposers
48. mesosome
49. Protista
50. middle lamella
51. phospholipid
52. pseudopods
53. polyribosomes
54. cisternae
55. lysosomes
56. catalase
57 vacuoles
58. stroma
59. ATP synthetase
60. proplastids
61. endosymbionts
62. matrix
63. Centrioles
64. axoneme
65. resolving power

CHAPTER 5 Cell Transport

Learning Objectives

After mastering the material covered in chapter 5, you should be able to confidently do the following tasks:

- Give the details of the fluid mosaic model for the structure of plasma membrane.
- Explain the known roles for membrane proteins, glycoproteins, and glycolipids.
- Show how gap junctions and plasmodesmata connect cells to one another.
- Demonstrate how diffusion can move molecules along a concentration gradient and explain how osmosis is involved in maintaining water balance in plant and animal cells.
- Describe the active transport mechanisms that can move molecules against a concentration gradient.
- Distinguish the mechanisms of endocytosis and exocytosis, and give examples of their activity.

Chapter Outline

I. The Plasma Membrane.

A. The fluid mosaic model.
   1. membranal proteins.
   2. glycoproteins and glycolipids.

B. Cell junctions.
   1. gap junctions.
   2. plasmodesmata.

II. Mechanisms of Transport.

A. Passive transport.
   1. diffusion.
   2. osmosis.
   3. facilitated diffusion and permeases.

B. Active transport.
   1. the work of membranal carriers.
      a. sodium/potassium ion exchange pump.
      b. other ion pumps.
   2. endocytosis and exocytosis.
      a. endocytosis.
      b. exocytosis.
      c. receptor-mediated endocytosis.

Key Words

plasma membrane
TEM
phospholipid
fluid mosaic model
choline
phosphatidylcholine
hydrophilic
hydrophobic
cholesterol
globular protein
linear protein
freeze fracture
transmembranal protein
peripheral protein
surface protein
glycocalyx
glycoprotein
glycolipid
gap junction
plasmodesma (-ata)
desmotubule
cytoplasmic streaming
permeability
selective permeability
passive transport
concentration gradient
diffusion
dynamic equilibrium
osmosis
semipermeable membrane
thistle tube
osmotic pressure
isotonic
hypotonic
hypertonic
plasmolysis
hydrostatic pressure
turgor
turgor pressure
wilting
water potential
adhesion
facilitated diffusion
carrier molecule
permease
active transport
cotransport carrier
sodium/potassium ion exchange pump
Na^+/Ca^+ pump
calcium ion pump
neurotransmitter
proton pump
electrochemical gradient
endocytosis
vacuole
phagocytosis
pinocytosis
exocytosis
receptor-mediated endocytosis
coated pit
coated vesicle
storage vesicle
low density lipoprotein (LDL)
LDL receptor
cholesterol
hypercholesterolemia
atherosclerosis

Exercises

1. Provide labels for this diagram showing the fluid-mosaic model of the plasma membrane. Use terms from the list below. Check your work with figure 5.1 in the text.

glycolipid
glycoprotein
hydrophilic "heads"
hydrophobic "tails"
inner surface
microfilament
microtubule
outer surface
peripheral protein
phospholipid bilayer
surface protein
transmembranal protein

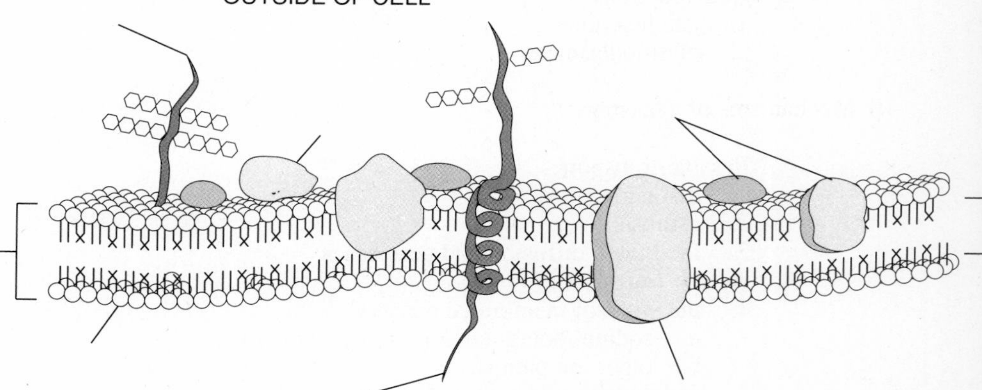

2. Osmotic relations are important for cells. For each of the cells below, show the movement of water into and out of the cells with arrows. Sketch a diagram to show any changes to the cells under the conditions indicated. Compare your work with figure 5.9 in the text.

a. This plant cell is placed in a 5% salt solution that is hypertonic to the cytoplasm.

Before: After:

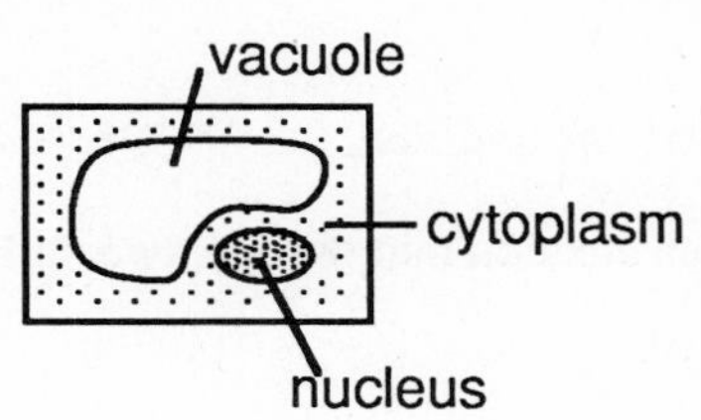

b. This plant cell is placed in distilled water. Is this hypertonic, hypotonic, or isotonic?

Before: After:

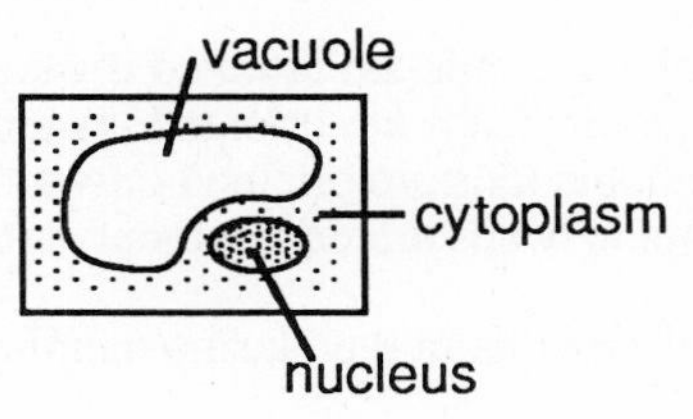

c. This animal cell is placed in an isotonic solution. What happens?

Before: After:

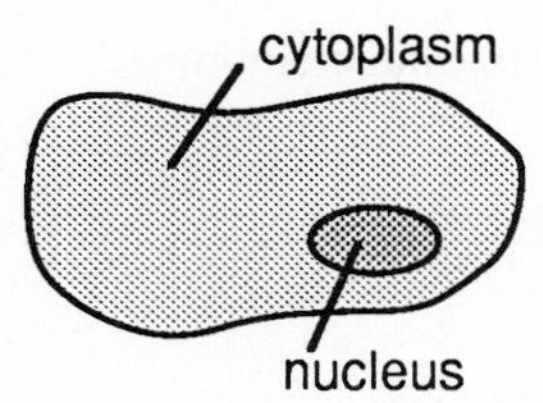

d. This animal cell is placed in a 7% salt solution. Is this hypotonic, hypertonic or isotonic?

Before: After:

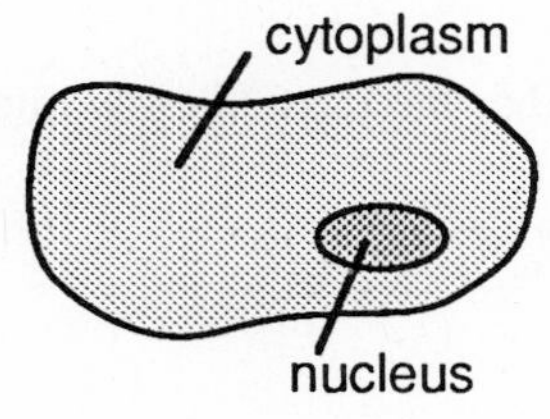

e. This animal cell is placed in distilled water. Is this hypotonic, hypertonic, or isotonic?

Before: After:

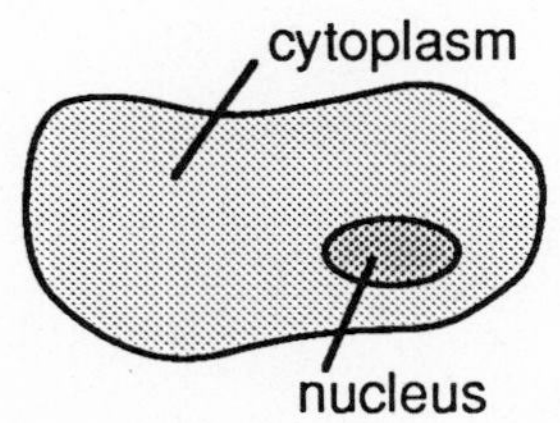

Self-Exam

You should be able to easily answer the following questions after learning the material in chapter 5. If you have difficulty with any question, study the appropriate section in the text and try again.

A. Multiple Choice Questions

Circle one alternative that best completes the statement or answers the question.

1. When plasma membrane is visualized edge-on by transmission electron microscopy, two dark lines are visible. These lines represent
   a. the phosphorus-rich nonpolar tails of the phospholipids.
   b. the phosphorus-rich polar heads of the phospholipids.
   c. the transmembrane proteins of the membrane.
   d. the sugar-rich glycoproteins of the plasma membrane.

2. In a phospholipid bilayer membrane,
   a. the hydrophilic heads are oriented outward and the hydrophobic tails are oriented inward.
   b. the hydrophilic tails are oriented outward and the hydrophobic heads are oriented inward.
   c. the hydrophilic heads are oriented inward and the hydrophobic tails are oriented outward.
   d. the hydrophilic tails are oriented inward and the hydrophobic heads are oriented outward.

3. Which of the following is *not* a widely recognized function of proteins in the plasma membrane?
   a. transport, functioning as a carrier molecule.
   b. enzymes, carrying out reactions in the membrane.
   c. support, forming bridges to the cytoskeleton.
   d. storage, providing energy reserves for the cell.

4. Which of the molecules below make up the glycocalyx region of the plasma membrane?
   a. glycogen and glucose.
   b. glycoproteins and glycolipids.
   c. cholesterol and phosphatidylcholine.
   d. glycocorolla and glycocorona.

5. Cell recognition is sometimes a function of which type of membrane molecule?
   a. phospholipid.
   b. cholesterol.
   c. phosphatidylcholine.
   d. glycoprotein.

6. Molecules can pass from cell to cell via gap junctions. The largest that can pass this way might be
   a. about 10 mm diameter.
   b. about 10 daltons.
   c. about 1000 mm diameter.
   d. about 1000 daltons.

7. Cytoplasmic connections between adjacent plant cells are termed
   a. microtubules.
   b. gap junctions.
   c. plasmodesmata.
   d. middle lamella.

8. We can characterize the plasma membrane that surrounds cells as
   a. completely permeable.
   b. completely impermeable.
   c. selectively permeable.
   d. completely waterproof.

9. Which of these transport mechanisms is an active transport mechanism?
   a. diffusion.
   b. osmosis.
   c. facilitated diffusion.
   d. exocytosis.

10. The energy that leads to diffusion of molecules is energy from
    a. the sun.
    b. the motion of the molecules themselves.
    c. active transport.
    d. ATP.

11. If we examine diffusion of molecules, we find that at dynamic equilibrium
    a. molecular movement ceases, and there is no net movement in any direction.
    b. there is no *net* movement of molecules in any direction, although molecular movement continues.
    c. there is a *net* movement of molecules, although molecular movement has ceased.
    d. there is a net movement of molecules in one direction, because molecular movement continues.

12. In osmosis, the net movement of water across a membrane is
    a. from a region of lower solute concentration into a region of higher solute concentration.
    b. from a region of higher solute concentration into a region of lower solute concentration.
    c. from a region of lower water concentration to a region of higher water concentration.
    d. by an active transport mechanism.

13. If two aqueous solutions are separated by a membrane. osmotic pressure is least when
    a. the two solutions are very different in concentration.
    b. the two solutions are only slightly different in concentration.
    c. the two solutions have the same concentration.
    d. one solution is really pure water.

14. If human red blood cells are placed in a hypotonic solution,
    a. the cells will gain water through osmosis until they burst.
    b. the cells will lose water through osmosis until they shrink.
    c. the cells will neither gain nor lose water.
    d. some cells will swell, while others shrink, until equilibrium is reached.

15. A host making salad discovers his salad greens slightly wilted; he can still have crisp salad for his guests by
    a. immersing the greens in chilled salt water.
    b. immersing the greens in chilled pure water.
    c. sprinkling monosodium glutamate on the wilted greens.
    d. adding sulfites as is the practice at some salad bars.

16. The process of facilitated diffusion is
    a. an active transport process that employs permeases.
    b. a passive transport mechanism that utilizes carrier molecules.
    c. synonymous with osmosis.
    d. a process that occurs only within cells.

17. The chemical energy that is used for active transport mechanisms is typically stored by cells in the form of
    a. ATP.
    b. membrane carrier proteins.
    c. DNA.
    d. membrane phosphatidylcholine.

18. One of the most common active transport mechanisms is the sodium/potassium ion exchange pump, and
    a. it transports sodium ions into the cell and transports potassium ions out of the cell.
    b. it transports sodium ions out of the cell and transports potassium ions into the cell.

c. it transports both sodium ions and potassium ions out of the cell.
d. it transports both sodium ions and potassium ions into the cell.

19. Which of the following does *not* involve an active transport mechanism?
    a. restoring sodium and potassium conditions in a neuron after an impulse.
    b. producing an electrochemical gradient to produce ATP.
    c. maintaining turgor pressure in plant cells.
    d. maintaining proper calcium ion concentration in muscle cells.

20. A bacterium can be engulfed by a cell and end up in a vacuole via
    a. exocytosis.
    b. phagocytosis.
    c. pinocytosis.
    d. osmosis.

21. Engulfed food in a food vacuole is often digested by
    a. enzymes contained in a lysosome.
    b. enzymes embedded in the plasma membrane.
    c. active transport mechanisms.
    d. exocytosis.

22. In receptor-mediated endocytosis, after many receptor molecules are bound to molecules of the appropriate external substance, internal inpouching leads to formation of
    a. a lysosome.
    b. a coated vesicle.
    c. a pitted vesicle.
    d. atherosclerosis.

B. True or False Questions

Mark the following statements either T (True) or F (False).

_____ 23. The phospholipids of the plasma membrane form a rigid framework in which proteins and other membrane components are embedded.

_____ 24. .Cholesterol is often widely distributed in plasma membrane of animal cells.

_____ 25. Transmembranal proteins extend completely through the plasma membrane.

_____ 26. Glycolipids have long fatty acid chains that project from the surface of the plasma membrane.

_____ 27. Cell surface proteins are important in the functioning of the immune system.

_____ 28. Gap junctions are normally blocked in cells that have very low concentrations of calcium ions.

_____ 29. Gap junctions and plasmodesmata can limit to some extent what passes between cells.

_____ 30. Both water and sugar can easily diffuse through a typical cell's plasma membrane.

_____ 31. Passive transport mechanisms are generally responses to concentration gradients of molecules.

_____ 32. In general, phospholipid bilayer membranes are more permeable to ions than to oxygen or water.

_____ 33. In the diffusion process, all molecular motion stops when equilibrium is reached.

_____ 34. Wilted celery can be made crisp by osmotic uptake of water.

_____ 35. A thistle tube containing distilled water, its large end covered by a semipermeable membrane, is placed in a beaker of 3% sugar solution; water will flow into the thistle tube by osmosis.

_____ 36. The greater the concentration difference between two solutions, the greater the osmotic pressure.

_____ 37. The physiological saline solutions used for intravenous (IV) injections in human medicine should be isotonic with human blood.

_____ 38. Watering plants with sea water does no harm, since the plants take up only water, not the solutes.

_____ 39. The internal water pressure is uniform in plants, from roots to stems and leaves.

_____ 40. Although membrane-embedded proteins called permeases are involved, facilitated diffusion is a passive transport process.

_____ 41. Carrier proteins are believed to function by undergoing conformational changes.

_____ 42. The sodium/potassium ion exchange pump involves cotransport of sodium ions and potassium ions in the same direction.

_____ 43. Calcium ion pumps are involved in correct orientation of root growth in plants.

C. Fill in the Blanks

Answer the question or complete the statement by filling in the blanks with the correct word or words.

44. The fluid mosaic model of membrane structure is based on a bilayer of _______________ molecules.

45. The glycocalyx of the plasma membrane is made up of _______________ and _______________.

46. Cell surface molecules can function in cell recognition or act as _______________ for hormones.

47. Elements of the endoplasmic reticulum passing between cells through plasmodesmata form slender channels called _______________.

48. In diffusion, atoms, molecules, or ions follow a _______________.

49. Oxygen and carbon dioxide enter and leave cells by _______________.

50. The diffusion of water molecules across a membrane is _______________.

51. A 3% sucrose solution is _______________-tonic with respect to pure water.

52. Nonwoody plants can remain erect because of the _______________ of their cells.

53. A membrane-embedded carrier molecule that facilitates diffusion may be called a _______________.

54. A common energy source for active transport mechanisms is the molecule _______________.

55. In one of the most common active transport mechanisms, sodium ions are cotransported with _______________, in opposite directions.

56. In plants, _______________ molecules are actively transported from leaf cells, where they are produced, into the plants transport system.

57. In chloroplasts and mitochondria, proton pumps establish ________________, which provide energy for the production of ATP.

58. Liquids can be taken into cells by the endocytotic process of ________________.

59. Golgi vesicles of secretory cells expel their contents via ________________.

60. When receptor-mediated endocytosis is occurring, a membrane surface indentation termed a ________________ is formed as receptor sites are filled.

Questions for Discussion

1. Thinking about osmosis, discuss the following situations. We water garden plants with pure water, and they take it into their cells; why do the cells not swell and rupture? When a person takes a bath, she is immersed in pure water; is there any osmotic risk to bathing? Can fish that live in the oceans also survive in fresh water? What are the difficulties they encounter? Do you think seawater could quench your thirst?

2. One of the common features of all cells is the plasma membrane. Apparently it is very important, since it is so universal. Outline the important functions of the plasma membrane.

3. What are some problems that cells would have if the only transport mechanisms were *passive* transport? List some of the processes that would fail without active transport mechanisms.

Answers to Self Exam

Multiple Choice Questions

1. b
2. a
3. d
4. b
5. d
6. d
7. c
8. c
9. d
10. b
11. b
12. a
13. c
14. a
15. b
16. b
17. a
18. b
19. c
20. b
21. a
22. b

True or False Questions

23. F
24. T
25. T
26. F
27. T
28. F
29. T
30. F
31. T
32. F
33. F
34. T
35. F
36. T
37. T
38. F
39. F
40. T
41. T
42. F
43. T

Fill in the Blank Questions

44. phospholipid
45. glycoproteins, glycolipids
46. receptors
47. desmotubules
48. concentration gradient
49. diffusion
50. osmosis
51. hyper-
52. turgor pressure
53. permease
54. ATP
55 potassium ions
56. sugar
57. electrochemical gradients
58. pinocytosis
59. exocytosis
60. coated pit

CHAPTER 6 Cell Energetics

Learning Objectives

After mastering the material covered in chapter 6, you should be able to confidently do the following tasks:

- List the various forms energy may take, and give the basic laws of thermodynamics that govern all chemical reactions, including those in living systems.
- Explain the involvement of free energy, energy of activation, and catalysts in determining the rates and directions of biological chemical reactions.
- Describe the ways that enzymes are thought to work, and list the factors that influence enzyme activity.
- Show the basic structure of the important energy-carrying molecule ATP and the nucleotide coenzymes NAD, NADP, and FAD.
- Outline the energetic basis for ATP production via chemiosmotic phosphorylation.

Chapter Outline

I. Energy.

A. Forms of energy.
B. Laws of thermodynamics.

II. Energy in Living Systems.

A. Free energy.
B. Exergonic and endergonic reactions.
C. Chemical reactions, free energy, and equilibrium.
D. Coupled reactions.
E. Energy of activation.
F. Catalysts.

III. Enzymes: Biological Catalysts.

A. Cofactors.
B. Enzyme-substrate complex.
C. Influences on the rates of enzyme action.
D. Mechanisms of enzyme control.
1. enzyme-activating mechanisms.
2. enzyme-inhibiting mechanisms.
3. allosteric control.

IV. ATP: The Energy Currency of the Cell.

A. Molecular structure of ATP.
B. Characteristics of ATP.
C. Cycling of ATP.

V. The Coenzymes: NAD, NADP, and FAD.

A. Redox reactions: Oxidation and reduction.
B. Coenzymes in action.

VI. Chemiosmotic Phosphorylation.

A. Charging the system.
B. The proton gradient.
C. Tapping the free energy of the gradient.

Key Words

energy
work
potential energy
kinetic energy
laws of thermodynamics
first law of thermodynamics
law of conservation of energy
isolated system
energy transformation
second law of thermodynamics
law of entropy
entropy
free energy
Gibbs free energy
ΔG
exergonic reaction
endergonic reaction
reactant
product
equilibrium
photosynthesis
respiration
coupled reactions
calorimeter
energy of activation
catalyst
enzyme
substrate
cofactor
vitamin
mineral
coenzyme
enzyme-substrate complex
active site
induced fit hypothesis
enzyme saturation
turnover time
metabolic pathway
enzyme activation
epinephrine
enzyme inhibition
competitive inhibition
noncompetitive inhibition
allosteric control
allosteric enzymes
modulators
stimulators
inhibitors
negative feedback inhibition
adenosine triphosphate
ATP
adenine base
ribose
phosphate group
phosphorylation
adenosine diphosphate
ADP
adenosine monophosphate
AMP
coenzymes
nicotinamide adenine dinucleotide
NAD
nicotinamide adenine dinucleotide phosphate
NADP
flavin adenine dinucleotide
FAD
nicotinic acid
riboflavin
oxidation
reduction
redox reactions
electron donor
reducing agent
electron acceptor
oxidizing agent
electron carriers
NAD^+
$NADP^+$
NADH
NADPH
$FADH_2$
reducing power
cytochrome
heme group
Fe^{2+}
Fe^{3+}
chemiosmotic phosphorylation
substrate-level phosphorylation
Peter Mitchell
proton concentration gradient
electron transport system
pH gradient
CF_0 - CF_1 complex
F_0 - F_1 complex
ATP synthetase

Exercises

1. Below are axes for plotting the free energy levels of reactants and products over the course of a reaction. Follow the instructions for parts 1a and 1b. Check your work with figure 6.4.

1a. Sketch a line showing the free energy levels as the reaction proceeds from left to right, when the reactants have G = a and the products have G = b. What is the change in free energy Δ G? Is it positive or negative? Show this in your sketch. Is this reaction endergonic or exergonic?

1a. Sketch a line showing the free energy levels as the reaction proceeds from left to right, when the products have G = a and the reactants have G = b. What is the change in free energy Δ G? Is it positive or negative? Show this in your sketch. Is this reaction exergonic or endergonic?

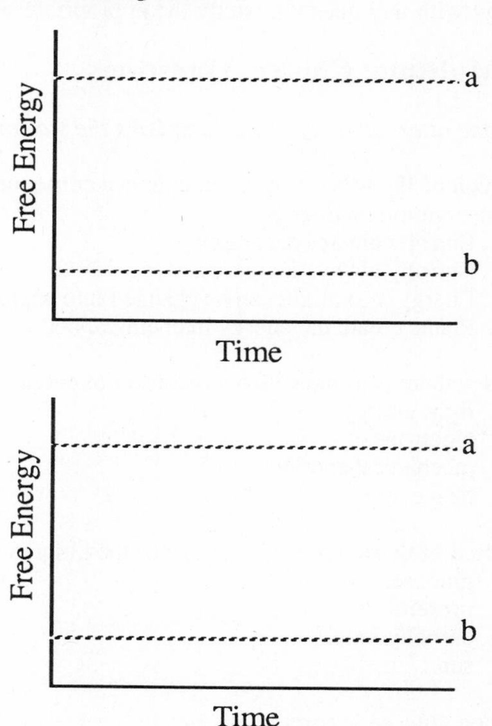

2. A number of factors influence the rate at which enzyme catalyzed reactions proceed. Among these factors are substrate concentration, temperature, and pH. For each of the situations below, sketch a curve suggesting the relative rates of reaction that might be observed. Check your work with figures 6.4 and 6.9.

2a. Let us imagine the same type of enzyme occurs both in human cells and in cells of bacteria that grow in the very hot springs in Yellowstone National Park. Draw and label curves comparing the enzyme activities versus temperature in the two cell types.

2b. Now, compare two different enzymes from the same organism — a human. Pepsin cleaves protein in the stomach, where HCl has been added, while trypsin cleaves protein in the small intestine after bicarbonate has been added. Draw and label curves comparing the activities of these enzymes versus pH.

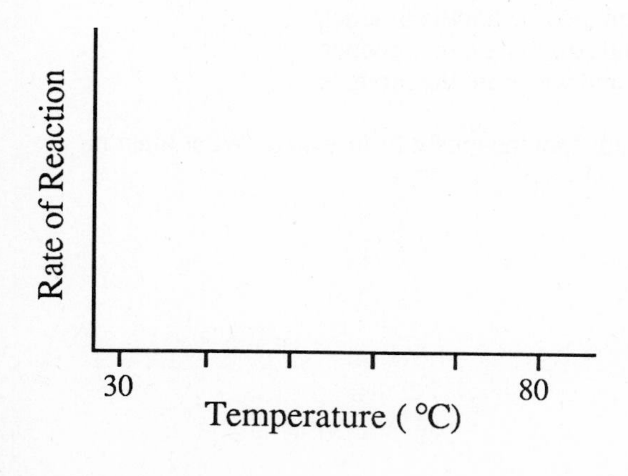

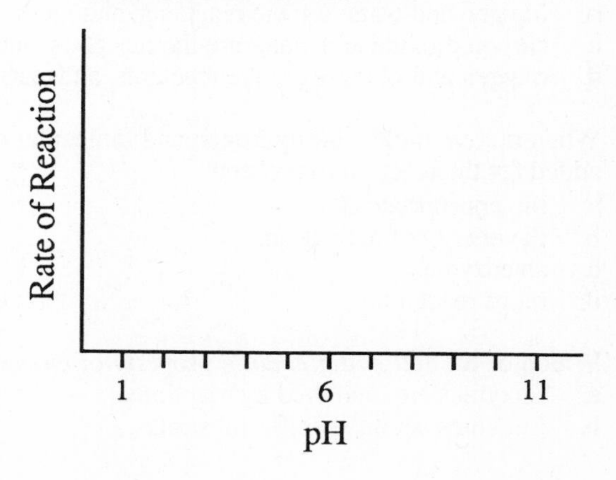

Self-Exam

You should be able to easily answer the following questions after learning the material in chapter 6. If you have difficulty with any question, study the appropriate section in the text and try again.

A. Multiple Choice Questions

Circle one alternative that best completes the statement or answers the question.

1. Which of the following statements is a consequence of the first law of thermodynamics, the "Law of Conservation of Energy"?
   a. Entropy always decreases.
   b. Entropy always increases.
   c. Energy in sunlight can be changed into energy in chemical bonds.
   d. Plants create the energy in chemical bonds, and animals destroy it.

2. When there is a change from one form of energy to another, some energy usually escapes as
   a. light energy.
   b. electrons.
   c. mechanical energy.
   d. heat energy.

3. Which of these items best describes the ultimate source of the chemical bond energy in ATP?
   a. glucose.
   b. protein.
   c. carbohydrate.
   d. sun.

4. When glucose is completely metabolized as in $C_6H_{12}O_6 + 6\ O_2 \rightarrow 6\ CO_2 + 6\ H_2O$, ΔG is
   a. -686 kcal per mole glucose.
   b. 686 kcal per mole glucose.
   c. -686 kcal per gram glucose.
   d. 686 kcal per gram glucose.

5. An exergonic reaction
   a. releases energy, as in the breakdown of glucose.
   b. requires energy input, as in the synthesis of glucose.
   c. releases energy, as in the synthesis of glucose.
   d. requires energy input, as in the breakdown of glucose.

6. In the overall chemical reaction of photosynthesis,
   a. oxygen and carbon dioxide are the reactants, and glucose and water are the products.
   b. oxygen and water are the reactants, and glucose and carbon dioxide are the products.
   c. carbon dioxide and water are the reactants, and glucose and oxygen are the products.
   d. oxygen and glucose are the reactants, and carbon dioxide and water are the products.

7. When mixed, molecular hydrogen and molecular oxygen do not spontaneously form water. What must be added for the reaction to proceed?
   a. the appropriate ΔG.
   b. the energy of activation.
   c. an enzyme.
   d. more reactants.

8. Which of the following is *not* a property of enzymes?
   a. Enzymes are unaltered by reactions.
   b. Enzymes act on specific substrates.

c. Enzymes lower the free energy change, ΔG, of a reaction.
d. Enzymes do not change the equilibrium point for a given reaction.

9. If there is plenty of enzyme for a given reaction, as the concentration of substrate increases from zero,
   a. the overall rate of reaction will increase until all enzyme is saturated.
   b. the overall rate of reaction will be constant, but at a faster rate than without enzyme.
   c. the overall rate of reaction will decrease as enzyme is used up.
   d. the overall rate of reaction will vary, but in which direction cannot be predicted.

10. As a generalization, the rate of an enzyme catalyzed reaction will double with
   a. an increase of one pH point.
   b. a doubling of the substrate concentration at the enzyme saturation point.
   c. an increase in temperature of 10° C.
   d. the addition of cofactors.

11. Which of the following agents leads to enzyme activation rather than to enzyme inhibition?
   a. epinephrine.
   b. DDT.
   c. cyanide.
   d. penicillin.

12. In negative feedback inhibition
   a. substrate molecules activate a metabolic pathway.
   b. substrate molecules block allosteric sites on enzyme molecules.
   c. reaction products activate a metabolic pathway.
   d. reaction products block allosteric sites on enzyme molecules.

13. Enzymes
   a. are generally proteins.
   b. often are nonspecific as to substrate.
   c. are slowly used up in the catalytic process.
   d. raise the activation energy for given reactions.

14. When the terminal phosphate bond on ATP is hydrolyzed, ΔG is
   a. -686 kcal/mole.
   b. 686 kcal/mole.
   c. -7.3 kcal/mole.
   d. 7.3 kcal/mole.

15. One of the most important functions of the hydrolysis of ATP is
   a. to provide heat to cells.
   b. to provide energy to coupled endergonic reactions in cells.
   c. to provide energy to coupled exergonic reactions in cells.
   d. to provide the basic energy input to cells.

16. NAD, NADP, and FAD are best described as
   a. enzymes.
   b. coenzymes.
   c. proteins.
   d. catalysts.

17. An oxidizing agent
   a. is an electron acceptor, and the substance oxidized becomes more electropositive.
   b. is an electron acceptor, and the substance oxidized becomes more electronegative.
   c. is an electron donor, and the substance oxidized becomes more electropositive.
   d. is an electron donor, and the substance oxidized becomes more electronegative.

18. During cell respiration, glucose is
    a. completely reduced in chemical reactions of a lengthy metabolic pathway.
    b. completely oxidized in chemical reactions of a lengthy metabolic pathway.
    c. completely reduced in a simple chemical reaction.
    d. completely oxidized in a simple chemical reaction.

19. In being electron carriers, NAD and FAD are similar to
    a. catalysts.
    b. cytochromes.
    c. ATP.
    d. glucose.

20. The important function of membrane electron transport systems is to
    a. reduce the cytochromes.
    b. oxidize the cytochromes.
    c. use free energy of electrons to pump protons across the membrane.
    d. hydrolyze ATP molecules.

B. True or False Questions

Mark the following statements either T (True) or F (False).

_____ 21. Chemical bond energy is interchangeable with the radiant energy of light.

_____ 22. A large stone falling through the air has only potential energy until it reaches the ground.

_____ 23. All biological phenomena are consistent with the laws of thermodynamics.

_____ 24. Because of the first law of thermodynamics, almost all the energy in glucose can be converted to chemical bond energy in ATP molecules.

_____ 25. Living things utilize about 40% of the light energy reaching earth from the sun.

_____ 26. Photosynthesis is, overall, an endergonic reaction, with $\Delta G = +686$ kcal/mole.

_____ 27. In weakly exergonic reactions at equilibrium, the concentrations of product and reactants remain equal.

_____ 28. The overall process of glucose metabolism during respiration is an endergonic process.

_____ 29. Many exergonic reactions will not proceed until the energy of activation is provided.

_____ 30. Enzymes act as catalysts by accelerating chemical reactions.

_____ 31. An atom such as zinc or copper that is needed for enzyme function is a coenzyme.

_____ 32. After enzyme saturation is reached, the rate of a reaction can be increased only by increasing the amount of substrate available for the reaction.

_____ 33. Increasing temperature can have two effects on enzyme-catalyzed reactions — the rate of reaction can increase, or the enzyme can be denatured.

_____ 34. Penicillin acts as a competitive inhibitor of enzymes important in bacterial cell wall synthesis.

_____ 35. Allosteric enzymes are proteins containing a single polypeptide chain.

_____ 36. The specificity between enzyme and substrate is often determined by the enzyme's active site.

_____ 37. ATP is structurally related to the polypeptides.

_____ 38. Hydrolysis of a phosphate from ATP is often coupled to phosphorylation of other molecules.

_____ 39. Important parts of NAD and FAD are provided by B vitamins.

_____ 40. A reducing agent serves as an electron acceptor in redox reactions.

C. Fill in the Blanks

Answer the question or complete the statement by filling in the blanks with the correct word or words.

41. The energy of matter in motion is _______________.

42. According to the second law of thermodynamics, _______________ always increases.

43. The chemical bond energy in glucose originates as _______________ energy.

44. ΔG refers to the change in free energy between _______________ and _______________.

45. A chemical reaction in which the products have less free energy than the reactants is an _______________ reaction.

46. Once the energy of activation is provided, the reaction $2H_2 + O_2 \leftrightarrow 2H_2O$ will proceed from _______________ to _______________.

47. The substance acted upon by an enzyme is its _______________.

48. Many vitamins are _______________ or their precursors.

49. An enzyme typically has an optimal _______________ and an optimal _______________.

50. A long sequence of related reactions is termed a _______________.

51. A non-substrate molecule binds and blocks the active site of an enzyme in _______________.

52. When the final product of a metabolic pathway slows or stops the action of the pathway, the regulation is called _______________.

53. An ATP molecule is composed of an _______________ base, a _______________ sugar, and three phosphate groups.

54. ATP becomes ADP + P_i when the terminal phosphate bond is broken by _______________.

55. NAD, NADP, and FAD are important energy carrying _______________.

56. When a substance loses electrons, it is _______________.

57. During respiration, glucose is completely _______________ in a lengthy metabolic pathway.

58. The reduced form of nicotinamide adenine dinucleotide is written _______________.

59. The heme group of a cytochrome can be in either a reduced or oxidized state; Fe^{2+} is the _______________ state and Fe^{3+} is the _______________ state.

60. In a chloroplast, the electrical potential across an inner membrane is due to a _______________ concentration gradient established by chemiosmosis.

Questions for Discussion

1. Organisms and all biological processes are governed by the basic laws of thermodynamics. The second law states that entropy always increases - that is, systems spontaneously tend to become more disordered. Yet a complex organism like an insect or an orchid or a human is very organized and complexly structured. Can you explain how the existence of such very orderly systems is consistent with the first and second laws?

2. You have probably tired of the old dilemma of "which came first — the chicken or the egg?" Consider a similar dilemma. Many chemical reactions of life are dependent on the activities of specific enzymes, yet the enzymes are neither reactants nor products in the overall process. Which do you think came first — the biological reaction, or the enzyme? Support your reasoning, drawing on your knowledge of cell energetics.

3. Enzymes are in a sense very delicate molecules. Their shapes and structures are very specific for the roles they play. Many conditions affect how well they do their jobs. Outline the factors that may influence the rate of enzyme-assisted reactions. Then describe the ways that enzyme activity can be controlled.

4. The presence of metabolically important coenzymes depends on an adequate supply of certain B vitamins in diet. What might be the consequences of a deficiency or absence of these vitamins. What are the sources of these vitamins in your diet?

Answers to Self-Exam

Multiple Choice Questions

1. c
2. d
3. d
4. a
5. a
6. c
7. b
8. c
9. a
10. c
11. a
12. d
13. a
14. d
15. b
16. b
17. a
18. b
19. b
20. c

True or False Questions

21. T
22. F
23. T
24. F
25. F
26. T
27. T
28. F
29. T
30. T
31. F
32. F
33. T
34. T
35. F
36. T
37. F
38. T
39. T
40. F

Fill in the Blank Questions

41. kinetic energy
42. entropy
43. sunlight
44. reactants, product
45. exergonic
46. left, right
47. substrate
48. coenzymes
49. pH, temperature
50. metabolic pathway
51. competitive inhibition
52 negative feedback inhibition
53. adenine, ribose
54. hydrolysis
55. coenzymes
56. oxidized
57. oxidized
58. NADH or NADH + H^+
59. reduced, oxidized
60. proton

CHAPTER 7 Photosynthesis

Learning Objectives

After mastering the material covered in chapter 7, you should be able to confidently do the following tasks:

- Present a comprehensive overview of the entire process of photosynthesis by describing events of the two important parts of the process, the light reactions and the light-independent reactions.
- Show how chloroplast structure and the light reactions are involved in producing ATP and NADPH.
- Outline the sequence of events in the light-independent reactions of the Calvin cycle.
- Describe alternative pathways found in photorespiration, C4 carbon fixation, and crassulacean acid metabolism.

Chapter Outline

I. Overview of Photosynthesis.

II. Chloroplasts.

A. Thylakoids.
1. the two photosystems.
a. chlorophyll and light-harvesting antennas.
b. reaction centers.
2. electron transport systems.
3. CF_0-CF_1 complexes.

III. Light Reactions of Photosynthesis.

A. Photosystems I, II, and the noncyclic events.
1. water and the reduction of P680.
2. electron energy and proton transport.
3. photosystem I: reduction of NADP.
B. Photosystem I and cyclic events.
C. Chemiosmotic phosphorylation.

IV. Light-independent Reactions of Photosynthesis.

A. Discovery of the Calvin cycle.
B. The Calvin cycle.
1. phosphorylation of ribulose phosphate.
2. carbon fixation.
3. synthesis of glyceraldehyde-3-phosphate.

V. Alternative Pathways.

A. Photorespiration.
B. The C4 pathway.
C. Crassulacean acid metabolism (CAM).

Key Words

photosynthesis
photons
electron transport system
chlorophyll
light reactions
light-independent reactions
chloroplasts
stroma
thylakoid membrane system
thylakoid
granum (-a)
granum thylakoid
stroma thylakoid
lumen
photosystem I (PS I)
photosystem II (PS II)
light-harvesting antenna
reaction center
pigments
chlorophyll *a*
chlorophyll *b*
carotenoid
absorption spectrum
action spectrum
light wavelength
visible light
accessory pigment
P700
P680
excited state
ground state
bioluminescence
CF_0 - CF_1 complex
ATP synthetase
noncyclic photophosphorylation
cyclic photophosphorylation
plastoquinone (PQ)
PQH
cytochrome
plastocyanin (PC)
Z-scheme
chemiosmotic gradient
pH gradient
Melvin Calvin
Calvin cycle
Chlorella
carbon 14 (^{14}C
carboxylation
ribulose -5-phosphate (RuP)
ribulose-1,5-bisphosphate (RuBP)
ribulose-1,5-bisphosphate carboxylase
rubisco
3-phosphoglyceric acid (3-PG)
phosphoglycerate
diphosphoglyceric acid
1,3-diphosphoglycerate (DPG)
glyceraldehyde-3-phosphate (G3P)
fructose-1,6-diphosphate
glucose-1-phosphate
thermal efficiency
photorespiration
glycolate
microbodies
C4 pathway
C3 plants
C4 plants
phosphoenolpyruvate (PEP)
phosphoenolpyruvate carboxylase (PEP carboxylase)
leaf mesophyll cells
oxaloacetate
malate
bundle sheath cells
pyruvate
crassulacean acid metabolism (CAM)
Crassulaceae
stoma (-ata)

Exercises

1. A summary equation for photosynthesis may be as simple as $6CO_2 + 6H_2O \rightarrow 6O_2 + C_6H_{12}O_6$. This summary obscures what actually happens in the light reactions and the light-independent reactions and overlooks many of the important intermediates in the process of turning carbon dioxide into organic molecules. Complete the "black box" diagram below to show the various inputs and outputs. Use terms from the following list; don't worry about balancing the reactions numerically.

ADP	light
ATP	$NADP^+$
$C_6H_{12}O_6$	NADPH + H^+
CO_2	O_2
H_2O	P_i

Light Reactions

Calvin Cycle

2. In the space below, sketch a careful diagram of the "Z-scheme" to illustrate electron flow during the noncyclic events of the photosynthetic light reactions. The photosystems are shown to get you started. Indicate the relative free energy levels of the various intermediates. Check your work with figures 7.8 and 7.9 in the text. You should include at least the following elements:

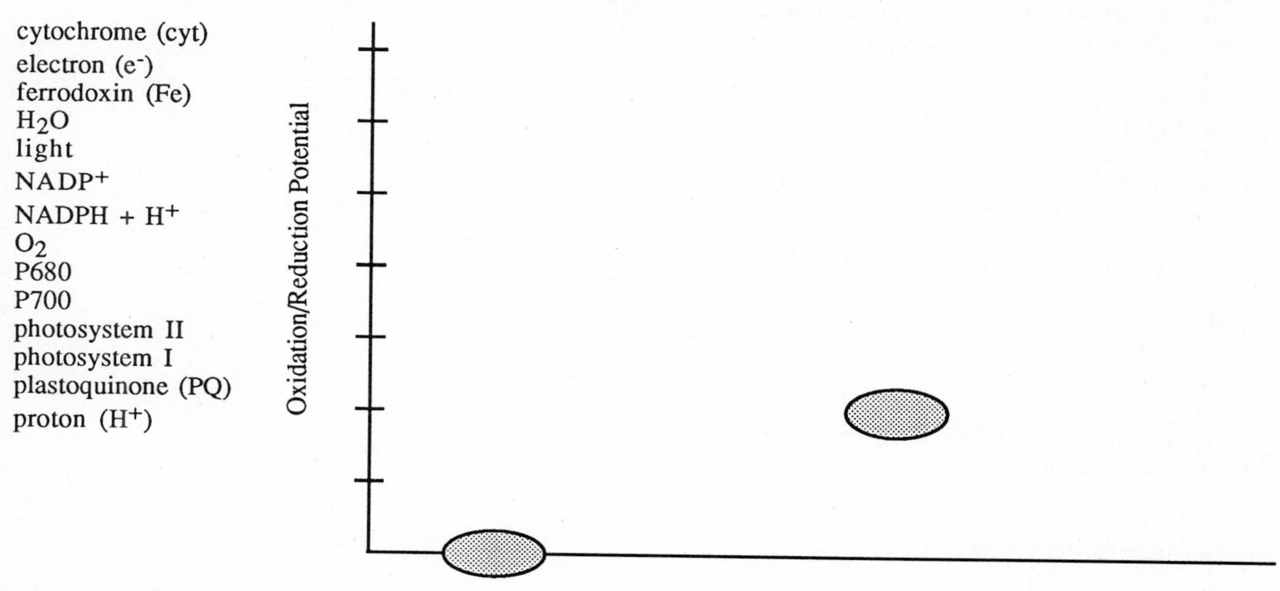

Self-Exam

You should be able to easily answer the following questions after learning the material in chapter 7. If you have difficulty with any question, study the appropriate section in the text and try again.

A. Multiple Choice Questions

Circle one alternative that best completes the statement or answers the question.

1. Which of the below is a general equation for photosynthesis?
   a. $6\ CO_2 + 12\ H_2O \rightarrow C_6H_{12}O_6 + 6\ O_2 + 6H_2O$.
   b. $C_6H_{12}O_6 + 6\ O_2 + 6H_2O \rightarrow 6\ CO_2 + 12\ H_2O$.
   c. $6\ CO_2 + 6\ O_2 + 12\ H_2O \rightarrow C_6H_{12}O_6 + 6H_2O$.
   d. $6\ CO_2 + C_6H_{12}O_6 + 12\ H_2O \rightarrow 6\ O_2 + 6H_2O$.

2. The oxygen produced during photosynthesis originally comes from which molecule?
   a. atmospheric carbon dioxide.
   b. atmospheric oxygen.
   c. water.
   d. glucose.

3. After the light reactions of photosynthesis, the energy of sunlight has been converted to
   a. electromagnetic energy of photons.
   b. chemical bond energy in carbohydrates.
   c. energy in ATP and NADPH.
   d. free energy in carbon dioxide.

4. After the light-independent reactions of photosynthesis, the energy captured from sunlight is found as
   a. electromagnetic energy of photons.
   b. chemical bond energy in carbohydrates.
   c. energy in ATP and NADPH
   d. free energy in carbon dioxide.

5. Which of these locations best describes the site of the light reactions of photosynthesis?
   a. the stroma of the chloroplasts.
   b. the thylakoid membranes of the chloroplast.
   c. cytoplasmic membranes adjacent to the chloroplast.
   d. the lumen of the thylakoids.

6. A proton gradient is important to photosynthesis; the proton concentration is higher
   a. in the cytoplasm.
   b. in the stroma.
   c. in the lumen of the thylakoid.
   d. in the matrix.

7. The reaction center of photosystem I absorbs light at
   a. 680 µm.
   b. 700 µm.
   c. 680 nm.
   d. 700 nm.

8. Chlorophylls α and *b* are most absorbent of light of which colors?
   a. reds and greens.
   b. reds and blues.
   c. blues and yellows.
   d. greens and yellows.

9. During active photosynthesis, a chlorophyll electron is "excited." What happens to this electron?
   a. It gives off light as bioluminescence.
   b. It returns to "ground state."
   c. It is captured by an electron transport system.
   d. It is captured by carbon dioxide.

10. The electron transport systems of photosynthesis need a supply of electrons. Where do they come from?
    a. water.
    b. oxygen.
    c. carbon dioxide.
    d. the sun.

11. Which carrier molecule plays a role as "proton shuttle," moving hydrogen ions to the thylakoid lumen?
    a. plastoquinone.
    b. plastocyanin.
    c. chlorophyll *a*.
    d. ferredoxin.

12. The final step in the noncyclic events of the photosystem I transport system is
    a. the oxidation of $NADP^+$ to NADPH.
    b. the reduction of $NADP^+$ to NADPH.
    c. the reduction of the P680 reaction center.
    d. the reduction of the P700 reaction center.

13. Which statement best describes the net result of the *cyclic* actions of photosystem I?
    a. Energy of P700 electrons is used to reduce NADP.
    b. Energy of P680 electrons is used to reduce P700 electrons.
    c. Energy of P700 electron is used to pump protons to the thylakoid lumen, increasing the proton gradient.
    d. Energy of NADPH is used to pump protons to the thylakoid lumen, increasing the proton gradient.

14. Which statement best describes the pH gradient is established in the chloroplast during active photosynthesis?
    a. The thylakoid lumen and stroma are both acid, the stroma being much more acidic.
    b. The thylakoid lumen and stroma are both alkaline, the stroma being much less alkaline.

c. The thylakoid lumen is acidic, and the stroma is alkaline.
d. The thylakoid lumen is alkaline, and the stroma is acidic.

15. The ATP-producing enzyme ATP synthetase is localized in chloroplasts
a. in the CF_0 complex.
b. in the CF_1 complex.
c. in the P680 reaction center.
d. in the P700 reaction center.

16. What happens during the Calvin cycle?
a. Photon energy is captured as energy in ATP and NADPH.
b. Six carbon dioxide molecules are joined to form glucose.
c. Light energy is converted to chemical energy.
d. Carbon dioxide and energy in ATP and NADPH are incorporated into organic molecules.

17. In his studies with *Chlorella*, what did Melvin Calvin use to trace the chemical pathways of carbohydrate synthesis in the light-independent reactions?
a. radioactively labeled chlorophyll *a*.
b. radioactively labeled water.
c. radioactively labeled carbon dioxide.
d. radioactively labeled molecular oxygen.

18. In the Calvin cycle, carbon dioxide is added to the 5-carbon molecule
a. ribulose-5-phosphate (RuP).
b. ribulose-1,5-bisphosphate (RuBP).
c. 3-phosphoglyceric acid ((3-PG).
d. 1,3 diphosphoglycerate (1,3-DPG).

19. The molecules resulting at the completion of the synthesizing portion of the Calvin cycle, after incorporation of carbon dioxide and energy from ATP and NADPH, are
a. ribulose-1,5-bisphosphates (RuBPs).
b. 3-phosphoglyceric acids (3-PGs).
c. 1,3 diphosphoglycerates (1,3-DPGs).
d. glyceraldehyde-3-phosphates (G-3-Ps).

20. Which of the following sequences gives the correct pathway to glucose from the Calvin cycle?
a. two glyceraldehyde-3-phosphates ⇒ fructose-1,6-diphosphate ⇒ glucose-1-phosphate ⇒ glucose.
b. two glyceraldehyde-3-phosphates ⇒ glucose-1-phosphate ⇒ fructose-1,6-diphosphate ⇒ glucose.
c. two glyceraldehyde-3-phosphates ⇒ glucose ⇒ glucose-1-phosphate ⇒ fructose-1,6-diphosphate.
d. two fructose-1,6-diphosphates ⇒ glyceraldehyde-3-phosphate ⇒ glucose-1-phosphate ⇒ glucose.

21. How many "turns" of the Calvin cycle are necessary to incorporate carbon for the equivalent of one glucose?
a. one.
b. two.
c. six.
d. twelve.

22. Which statement best describes photorespiration?
a. Light energy is used to extract energy from glucose.
b. The photosynthetic pathway operates in the reverse direction.
c. RuBP carboxylase outcompetes PEP carboxylase.
d. Oxygen successfully competes with carbon dioxide for RuBP carboxylase.

23. Which of the following plants would be most likely to use PEP carboxylase to capture carbon?
a. sugar cane in Hawaii.
b. Douglas fir in Washington.
c. a fern in Massachusetts.
d. a spruce in Siberia.

24. In C4 plants, the main site of PEP carboxylase activity is
    a. the leaf mesophyll cells.
    b. the bundle sheath cells.
    c. the cells of the vascular bundles.
    d. the guard cells of the stomata.

25. Compared with C3 plants, C4 plants can carry out photosynthesis
    a. at lower carbon dioxide concentrations, and by using less ATP.
    b. at lower carbon dioxide concentrations, but by using more ATP.
    c. at higher carbon dioxide concentrations, but by using less ATP.
    d. at higher carbon dioxide concentrations, and by using more ATP.

B. True or False Questions

Mark the following statements either T (True) or F (False).

_____ 26. The light-independent reactions of photosynthesis occur only in the dark.

_____ 27. The light reactions of photosynthesis result in the reduction of NADP and the formation of ATP.

_____ 28. Both the light reactions and the light-independent reactions take place in the chloroplast.

_____ 29. Chloroplasts have an inner thylakoid membrane system characterized by many small pores.

_____ 30. Accessory pigments such as carotenoids can absorb some green wavelengths of light.

_____ 31. Photosystem II is often referred to as P680.

_____ 32. The energy that transforms chlorophyll *a* to its excited state is provided by a photon.

_____ 33. The reduction of $NADP^+$ occurs in photosystem II.

_____ 34. The CF_0 complex of the chloroplast is found in its thylakoid membrane.

_____ 35. The oxygen atoms released during photosynthesis come from carbon dioxide.

_____ 36. The protons from the splitting of water are released into the thylakoid lumen.

_____ 37. During photosynthesis, electrons move from P700 to P680.

_____ 38. Because of the transport of protons, the thylakoid lumen is acidic compared to the stroma.

_____ 39. ATP is formed as protons flow through CF_0-CF_1 complexes from the stroma into the thylakoid.

_____ 40. In the Calvin cycle, carbon dioxide molecules are joined to one another.

_____ 41. The carbon-accepting molecule in the Calvin cycle is RuBP.

_____ 42. Glyceraldehyde-3-phosphate (G-3-P) molecules can be assembled into starch.

_____ 43. Each turn of the Calvin cycle produces a glucose molecule, or the equivalent.

_____ 44. Often, less than 1% of the light energy absorbed by a plant is converted into chemical bond energy.

_____ 45. After photorespiration, carbon dioxide is released.

_____ 46. All plants with the C4 pathway of photosynthesis also have the C3 pathway.

_____ 47. C4 carbon fixation results in the 4-carbon molecule oxaloacetate.

_____ 48. C3 plants can photosynthesize at lower CO_2 concentrations than can C4 plants.

_____ 49. Many members of the cactus family use the C4 pathway.

_____ 50. CAM plants tend to open their stomata in the daytime and close them at night.

C. Fill in the Blanks

Answer the question or complete the statement by filling in the blanks with the correct word or words.

51. The chemical reactants of the photosynthesis reaction are ________________.

52. Products of the light reactions are reduced NADP and ________________.

53. Stacks of thylakoid membranes are termed ________________.

54. The proton reservoir of the chloroplast is the ________________.

55. Chlorophylls and carotenoid pigments make up the ________________ of each photosystem.

56. Chlorophyll molecules reflect rather than absorb ________________ colored wavelengths.

57. The reaction center of photosystem II absorbs most strongly light with ________________nm wavelength.

58. After it has lost an electron, chlorophyll *a* regains an electron from ________________, via a manganese-containing protein.

59. The mobile carrier in PS II that shuttles protons to the thylakoid lumen is ________________.

60. On a bright, sunny day there are bubbles among the algal mats in a pond; the bubbles contain ________________.

61. Unlike the cyclic events in PS I, the noncyclic events in PS I produce ________________.

62. As negative chloride ions join protons in the thylakoid lumen, ________________ is formed.

63. The proton gradient in chloroplasts leads to ATP synthesis via ________________.

64. The free energy level of carbon dioxide is ________________ than that of glucose.

65. In his studies of chemical pathways of photosynthesis, Calvin labeled molecules with ________________.

66. The key enzyme in carbon fixation, probably the earth's most common protein, is ________________.

67. The energy that powers the Calvin cycle is in (RuBP) molecules.

68. At low CO_2 concentrations, RuBP carboxylase can join ________________ to RuBP.

69. The C4 pathway is one way plants avoid ________________ at low carbon dioxide concentrations.

70. The origin of all biomass on earth can be traced to ________________.

Questions for Discussion

1. We can make the assertion that all the biomass on earth today is derived from carbon assimilated in photosynthesis by plants and algae. Does this mean that photosynthesis is an ancient process? Do you think the very first organisms were photosynthetic? If not, where did their biomass come from? Can you think of evidence that sheds light on the antiquity of photosynthesis?

2. Some scientists think that global nuclear war might lead to conditions that would significantly reduce the amount of sunlight reaching the earth's surface. How might this affect global photosynthesis? Might biomass production be altered? Might temperature be altered, and, if so, would this affect photosynthesis? How might all of this affect us humans?

3. Tropical rain forests, sites of much photosynthesis, are being destroyed at an alarming rate, and this is cause for great concern. Is it plausible that cutting and burning tropical forests could alter the gas content of the earth's atmosphere? What changes might occur?

4. Home gardeners in the midwest U.S. know that bluegrass thrives in the cool weather of spring and early autumn, but in the hottest part of the summer, crabgrass will take over a lawn. Which is a C3 grass and which is a C4 grass? Support your conclusions.

Answers to Self-Exam

Multiple Choice Questions

1. a
2. c
3. c
4. b
5. b
6. c
7. d
8. b
9. c
10. a
11. a
12. b
13. c
14. c
15. b
16. d
17. c
18. b
19. d
20. a
21. c
22. d
23. a
24. a
25. b

True or False Questions

26. F
27. T
28. T
29. F
30. T
31. T
32. T
33. F
34. T
35. F
36. T
37. F
38. T
39. F
40. F
41. T
42. T
43. F
44. T
45. T
46. T
47. T
48. F
49. T
50. F

Fill in the Blank Questions

51. carbon dioxide and water
52. ATP
53. grana
54. thylakoid lumen
55. light-harvesting antenna
56. green
57. 680 nm
58. water
59. plastoquinone (PQ)
60. oxygen
61. NADPH
62 hydrochloric acid (HCl)
63. chemiosmotic phosphorylation
64. lower
65. radioactive carbon-14
66. ribulose-1,5-bisphosphate (RuBP) carboxylase
67. ATP and NADPH
68. oxygen
69. photorespiration
70. plants and algae

CHAPTER 8 Respiration

Learning Objectives

After mastering the material covered in chapter 8, you should be able to confidently do the following tasks:

- Explain the relationships among the three major parts of respiration — glycolysis, the citric acid cycle, and electron transport and chemiosmotic phosphorylation.

- Outline the pathway of glycolysis, including reactants, products, and energy yield.

- Show how the structure of the mitochondrion is important in aerobic respiration.

- Outline the citric acid cycle, including reactants, products, and energy yield.

- Explain how the electron transport system in the mitochondrial membrane establishes a chemiosmotic gradient, and show how this gradient is used in ATP synthesis.

- Describe anaerobic fermentation pathways that lead to alcohol and lactate.

- Show how fats and proteins can enter metabolic pathways.

Chapter Outline

I. Glucose Utilization.

A. Glycolysis.
B. Aerobic vs. anaerobic respiration.
C. Fermentation.

II. Glycolysis.

A. Reactions of glycolysis.
B. Energy yield of glycolysis.
C. Control of glycolysis.

III. Respiration in the Mitochondrion: Citric Acid Cycle.

A. Pyruvate to acetyl-CoA.
B. Reactions of the citric acid cycle.

IV. Electron Transport and Chemiosmotic Phosphorylation.

A. Mitochondrial structure.

B. Electron transport systems.
C. Chemiosmotic gradient and ATP synthesis.
D. Other uses for the proton gradient.

V. Fermentation Pathways.

A. Alcohol fermentation.
B. Lactate fermentation.

VI. Alternative Fuels for the Cell.

A. Fat metabolism.
B. Protein metabolism.
C. Biosynthetic pathways.

Key Words

metabolism
catabolism
endergonic
exergonic
aerobic respiration
anaerobic respiration
fermentation
glycolysis
phosphorylate
substrate level phosphorylation
chemiosmotic phosphorylation
sulfate ion
nitrate ion
nitrite
nitrous oxide
hydrogen sulfide
methane
glycolytic pathway
isopropyl alcohol
butyl alcohol
ethyl alcohol
acetic acid
lactic acid
propionic acid
formic acid
yeast
citric acid cycle
glucose
hexokinase
glucose-6-phosphate
phosphoglucoisomerase
fructose-6-phosphate
phosphofructokinase
fructose-1,6-diphosphate
aldolase
dihydroxyacetone phosphate
glyceraldehyde-3-phosphate
G-3-P dehydrogenase
3-phosphoglyceroyl phosphate
phosphoglyceroyl kinase
3-phosphoglycerate
phosphoglyceromutase
2-phosphoglycerate
enolase
phosphoenolpyruvate
pyruvic kinase
pyruvate
hydrolyze
kcal/mole
allosteric enzyme
activator
inhibitor
acetyl-coenzyme A
acetyl-CoA
Hans Krebs
Krebs cycle
citrate
aconitase
cis-aconitate
isocitrate
isocitrate dehydrogenase
α-ketoglutarate
α-ketoglutarate dehydrogenase
succinyl CoA
succinyl CoA synthetase
GDP
GTP
succinate
succinate dehydrogenase
fumarate
fumarase
malate
malate dehydrogenase
oxaloacetate
citrate synthetase
decarboxylation
NAD^+
FAD
NADH
$FADH_2$
mitochondrion
matrix
cristae
cytochrome
heme group
flavin mononucleotide (FMN)
iron-sulfur protein (FeS)
coenzyme Q (CoQ)
cytochrome b
cytochrome c_1
cytochrome c
cytochrome a
cytochrome c_3
NADH-CoQ reductase
$CoQH_2$-cytochrome c reductase
cytochrome c oxidase
F_0F_1 complex
electrochemical gradient
proton-motive force
adenine nucleotide translocase
brown fat
alcoholic fermentation
decarboxylase
acetaldehyde
alcohol dehydrogenase
lactate fermentation
Lactobacillus
Streptococcus
intermediary metabolism
palmitic acid
deamination
biosynthesis

Exercises

1. Below is a diagram outlining the relationships between important parts of aerobic respiration. Complete the diagram, using terms from the following list: ATP, CO_2, CoA, FAD, $FADH_2$, glucose, NAD^+, NADH + H^+, oxygen, and water. Check your work with figures 8.2, 8.7, and 8.10 in the text.

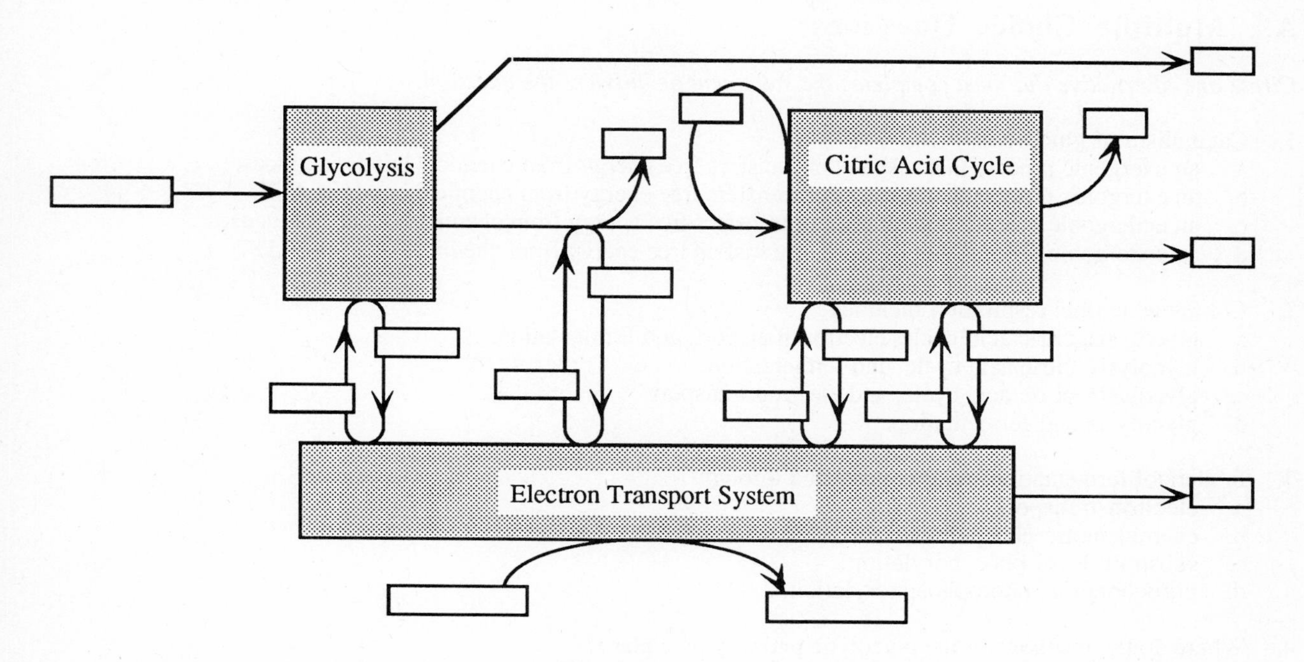

2. The caloric content of glucose is 686 kcal/mole. The reactions of respiration convert some of this to bond energy in ATP. Fill in the table below to outline the energy yields in respiratory processes.

Glycolysis:

Substrate-level phosphorylation			_______ ATPs	_______ kcal/mole
Chemiosmotic phosphorylation	_______ NADH	→	_______ ATPs	_______ kcal/mole

Pyruvate to Acetyl-CoA:

Chemiosmotic phosphorylation	_______ NADH	→	_______ ATPs	_______ kcal/mole

Citric Acid Cycle:

Substrate-level phosphorylation			_______ ATPs	_______ kcal/mole
Chemiosmotic phosphorylation	_______ NADH	→	_______ ATPs	_______ kcal/mole
	_______ $FADH_2$	→	_______ ATPs	_______ kcal/mole

Total yield per mole of glucose: _______ ATPs _______ kcal/mole

Efficiency of glycolysis alone = _______ kcal/mole ÷ _______ kcal/mole = ___________%

Efficiency of complete aerobic respiration = _______ kcal/mole ÷ _______ kcal/mole = ___________%

Self-Exam

You should be able to easily answer the following questions after learning the material in chapter 8. If you have difficulty with any question, study the appropriate section in the text and try again.

A. Multiple Choice Questions

Circle one alternative that best completes the statement or answers the question.

1. Catabolism of glucose is
   a. an exergonic process that releases or transfers free energy from chemical bonds in glucose.
   b. an exergonic process that releases or transfers free energy from chemical bonds in ATP.
   c. an endergonic process that releases or transfers free energy from chemical bonds in glucose.
   d. an endergonic process that releases or transfers free energy from chemical bonds in ATP.

2. Complete aerobic respiration includes
   a. glycolysis, citric acid cycle, electron transport, and fermentation.
   b. glycolysis, citric acid cycle, and fermentation.
   c. glycolysis, citric acid cycle, and electron transport.
   d. glycolysis and fermentation.

3. In alcohol fermentation, ATP is produced through
   a. electron transport.
   b. chemiosmotic phosphorylation.
   c. substrate level phosphorylation.
   d. mitochondrial photophosphorylation.

4. Where do the reactions of the glycolytic pathway take place?
   a. in the endoplasmic reticulum.
   b. in the cytoplasmic fluids.
   c. in the mitochondria.
   d. in the chloroplasts.

5. The glycolytic pathway begins with and ends with which molecules, respectively?
   a. begins with carbon dioxide and oxygen and ends with glucose.
   b. begins with glucose and ends with carbon dioxide.
   c. begins with glucose and ends with coenzyme A.
   d. begins with glucose and ends with pyruvate.

6. How many molecule of ATP are produced during glycolysis of one glucose molecule?
   a. one.
   b. two.
   c. four.
   d. thirty-six.

7. Which is the six-carbon molecule that is split into two three-carbon molecules during glycolysis?
   a. glucose.
   b. glucose-6-phosphate.
   c. fructose-1,6-diphosphate.
   d. phosphoenolpyruvate.

8. What is the difference in free energy per mole between ADP molecules and ATP molecules?
   a. 7.3 kcal/mole.
   b. 14.6 kcal/mole.
   c. 263 kcal/mole.
   d. 686 kcal/mole.

9. Which enzyme is thought to be involved in regulation of the rate of glycolysis?
   a. amylase.
   b. phosphofructokinase.
   c. decarboxylase.
   d. cytochrome oxidase.

10. The metabolic steps between glycolysis and the citric acid cycle are best represented by
    a. pyruvate + NAD^+ + CoA $\rightarrow$ acetyl CoA + NADH + H^+ + CO_2.
    b. pyruvate + NADH + H^+ + CoA $\rightarrow$ acetyl CoA + NAD^+ + CO_2.
    c. pyruvate + NAD^+ + CO_2 $\rightarrow$ acetyl CoA + NADH + H^+ + CoA.
    d. pyruvate + NADH + H^+ + acetyl CoA $\rightarrow$ CoA + NAD^+ + CO_2.

11. Substrate level phosphorylation occurs
    a. only in glycolysis.
    b. in glycolysis and in the citric acid cycle.
    c. only in the citric acid cycle.
    d. in glycolysis, in the citric acid cycle, and in the electron transport system.

12. In the citric acid cycle
    a. fuel molecules are oxidized and the coenzymes FAD and NAD^+ are oxidized.
    b. fuel molecules are oxidized and the coenzymes FAD and NAD^+ are reduced.
    c. fuel molecules are reduced and the coenzymes FAD and NAD^+ are oxidized.
    d. fuel molecules are reduced and the coenzymes FAD and NAD^+ are reduced .

13. The four-carbon molecule to which two carbons are added to begin the citric acid cycle is
    a. citric acid.
    b. acetyl-CoA.
    c. pyruvate.
    d. oxaloacetate.

14. Which of the following best describes the key reactions of the citric acid cycle?
    a. decarboxylations and oxidation/reductions, but not substrate level phosphorylations.
    b. substrate level phosphorylations and oxidation/reductions, but not decarboxylations .
    c. oxidation/reductions, but not decarboxylations or substrate level phosphorylations.
    d. decarboxylations, oxidation/reductions, and substrate level phosphorylations.

15. The carriers of the respiratory electron transport system are located
    a. in the mitochondrial matrix.
    b. in the mitochondrial inner membrane.
    c. in the mitochondrial outer membrane.
    d. in the cytoplasm of the cell.

16. The reactions that occur as electrons are passed from one carrier to the next are
    a. oxidations and reductions.
    b. decarboxylations.
    c. phosphorylations.
    d. dehydrations.

17. The site of chemiosmotic phosphorylation of ADP to ATP is
    a. the NADH-CoQ reductase complex.
    b. the $CoQH_2$-cytochrome *c* reductase complex.
    c. the cytochrome *c* oxidase complex.
    d. the F_0F_1 complex.

18. In the electron transport system, a proton pump such as FMN transports how many protons at a time?
    a. one.
    b. two.

c. four.
d. six.

19. If a pair of electrons from NADH is passed completely through the electron transport system, what is the total number of protons added to the chemiosmotic gradient?
a. one.
b. two.
c. four.
d. six.

20. If a pair of electrons from $FADH_2$ is passed completely through the electron transport system, what is the total number of protons added to the chemiosmotic gradient?
a. one.
b. two.
c. four.
d. six.

21. Which statement best describes the role of oxygen (O_2) in respiration?
a. It provides the oxygen in carbon dioxide, which is given off.
b. It reduces the fuel molecules as they are metabolized.
c. It is the final electron acceptor in the electron transport system.
d. It converts ADP to ATP at the F_0F_1 complex.

22. The coenzyme NAD^+ is reduced in reactions
a. during glycolysis and the citric acid cycle.
b. during conversion of pyruvate to acetyl CoA and during the citric acid cycle.
c. during glycolysis, during conversion of pyruvate to acetyl CoA, and during the citric acid cycle.
d. during electron transport.

23. When glucose is respired, about what fraction of its bond energy is transferred to ATP?
a. about 2% in glycolysis alone, and about 38-40% in complete aerobic respiration.
b. about 10% in glycolysis alone, and about 38-40% in complete aerobic respiration.
c. about 2% in glycolysis alone, and about 53-55% in complete aerobic respiration.
d. about 38-40% in glycolysis alone, and about 53-55% in complete aerobic respiration.

24. The energy of a proton gradient can be used for all of the following *except*
a. substrate level phosphorylation.
b. transporting calcium ions across membranes.
c. generating additional body heat.
d. spinning a prokaryotic flagellum.

25. In fermentation by yeast, the final energy-rich organic molecule is
a. pyruvate.
b. acetaldehyde.
c. ethanol.
d. carbon dioxide.

26. In the lactate fermentation pathway, as pyruvate is reduced to lactate, what is produced?
a. NADH + H^+.
b. NADP.
c. NAD^+.
d. FAD.

27. For a long fatty acid to be metabolized it is converted to
a. glucose subunits.
b. starch.
c. two-carbon acetyl-CoA units.
d. citric acid.

28. Which must happen *first* if a protein is to be metabolized as an energy source?
    a. It must be deaminated.
    b. It must be hydrolyzed into amino acids.
    c. It must be converted to acetyl-CoA.
    d. It must be converted to pyruvate.

B. True or False Questions

Mark the following statements either T (True) or F (False).

_____ 29. In catabolism of glucose, the energy is primarily in the bonds linking hydrogen to carbon.

_____ 30. Overall, the breakdown of glucose is endergonic.

_____ 31. Glycolysis produces ATP directly via substrate level phosphorylation.

_____ 32. In aerobic respiration, the final electron acceptor is oxygen.

_____ 33. Chemiosmotic phosphorylation never occurs in anaerobic respiration in bacteria.

_____ 34. Fermentation does not require oxygen.

_____ 35. Glycolysis occurs in the cytoplasm, while the reactions of the citric acid cycle are in the mitochondria.

_____ 36. The glycolytic pathway begins with glucose and ends with carbon dioxide.

_____ 37. During glycolysis of one glucose molecule, four ATPs are produced, but two ATPs are expended.

_____ 38. When ATP is hydrolyzed to ADP, $\Delta G = -7.3$ kcal/mole.

_____ 39. In cells, the rate of glycolysis is independent of the concentration of ATP, ADP, and AMP.

_____ 40. Chemically speaking, NAD^+ and FAD are nucleotides.

_____ 41. Substrate level phosphorylation during the Krebs cycle involves conversion of GDP to GTP.

_____ 42. Complete respiration of one molecule of glucose requires six turns of the citric acid cycle.

_____ 43. The cytochromes have iron-containing heme groups.

_____ 44. Electrons travel in groups of two in the electron transport system.

_____ 45. The oxygen required for aerobic respiration ends up as an oxygen atom in carbon dioxide.

_____ 46. It appears that two protons passing through an F_0F_1 complex generates one ATP.

_____ 47. We can say that aerobic respiration is more energy efficient than an automobile engine.

_____ 48. In alcohol fermentation, pyruvate is oxidized by NAD^+ to form alcohol.

_____ 49. Gram - for - gram, fats yield more energy in metabolic breakdown than does glucose.

_____ 50. Proteins are important a building blocks, but they cannot be broken down for energy in cells.

C. Fill in the Blanks

Answer the question or complete the statement by filling in the blanks with the correct word or words.

51. Fermentation is a type of ________________ respiration.

52. The bubbles in beer contain the gas ________________.

53. Glycolysis of one glucose molecule results in two ________________.

54. The reactions of the citric acid cycle occur in the ________________.

55. Formation of ATP during glycolysis is an example of ________________ phosphorylation.

56. The total free energy of one mole of glucose is about ________________.

57. An allosteric enzyme involved in regulation of the rate of glycolysis is ________________.

58. The molecule necessary for carbon to enter the Krebs cycle is ________________.

59. The so-called "resident molecule" of the citric acid cycle is ________________.

60. The major reactions of the Krebs cycle are ________________ of the fuel molecules.

61. Each turn of the citric acid cycle produces two ________________ molecules as final products.

62. The inner membrane folds within the mitochondrion are termed the ________________.

63. In electron transport, NADH first passes two electron and two protons to ________________.

64. The final electron acceptor in electron transport is ________________.

65. The "waste" products of complete aerobic respiration are ________________.

66. Anaerobic respiration in vertebrate muscle leads to an accumulation of ________________.

67. A process of intermediary metabolism that builds new molecules is termed ________________.

Questions for Discussion

1. Yeasts are sometimes termed facultative anaerobes — depending on conditions, they can carry out either fermentation or complete aerobic respiration. What types of conditions can you think of that would lead to one pathway or the other? What is the advantage of this metabolic flexibility? Do humans have alternative pathways for respiration? Explain your answer.

2. In humans, lack of oxygen can lead to cell death and eventually to death of the complete individual. How does the lack of oxygen interfere with cellular processes? Explain why the metabolic machinery grinds to a halt.

3. At many groceries, one can buy milk labeled "whole milk," "2%," "1/2%," etc. What is the difference in the composition of these milks? Compare their protein contents, their carbohydrate contents, and their fat contents. What does this say about the energy (caloric) contributions of the components?

4. What happens to mitochondrial function if the mitochondrial membranes are disrupted? How is mitochondrial structure important for mitochondrial function?

Answers to Self-Exam

Multiple Choice Questions

1. a
2. c
3. c
4. b
5. d
6. c
7. c
8. a
9. b
10. a
11. b
12. b
13. d
14. d
15. b
16. a
17. d
18. b
19. d
20. c
21. c
22. c
23. a
24. a
25. c
26. c
27. c
28. b

True or False Questions

29. T
30. F
31. T
32. T
33. F
34. T
35. T
36. F
37. T
38. T
39. F
40. T
41. T
42. F
43. T
44. T
45. F
46. T
47. T
48. F
49. T
50. F

Fill in the Blank Questions

51. anaerobic
52. carbon dioxide
53. pyruvates
54. mitochondrion
55. substrate level
56. 686 kcal
57. phosphofructokinase
58. coenzyme A (CoA)
59. oxaloacetate
60. oxidations
61. carbon dioxide
62. cristae
63. flavin mononucleotide (FMN)
64. oxygen
65. carbon dioxide and water
66. lactate
67. biosynthesis

CHAPTER 9 Eukaryotic Cell Reproduction

Learning Objectives

After mastering the material covered in chapter 9, you should be able to confidently do the following tasks:

- Explain how DNA is packaged into eukaryotic chromosomes.
- Outline the stages of a typical cell cycle.
- Describe the sequence of events during mitosis.
- Distinguish cell division in plants and animals.
- Detail the sequence of events in meiosis.
- Demonstrate the significance of meiosis.

Chapter Outline

I. The Nucleus.

A. DNA, chromatin, and the eukaryotic chromosomes.
B. The cell cycle: growth, replication, and division stages in the cell cycle.

II. Mitosis and Cell Division.

A. Prophase.
B. Metaphase.
C. Anaphase.
D. Telophase.
E. Cytokinesis.
  1. animals.
  2. plants.
  3. mitosis without cell division.
F. The adaptive advantages of mitosis.

III. Meiosis.

A. Homologous chromosomes.
B. An overview of meiosis.
C. Meiosis I: the first meiotic division.
  1. prophase I; meiotic prophase and its substages.
    a. a closer look at pachytene and crossing over.

2. metaphase I.
3. anaphase I.
4. telophase I and meiotic interphase.

D. Meiosis II: the second meiotic division.
1. prophase II.
2. metaphase II ans anaphase II.
3. telophase II.

E. Summing up meiosis.

F. Where meiosis takes place.
1. meiosis in humans.
a. meiosis in females, a special case.

G. Why meiosis?

Key Words

anaphase
aster
basal bodies
cell cycle
cell plate
centrioles
centromere
chiasma (pl, chiasmata)
chromatids
chromatin
chromatin net
chromosomes
cleavage
crossing over
cytokinesis
daughter cell
diakinesis
diploid
diplotene
DNA replication
Down syndrome
equatorial furrow
G_1 phase
G_2 phase
gametes
gametogenesis
germinal epithelium
germinal tissues
haploid
helix
histones
homologues
interphase
karyotype
kinetochore
leptotene
M phase
meiosis I
meiosis II
metaphase
metaphase plate
microtubules
microtubule organizing centers
middle lamella
mitosis
nondisjunction
nonhistone chromosomal proteins
nuclear envelope
nucleolus
nucleoplasm
nucleosomes
oocytes
ovulation
ovum
pachytene
perinuclear space
polar body
primary oocyte
primary spermatocyte
prophase
recombination
reduction division
replication
S phase
secondary oocyte
seminiferous tubules
sister chromatids
sperm
spermatocyte
spindle poles
spindle
spores
supercoil
synapsis
synaptonemal complex
telophase
template
tetrad
trisomy-21
tubulin
zygote
zygotene

Exercises

1. Below are listed the important periods of mitosis and cell division. Next to each of these, list the important events that occur during that period. Check your completeness with table 9.1 in the text.

A. Premitotic interphase:

B. Prophase

C. Metaphase

D. Anaphase

E. Telophase

F. Postmitotic interphase

G. Cytokinesis

2. The diploid chromosome number for the fruit fly, *Drosophila melanogaster*, is 2N = 8. Complete the following table with the appropriate numbers.

Cell type	Number of chromosomal DNA molecules per cell	Number of centromeres (= number of chromosomes) per cell
Newly formed zygote	________	________
Muscle cell in G_1	________	________
Muscle cell in G_2	________	________
Spermatocyte, prophase I	________	________
Spermatocyte, prophase II	________	________
Muscle cell, prophase	________	________
Sperm cell	________	________
Larval salivary gland cell	________	________

Self-Exam

You should be able to easily answer the following questions after learning the material in chapter 9. If you have difficulty with any question, study the appropriate section in the text and try again.

A. Multiple Choice Questions

Circle one alternative that best completes the statement or answers the question.

1. The fluid interior of the cell nucleus is the
   a. nucleoplasm.
   b. perinuclear space.
   c. nucleolus.
   d. nucleosome.

2. Chromosomes are composed of a DNA and protein complex called
   a. nucleoplasm.
   b. chromatin.
   c. chromatids.
   d. microtubules.

3. The tiny, beadlike globules formed by DNA in association with histone molecules are called
   a. nucleoli.
   b. ribosomes.
   c. nucleosomes.
   d. centromeres.

4. The period of actual cell division during the cell cycle is
   a. anaphase.
   b. cytokinesis.
   c. S phase.
   d. mitosis.

5. The interphase period of the cell cycle includes the phase(s)
   a. M.
   b. M and S.
   c. G_1, S, and G_2.
   d. S.

6. By G_2, each chromosome has how many chromatids?
   a. one.
   b. two.
   c. three.
   d. none.

7. Which is the correct sequence for the stages of mitosis?
   a. metaphase, prophase, anaphase, telophase.
   b. prophase, anaphase, metaphase, telophase.
   c. prophase, metaphase, telophase, anaphase.
   d. prophase, metaphase, anaphase, telophase.

8. The mitotic spindle apparatus forms during
   a. anaphase.
   b. metaphase.
   c. prophase.
   d. telophase.

9. The chromosomes are arranged in the middle of the mitotic spindle during
   a. anaphase.
   b. metaphase.
   c. prophase.
   d. telophase.

10. The mitotic apparatus is broken down and the nuclear envelope is reassembled during
   a. anaphase.
   b. metaphase.
   c. prophase.
   d. telophase.

11. Chromosomes move to opposite poles of the mitotic cell during
   a. anaphase.
   b. metaphase.
   c. prophase.
   d. telophase.

12. Some spindle fibers attach to each chromosome at sites called
   a. centrosomes.
   b. centromeres.
   c. chromomeres.
   d. centrioles.

13. If there are eight centromeres in a cell during mitotic prophase, how many chromatids are present in the cell?
   a. four.
   b. eight.
   c. sixteen.
   d. thirty-two.

14. Spindle fibers are composed of
   a. microtubules.
   b. DNA.
   c. histone proteins.
   d. chromatin.

15. The type of cytokinesis that is called cleavage typically occurs in
   a. plant cells.
   b. animal cells.
   c. plant and animal cells.
   d. all cells.

16. Cell plate formation is typical of cytokinesis in
   a. plant cells.
   b. animal cells.
   c. plant and animal cells.
   d. all cells.

17. A cell that will undergo a meiotic division is typically
   a. haploid.
   b. diploid.
   c. a gamete.
   d. a spore.

18. A human sperm cell will have how many chromosomes?
   a. 92.
   b. 46.

c. 23.
d. 12.

19. In meiosis, synapsis occurs between
 a. gametes.
 b. meiocytes.
 c. centromeres.
 d. homologues.

20. In humans, the chromosome number is reduced from 46 to 23 during
 a. mitosis.
 b. meiosis I.
 c. meiosis II.
 d. fertilization.

21. During meiotic prophase, chiasmata are first clearly visible during
 a. diplotene.
 b. leptotene.
 c. pachytene.
 d. zygotene.

22. Chromosomes are first visible in meiotic cells during
 a. diplotene.
 b. leptotene.
 c. pachytene.
 d. zygotene.

23. The synaptonemal complex is completed and homologues are synapsed during
 a. diplotene.
 b. leptotene.
 c. pachytene.
 d. zygotene.

24. Genetic crossing over occurs during
 a. diplotene.
 b. leptotene.
 c. pachytene.
 d. zygotene.

25. Homologous chromosomes are separated from one another during
 a. metaphase I.
 b. anaphase I.
 c. anaphase II.
 d. telophase II.

26. Sister chromatids are separated from each other during meiotic
 a. metaphase I.
 b. anaphase I.
 c. anaphase II.
 d. telophase II.

27. The number of cells resulting from a mitotic division and a meiotic division are, respectively
 a. 4 and 2.
 b. 4 and 4.
 c. 2 and 2.
 d. 2 and 4.

28. In higher organisms, the numbers of mature functional products of meiosis resulting from spermatogenesis and oogenesis are, respectively,
    a. 2 and 2.
    b. 4 and 4.
    c. 4 and 1.
    d. 1 and 4.

29. A polar body and a secondary oocyte result from
    a. mitosis.
    b. meiosis I.
    c. meiosis II.
    d. fertilization.

30. Abnormalities in chromosome number, such as in Down syndrome, frequently occur as a result of
    a. nondisjunction.
    b. gene mutation.
    c. crossing over.
    d. mitosis.

B. True or False Questions

Mark the following statements either T (True) or F (False).

_____ 31. Chromatin is made up of DNA and tubulin protein

_____ 32. The sequence in the mitotic cell cycle is M phase, G_1 phase, S phase, G_2 phase, M phase.

_____ 33. DNA and chromosomal proteins are synthesized during mitotic prophase.

_____ 34. A typical human cell has during mitotic prophase 46 chromosomes and 92 chromatids total.

_____ 35. The mitotic spindle is a delicate structure composed of microtubules.

_____ 36. The nucleolus disappears during mitotic prophase and reappears during telophase.

_____ 37. Cytokinesis in plants is ordinarily by a process called cleavage.

_____ 38. Crossing over occurs during diakinesis.

_____ 39. DNA content is doubled during an S period of interphase between meiosis I and meiosis II.

_____ 40. In the males of higher organisms, germinal epithelium gives rise to primary spermatocytes via mitotic divisions.

C. Fill in the Blanks

Answer the question or complete the statement by filling in the blanks with the correct word or words.

41. The complex of DNA, histones, and nonhistone chromosomal proteins is called _________________.

42. In most cells, mitosis is followed by _________________, the division of the cytoplasm.

43. The actual sites of spindle fiber attachment to centromeres are called _________________.

44. The structures found at the poles of animal cell spindles, but not those of higher plants, are called ________________.

45. After cell plate formation between plant cells, the gummy layer that lies between the cell walls of adjacent cells is the ________________.

46. A graphic representation of an organism's chromosomes, arranged according to size and shape, is a ________________.

47. The structure made of RNA and protein that holds homologues together during prophase I is the ________________.

48. Genetic recombination between homologous chromosomes is the result of ________________.

49. The cells resulting from the first meiotic division of a primary oocyte are the secondary oocyte and ________________.

50. The X-shaped arrangement of chromatids resulting from crossing over in prophase I is referred to as a ________________.

Questions for Discussion

1. The last section of chapter 9 asks the question "Why meiosis?" We could just as easily ask the question "Why sexual reproduction?" What is the relation between sexual reproduction and meiosis? Do you think we can know which came first?

2. It is noted that the centriole is not necessary for mitosis to proceed. What evidence supports this conclusion? If it is not necessary for mitosis, why is it sometimes there?

3. Why do chromosomes typically become short and very condensed as they enter mitosis and subsequently become uncoiled and diffuse during interphase?

4. Does reproduction of bacterial cells follow the patterns we have seen in mitosis and meiosis? What does this tell us about the evolutionary relationship between bacteria and eukaryotic organisms? Can bacteria have sexual reproduction of the same type we observe in higher organisms?

Answers to Self-Exam

Multiple Choice Questions

1. a
2. b
3. c
4. b
5. c
6. b
7. d
8. c
9. b
10. d
11. a
12. b
13. c
14. a
15. b
16. a
17. b
18. c
19. d
20. b
21. a
22. b
23. d
24. c
25. b
26. c
27. d
28. c
29. b
30. a

True or False Questions

31. F
32. T
33. F
34. T
35. T
36. T
37. F
38. F
39. F
40. T

Fill in the Blank Questions

41. chromatin
42. cytokinesis
43. kinetochores
44. centrioles
45. middle lamella
46. karyotype
47. synaptonemal complex
48. crossing over
49. polar body
50. chiasma

CHAPTER 10 Mendelian Genetics

Learning Objectives

After mastering the material covered in chapter 10, you should be able to confidently do the following tasks:

- Outline Gregor Mendel's important contribution to our present knowledge of mechanisms of heredity.
- State and explain Mendel's first and second laws of inheritance.
- Predict the results of monohybrid crosses and dihybrid crosses involving simple Mendelian traits.
- Show how test crosses can reveal the genotypes that underlie dominant phenotypes.
- Construct a human pedigree and draw conclusions about the inheritance of traits by pedigree analysis.
- Explain why Mendelian inheritance is important in understanding Darwinian natural selection.

Chapter Outline

I. Theory of Natural Selection and Heredity: When Darwin Met Mendel (almost).

II. Mendel's Crosses.

A. The principle of dominance.

III. Mendel's First Law: Segregation of Alternate Alleles.

A. Segregation and probability.

IV. Mendel's Second Law: Independent Assortment.

A. Mendel's test crosses.

V. The Chromosomal Basis for Mendel's Laws.

VI. Pedigrees and Human Genetics.

VII. The Decline and Rise of Mendelian Genetics.

Key Words

Charles Darwin
blending inheritance
natural selection
inheritance of acquired traits
Jean Baptiste Lamarck
Gregor Mendel
peloric flowers
dominance
true-breeding
genotype
phenotype
self-pollinate
cross-pollinate
P_1 generation
F_1 generation
F_2 generation
dominant trait
recessive trait
hereditary factor
gene
allele
homozygous
heterozygous
progeny testing
Mendel's first law
law of segregation
multiplicative law
additive law
Punnett square
model
Mendel's second law
law of independent assortment
locus (-i)
dihybrid cross
testcross
Theodore Boveri
Walter Sutton
heteromorphic chromosome pairs
sex chromosomes
gene linkage
pedigree
sampling error
pedigree analysis

Exercises

1. An isolated group of rare flightless birds lives on a yet undiscovered Florida key. They are nondescript and uniform in appearance except for two traits. The birds are either tall or short, and they are either fat or skinny. An anonymous scientist studied these birds and concluded the traits were independently inherited Mendelian traits. Tallness (T) is dominant to shortness (t), and Fatness (F) is dominant to skinniness (f).

A. A true-breeding short fat female bird (ttFF) is mated to a true-breeding tall skinny male bird (TTff) to produce an F_1 progeny. Can you provide the following information?

Genotype(s) of female gametes ________________ Genotype(s) of male gametes ______________

F_1 Phenotype(s) ____________________________ F_1 Genotype(s) ___________________________

B. If F_1 progeny from the first mating are mated to one another, make predictions about the F_2 progeny. Complete the Punnett square by filling in the genotypes of the gametes produced by the F_1 males and females. Then fill in the squares for the F_2 zygotic combinations.

Female Gametes

Male Gametes	TF			
TF	TTFF			

F_2 Phenotype	Proportion
______________	______/16
______________	______/16
______________	______/16
______________	______/16

2. Below is a pedigree for a family in which a rare form of genetically determined albinism occurs. Affected individuals are homozygous for the responsible allele *a*, while individuals with the dominant allele *A* are normally pigmented.. Label each individual with the appropriate genotype, if you can. If you are unable to assign a genotype, give the probability that an individual is a carrier.

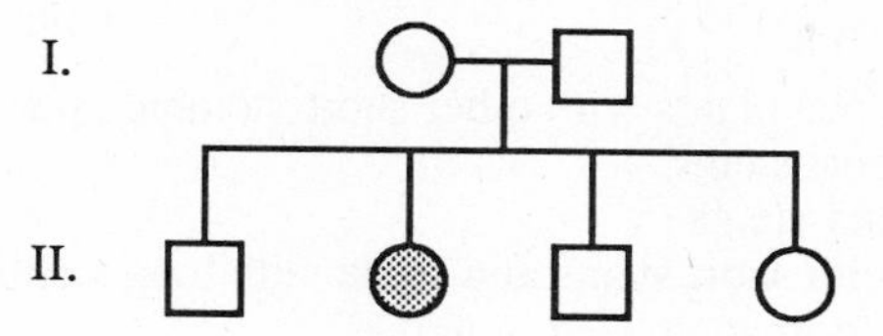

3. Below is a pedigree showing the inheritance of a rare recessive blood disorder in four generations of a family. Affected individuals have filled-in symbols. Individuals marrying into the family are assumed to be homozygous normal, unless there is evidence to the contrary. Using the symbol *B* for the dominant allele and the symbol *b* for the recessive allele, assign genotypes to as many individuals as you can.

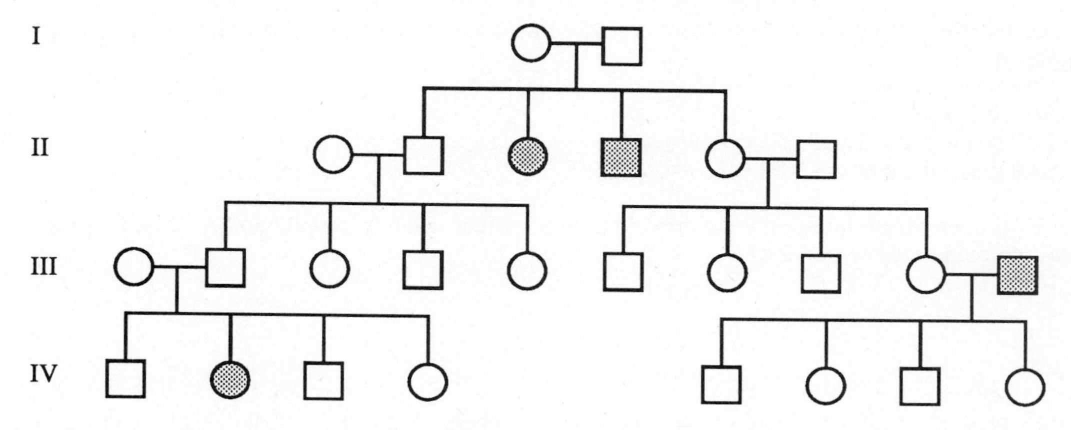

Self-Exam

You should be able to easily answer the following questions after learning the material in chapter 10. If you have difficulty with any question, study the appropriate section in the text and try again.

A. Multiple Choice Questions

Circle one alternative that best completes the statement or answers the question.

1. Which of these statements about blending inheritance is most accurate?
   a. There has never been any evidence to support the idea.
   b. Darwin demonstrated that this was the best genetic theory.
   c. Blending inheritance is suggested by the intermediate offspring of parents that differ in some trait.
   d. Mendel showed this with his pea plants, as well as other plants.

2. Which of these statements about the inheritance of acquired traits is most accurate?
   a. Evidence shows that use and disuse of body parts can lead to inherited change in those parts.
   b. Mendel provided evidence to support theories of the inheritance of acquired traits.
   c. Darwin disproved the theory of the inheritance of acquired traits.
   d. Lamarck's ideas of the inheritance of acquired traits are not generally accepted today.

3. If true-breeding strains of flowers that differ in color are crossed, the one color that appears in the progeny is
   a. recessive.
   b. dominant.
   c. preeminent.
   d. peloric.

4. Mendel crossed short-stemmed pea plants with other short-stemmed pea plants. What did he observe?
   a. Progeny pea plants with short stems.
   b. Progeny pea plants with long stems.
   c. Some progeny pea plants with short stems and some with long stems.
   d. Progeny pea plants with intermediate stem lengths.

5. Some of Mendel's round-seeded pea plants, when self-pollinated, produced only round-seeded progeny. Why?
   a. They were homozygous recessive.
   b. They were true-breeding.
   c. They had no hidden dominant alleles.
   d. Mendel had very good luck with his research.

6. If a true-breeding green-podded pea plant is crossed with a true-breeding yellow podded pea plant, the F_1 progeny will be
   a. All green-podded.
   b. All yellow-podded.
   c. 75% green-podded and 25% yellow-podded.
   d. 50% green-podded and 50% yellow-podded.

7. If the F_1 progeny from question 6 are crossed to one another, what is the probability of an F_2 progeny individual being yellow-podded?
   a. 100%.
   b. 75%.
   c. 25%.
   d. 0%.

8. If the F_1 progeny from question 6 are crossed to one another, what is the probability that a green-podded F_2 progeny individual is heterozygous for pod color?
   a. 1.
   b. 3/4.
   c. 2/3.
   d. 1/2.

9. Huntington disease is a lethal central nervous system disorder resulting from inheritance of a dominant allele. If a man's father has Huntington disease, what is the chance that he too will have the disease?
   a. 100%.
   b. 75%.
   c. 50%.
   d. 25%.

10. A pair of alleles determines a trait in an individual; the conclusion that these alleles separate from one another during the formation of gametes is
    a. Mendel's law of dominance.
    b. Mendel's law of segregation.
    c. Mendel's law of independent assortment.
    d. Mendel's principle of acquired traits.

11. Individuals heterozygous for a hair color gene (Yy) are crossed with one another. What is the probability that any one sperm or egg carries the recessive (y) allele?
    a. 0.
    b. 1/4.

c. 1/2.
d. 1.

12. What does the multiplicative law of probability tell us concerning the chances of a homozygous recessive offspring from the cross in question 11?
 a. The chance of yy is 1/4 x 1/4 = 1/16.
 b. The chance of yy is 1/2 x 1/2 = 1/4.
 c. The chance of yy is 1/4 + 1/4 = 1/2.
 d. The chance of yy is 1 x 1/4 = 1/4.

13. In crosses as in question 11, half of the progeny will be heterozygotes (Yy). Which of the following statements shows how the additive law of probability explains this?
 a. The overall chance of Yy is 1 x 1/2 = 1/2.
 b. The overall chance of Yy is (1/2 x 1/2) + (1/2 x 1/2) = 1/2.
 c. The overall chance of Yy is (1/4 + 1/4) x (1/4 + 1/4) = 1/2.
 d. The overall chance of Yy is 2 x 1/2 ÷ 2 = 1/2.

14. When Mendel crossed true-breeding peas having round, green peas with true-breeding peas having wrinkled, yellow peas, he obtained many seeds that were grown into F_1 plants. How are these F_1's best described?
 a. All produced round yellow seeds.
 b. All produced round green seeds.
 c. All produced wrinkled yellow seeds.
 d. The F_1's produced both round green seeds and wrinkled yellow seeds.

15. When Mendel crossed the F_1 plants resulting from a cross such as in question 14, many seeds were produced and grown into F_2 plants. What best describes the F_2 progeny?
 a. 3/4 had round green peas and 1/4 had wrinkled yellow peas.
 b. 1/4 had round green peas, 1/4 round yellow, 1/4 wrinkled green, and 1/4 wrinkled yellow peas.
 c. 9/16 had round green peas, 3/16 round yellow, 3/16 wrinkled green, and 1/16 wrinkled yellow peas.
 d. 3/16 had round green peas, 9/16 round yellow, 1/16 wrinkled green, and 3/16 wrinkled yellow peas.

16. Some guinea pigs can have black fur (BB or Bb), while others have recessive brown fur (bb). Which cross would be best to determine whether a particular black-furred animal is homozygous or heterozygous?
 a. Cross it with another black-furred animal.
 b. Cross it with one of its siblings.
 c. Cross it with a true-breeding black-furred animal.
 d. Cross it with a true-breeding brown-furred animal.

17. Suppose guinea pigs can also have curly hair (CC or Cc) or straight hair (cc). Which of the following crosses would be likely to produce the most progeny with black curly hair?
 a. BB CC x bb cc.
 b. Bb Cc x Bb Cc.
 c. Bb Cc x bb cc.
 d. BB Cc x BB Cc.

18. Guinea pigs heterozygous for both hair color and curliness are crossed with one another (Bb Cc x Bb Cc). When many progeny from numerous crosses are counted, there are three times as many black progeny as there are brown, and three times as many have curly hair as have straight. Yet the phenotypic combinations of black curly, black straight, brown curly, and brown straight show the 9:3:3:1 ratio. This ratio demonstrates
 a. the law of dominance.
 b. the law of segregation.
 c. the law of independent assortment.
 d. the principle of acquired traits.

19. When a pea plant heterozygous for pea shape and pea color (Rr Yy) produces pollen,
 a. 1/2 the pollen is Rr and 1/2 is Yy.
 b. 1/2 the pollen is RY and 1/2 is ry.

c. 1/4 the pollen is Ry, 1/4 is rY, 1/4 is RY, and 1/4 is ry.
d. 1/4 the pollen is RR. 1/4 is rr, 1/4 is YY, and 1/4 is yy.

20. A guinea pig, such as in questions 16-18, was testcrossed repeatedly, and the progeny all had either curly black hair or straight black hair. What is that animal's genotype?
a. BB CC.
b. Bb Cc.
c. BB Cc.
d. Bb CC.

21. Two people, each heterozygous for albinism, marry. They have four children. Which is the best prediction about their children's phenotypes?
a. There will be one albino and three normally pigmented children.
b. There will be two boys and two girls.
c. Among two boys and two girls there will be one albino and three normally pigmented children.
d. Several combinations of gender and pigmentation are possible, but the only certainty is four children.

B. True or False Questions

Mark the following statements either T (True) or F (False).

_____ 22. Darwin studied Mendel's work and rejected it in favor of the theory of blending inheritance.

_____ 23. Lamarck is well known as a proponent of the inheritance of acquired characteristics.

_____ 24. Mendel's first results were from studies of snapdragons with peloric vs. normal flowers.

_____ 25. Mendel worked with pea plants because they would not self-fertilize.

_____ 26. Plants designated "F_2" are the offspring of the original parental plants.

_____ 27. In Mendel' crosses, the recessive trait would appear in F_2 individuals.

_____ 28. A sperm cell carries two alleles for each gene.

_____ 29. When Mendel studied F_2 plants with the dominant phenotype (e.g., yellow peas) he found that 1/3 of these bred true, while 2/3 of these did not breed true.

_____ 30. Mendel's work was guided by his understanding of the mechanism of meiosis.

_____ 31. Suppose a pea plant is heterozygous for pea shape; the chance that a pollen grain will carry the recessive allele is 1/2.

_____ 32. Suppose two pea plants heterozygous for pea shape are crossed; the chance that a pollen grain and an ovule will both carry recessive alleles and form a homozygous recessive zygote is 1/2.

_____ 33. A dihybrid cross involves alleles at a single genetic locus.

_____ 34. Mendel's studies with peas involved alleles at seven different genetic loci altogether.

_____ 35. If a dihybrid cross is carried through to the F_2 generation, the frequency of F_2 individuals recessive for both traits should be about 1/16.

_____ 36. Mendel's law of independent assortment does not apply to gene loci that are close together on the same chromosome.

_____ 37. In a test cross, an individual with dominant phenotype is crossed to a known heterozygote.

_____ 38. The cross Rr x rr should give 1/2 dominant phenotype and 1/2 recessive phenotype progeny.

_____ 39. We would find no evidence for Mendelian inheritance in organisms lacking meiosis.

_____ 40. One of the first observations of sex chromosomes was in grasshopper spermatogenesis.

_____ 41. Assume each of a married couple is heterozygous for Tay-Sachs disease, a lethal recessive disorder. If they have four children, we know that only one will develop the disease.

_____ 42. Marriages between first cousins are considered consanguineous.

C. Fill in the Blanks

Answer the question or complete the statement by filling in the blanks with the correct word or words.

43. One of two or more alternative forms of a gene is termed an ________________.

44. A grid used to predict possible genotypes in offspring of a cross is a ________________.

45. The tendency for only one of two different alleles to be expressed in phenotype is ________________.

46. An individual with two identical alleles for a gene is ________________ for that locus.

47. Traits that are expressed only when homozygous are ________________ traits.

48. A cross of an unknown dominant phenotype to a homozygous recessive is a ________________.

49. A red-flowered plant is crossed with a white-flowered plant, and the offspring are pink-flowered. This seems to be an example of ________________.

50. Mendel's ________________ states that alleles separate from each other during gamete formation.

51. In probability, the ________________ law states that the chance of two independent events occurring together is the product of their independent probabilities.

52. A biological ________________ is an imaginary biological system that is consistent with past observations and can make predictions for outcomes of new situations.

53. In the F_2 of a typical dihybrid cross like one of Mendel's, there are ________________ different genotypes and ________________ different phenotypes.

54. The observation that two unlinked traits have no influence on each other in the results of a cross is the basis for Mendel's ________________.

55. Inheritance in humans is often studied by ________________ analysis, since controlled crosses are not practical and sample size is often too small.

56. A man has a rare hereditary trait, but his wife and her family have never been affected. Three of the couple's six children also show the trait. The allele responsible is most likely a ________________ allele.

Questions for Discussion

1. Albinism occurs in many different types of plants and animals and is usually due to a genetically determined failure in pigment production. Mutations to many different genes might lead to albinism, but it is almost

always due to homozygosity for a recessive allele. Two unrelated albinos marry, and they have four children, all with normal pigmentation. Is this consistent with Mendelian inheritance? How can you explain the phenotypes in this family?

2. One of the qualities that Mendel must have had to study inheritance in peas was patience. Results of crosses were not immediate. Suppose one spring you used pollen from a dominant round-seeded plant to pollinate a recessive wrinkled-seeded plant. What shape would the seeds be? Is this the F_1 phenotype? When would you be able to tell whether round or wrinkled were dominant, if you didn't already know? How many seasons would be necessary to carry the round vs. wrinkled cross through to determine the F_2 phenotypes?

3. The chromosome theory of inheritance that Sutton gave us has received considerable support over the years. Explain how the behavior of chromosomes during meiosis accounts for results of dihybrid crosses like the ones that Mendel did. You may find it useful to use diagrams of chromosomes during meiosis.

4. Cystic fibrosis is one of the more common recessive hereditary diseases affecting some human populations. An unaffected couple has a child with cystic fibrosis. What is the chance that their next child will also have the disease? What is the chance that the wife's unmarried sister is also a carrier for the recessive allele? Should she be interested in the health history of a prospective husband's family?

Answers to Self-Exam

Multiple Choice Questions

1. c
2. d
3. b
4. a
5. b
6. a
7. c
8. c
9. c
10. b
11. a
12. b
13. b
14. a
15. d
16. d
17. a
18. c
19. c
20. c
21. d

True or False Questions

22. F
23. T
24. F
25. F
26. F
27. T
28. F
29. T
30. F
31. T
32. F
33. F
34. T
35. T
36. T
37. F
38. T
39. T
40. T
41. F
42. T

Fill in the Blank Questions

43. allele
44. Punnett square
45. dominance
46. homozygous
47. recessive
48. test cross
49. blending inheritance
50. law of segregation
51. multiplicative
52. model
53. nine, four
54. law of independent assortment
55. pedigree
56. dominant

CHAPTER 11 Going Beyond Mendel

Learning Objectives

After mastering the material covered in chapter 11, you should be able to confidently do the following tasks:

- Explain exactly what genetic dominance is and the biological basis for dominance.
- Show how dominance relations are altered or modified by partial dominance. codominance, and lethal alleles.
- Explain the consequences of multiple alleles for a genetic locus, and use a human blood group system to illustrate a multiple allele locus.
- Demonstrate the ways that more than one genetic locus can interact to influence phenotype.
- Describe the ways in which expression of a gene can be influences by environment, by incomplete penetrance, or by variation in expressivity.
- Distinguish among sex limited, sex influenced, and sex linked traits.
- Show how apparently continuous variation can be the result of several genes affecting the same trait.
- Explain the chromosomal basis for genetic linkage, and demonstrate how the chromosomal locations of linked genes can be mapped.
- Identify the specific patterns of inheritance that are characteristic of sex-linked traits.

Chapter Outline

I. What Causes Dominance?

II. Other Dominance Relationships.

A. Partial dominance.
B. Codominance.
C. Recessive lethals.
D. Dominant lethals.
E. Other lethals.

III. Multiple Alleles.

A. Blood groups.
1. ABO blood groups.
2. Rh factor.

IV. Gene Interactions.

V. Conditional Gene Expression.

A. Environmental effects.
B. Incomplete penetrance and variable expressivity.
C. Sex limited and sex influenced effects.
D. Variable onset of gene action.
E. Pleiotropy.

VI. Continuous Variation and Polygenic Inheritance.

VII. Linkage.

A. Crossing over and recombination.
B. Mapping genes.
C. Chromosomes and sex.
   1. sex chromosomes in *Drosophila*.
   2. human sex chromosomes.

Key Words

dominance relationships
dominant inhibition
partial dominance
incomplete dominance
codominance
MN blood group system
antiserum (-a)
agglutinate
recessive lethal allele
dominant lethal allele
Huntington's chorea
achondroplastic dwarfism
multiple alleles
Drosophila melanogaster
fruit fly
albino
agouti
himalayan
ABO blood group system
antigen
antibody
cell-surface antigen
isoagglutinogen
type A
type B
type AB
type O
Rh blood group system
Rhesus
Rh positive
Rh negative
erythroblastosis fetalis
Rh incompatibility
hemolytic disease of the newborn
anti-Rh serum
gene interactions
epistasis
coat color B locus
coat color C locus
9:3:4 F_2 ratio
9:7 F_2 ratio
conditional gene expression
environmental influences
incomplete penetrance
tongue-rolling
variable expressivity
polydactyly
sex-limited trait
sex-influenced trait
pleiotropy
blue sclera-brittle bone disease
phenylketonuria (PKU)
continuous variation
polygenic inheritance
polygenic trait
linkage
linkage group
genetic recombination
crossing over
independent assortment
tightly linked
loosely linked
recombinants
nonrecombinants
frequency of recombination
Thomas Morgan
Alfred Sturtevant
map units
sex linkage
sex chromosome
white-eyed *Drosophila*
mutant allele
wild type allele
red-eyed *Drosophila*
X chromosome
Y chromosome
hemizygous
sex determining mechanism
testis determination factor (TDF)
Barr body
drumstick
Mary Lyon
Lyon effect
color-blindness
hemophilia
muscular dystrophy
Turner syndrome
Klinefelter syndrome
XXY syndrome
trisomy-X

Exercises

1. White leghorn chickens are white because they have a dominant allele *I* that inhibits color. Other strains of chicken may be white for other reasons. For example white Plymouth Rock chickens are homozygous for a recessive allele *cc*, lacking the dominant allele *C* that leads to colored feathers.

A. Suppose we start a series of crosses with true breeding white leghorn hens (II CC) and true breeding white Plymouth Rocks roosters (ii cc). Fill in the below information:

Genotype(s) of female gametes ________________ Genotype(s) of male gametes ____________

F_1 Phenotype(s) ________________________ F_1 Genotype(s) ____________________

B. If F_1 chickens from the first mating are mated to one another, make predictions about the F_2 progeny. Complete the Punnett square by filling in the genotypes of the gametes produced by the F_1 males and females. Then fill in the squares for the F_2 zygotic combinations. (Hint: when you fill in the F_2 phenotypes, you may not need all four lines.)

Female Gametes

Male Gametes	IC			
IC	IICC			

F_2 Phenotype	Proportion
____________	_____/16
____________	_____/16
____________	_____/16
____________	_____/16

2. The ABO blood group system was the first discovered in humans. It explains why transfusions of incompatible blood lead to clumping or agglutination of blood cells. The alleles for the ABO genetic locus can be designated I^A I^B, and i^O. Red blood cells may have A antigens or B antigens, and the serum may have anti-A or anti-B antibodies. Fill in the information required in the below table of characteristics of the various ABO phenotypes.

Blood Type	Genotype(s)	Antigens Present	Antibodies in serum	Does this blood agglutinate when mixed with Type A?	Type B?	Type AB?	Type O?
Type A	________	________	________	________	________	________	________
Type B	________	________	________	________	________	________	________
Type AB	________	________	________	________	________	________	________
Type O	________	________	________	________	________	________	________

Self-Exam

You should be able to easily answer the following questions after learning the material in chapter 11. If you have difficulty with any question, study the appropriate section in the text and try again.

A. Multiple Choice Questions

Circle one alternative that best completes the statement or answers the question.

1. A recessive mutant such as a genetically albino flower might lack colored pigment because
   a. the plant did not grow in the sun.
   b. the plant was deficient in some essential nutrient that it needed.
   c. the plant lacked an enzyme necessary to produce pigment.
   d. the temperature level was incorrect for the production of pigment.

2. A *Drosophila* fly that is heterozygous for eye color w^+/w (one wild-type allele and one mutant white allele)
   a. will have red eyes.
   b. will have white eyes.
   c. will have pink eyes.
   d. will have red and white eyes.

3. Two white sheep are mated, and of several lambs produced, one is black. What can you conclude?
   a. White fleece is recessive to black, and the black lamb is heterozygous.
   b. White fleece is recessive to black, and the parents are homozygous.
   c. Black fleece is recessive to white, and the black lamb is heterozygous.
   d. Black fleece is recessive to white, and the parents are heterozygous.

4. Snapdragons show partial dominance in flower color. Two pink snapdragons are crossed. You observe
   a. that all of the progeny bear pink flowers.
   b. that all of the progeny have red flowers.
   c. that there are three times as many red-flowered progeny as white-flowered progeny.
   d. that there are roughly equal numbers of red- and white-flowered progeny, and twice as many pink-flowered.

5. The alleles of the MN blood group system provide a good example of
   a. partial dominance.
   b. codominance.
   c. recessive lethals.
   d. dominant lethals.

6. For biology labs, corn seeds are produced that give a 3:1 ratio of normal green leaves to mutant white leaves when the seeds are planted. The seeds result from a cross between heterozygotes, Ww x Ww. What cross produces these parental heterozygotes?
   a. WW x ww.
   b. Ww x ww.
   c. Ww x Ww.
   d. ww x WW.

7. Tay-Sachs disease is a lysosomal storage disease that causes the death of affected individuals in the first few years of life. This disease must be due to
   a. partial dominance.
   b. codominance.
   c. recessive lethals.
   d. dominant lethals.

8. There are three alleles in the ABO blood group system. How many genotypes are possible?
   a. Three.
   b. Four.

c. Six.
d. Nine.

9. An individual with genotype $I^A I^B$ produces both A and B antigens. This is an example of
   a. partial dominance.
   b. codominance.
   c. recessive lethals.
   d. dominant lethals.

10. A person whose blood has neither anti-A nor anti-B antibodies has which blood type?
   a. Type A
   b. Type B.
   c. Type AB.
   d. Type O.

11. Which of the following situations poses the greatest health risk for a newborn?
   a. An Rh positive woman bears her first child, who is Rh negative.
   b. An Rh positive woman bears an Rh negative child after having already delivered an Rh negative child.
   c. An Rh negative woman bears her first child, who is Rh positive
   d. An Rh negative woman bears an Rh positive child after having already delivered an Rh positive child.

12. In which of the following marriages should couples be most concerned of the risks of erythroblastosis fetalis?
   a. An Rh negative woman marries an Rh positive man.
   b. An Rh negative woman marries an Rh negative man.
   c. An Rh positive woman marries an Rh negative man.
   d. An Rh positive woman marries an Rh positive man.

13. In mice, the allele for black hair (B) is dominant to the allele for brown hair (b), but albinos (cc) produce no pigment in their hair. Two black mice mate, and their litter includes black, brown, and white pups. What are the parental genotypes?
   a. BB cc x bb CC.
   b. Bb Cc x Bb Cc.
   c. Bb cc x Bb cc.
   d. BB Cc x bb Cc.

14. What would be the fastest way to develop a siamese cat with a dark "saddle" on its back?
   a. Cross standard Siamese cats with piebald cats that have large patches, and crossbreed the kittens.
   b. Clip the fur on the cat's saddle region to allow exposure to light.
   c. Tie an icebag on the cat's back, like a saddle.
   d. Threaten to use shoe polish, so the cat will voluntarily grow dark hair anywhere you ask.

15. A young man has one dominant allele at the tongue-rolling genetic locus. What can we say with certainty?
   a. The young man will be skilled at tongue-rolling.
   b. The young man will be unable to roll his tongue.
   c. His children have a 50% chance of inheriting the dominant allele from him.
   d. His children have less than a 50% chance of inheriting the dominant allele from him.

16. Pattern baldness in males is inherited as
   a. a sex-influenced trait.
   b. a sex-limited trait.
   c. a sex-linked dominant trait.
   d. a sex-linked recessive trait.

17. If we measure a large group of 18-year old army recruits, we find much variation in height. This is due to
   a. a large number of multiple alleles at a single locus.
   b. differing alleles at a number of gene loci affecting height.
   c. pleiotropic effects of a single gene.
   d. an artifact of measuring soldiers — they are all actually the same height.

18. Armadillos give birth to four genetically identical babies, derived from a single egg. Suppose we weigh four newborn siblings and find their weights to be nearly identical, but when captured and weighed one year later they differ substantially. What is this variation due to?
    a. A large number of multiple alleles at a single locus influences weight.
    b. Differing alleles at a number of gene loci affect the final weight.
    c. Environmental effects have introduced nongenetic variation during growth.
    d. We are seeing the effects of epistasis and pleiotropy.

19. A general statement about linked genes would be
    a. they do not conform to Mendel's law of dominance.
    b. they do not conform to Mendel's law of segregation.
    c. they do not conform to Mendel's law of independent assortment.
    d. they do not conform to Mendel's laws at all.

20. To oversimplify, assume brown hair is dominant to blond hair and brown eyes are dominant to blue eyes, and assume hair color and eye color are genetically linked. Which of the following statements is accurate?
    a. Brown-haired brown-eyed parents cannot have blue-eyed blond children.
    b. A brown-haired blue-eyed father and a brown-eyed blond mother can have a blue-eyed blond child.
    c. Blue-eyed parents cannot have brown haired children.
    d. Blond-haired parents always have blue-eyed children.

21. Genes P and Q are linked. A plant has the genotype Pp Qq. Assume a single crossover occurs between P and Q in exactly one-half of the meioses. What fraction of the gametes are recombinants ?
    a. 12.5%.
    b. 25%.
    c. 50%.
    d, 100%.

22. *Drosophila* with genotypes AA BB and aa bb are crossed. The Aa Bb progeny are then crossed with aa bb individuals, and we observe 25% Aa Bb, 25% aa Bb, 25% Aa bb, and 25% aa bb. We can conclude
    a. that genes A and B are linked, and 50 map units apart.
    b. that genes A and B are linked, and 25 map units apart.
    c. that genes A and B are linked and 12.5 map units apart.
    d. that genes A and B are assorting independently.

23. Snapdragons with genotypes CC DD and cc dd are crossed. The Cc Dd progeny are then crossed with cc dd plants, and we obtain 30% Cc Dd, 20% cc Dd, 20% Cc dd, and 30% cc dd offspring. We conclude.
    a. Genes C and D are linked, separated by 20 map units.
    b. Genes C and D are linked, separated by 30 map units.
    c. Genes C and D are linked, separated by 40 map units.
    d. Genes C and D are unlinked.

24. Snapdragons with genotypes DD EE and dd ee are crossed. The Dd Ee progeny are then crossed with dd ee plants, and we obtain 45% Dd Ee, 5% dd Ee, 5% Dd ee, and 45% dd ee. How far apart are genes D and E?
    a. 5 map units.
    b. 10 map units.
    c. 45 map units.
    d. Genes D and E are not linked.

25. Snapdragons with genotypes CC EE and cc ee are crossed. The Cc Ee progeny are then crossed with cc ee plants, and we obtain 35% Cc Ee, 15% cc Ee, 15% Cc ee, and 35% cc ee. Referring to questions 23 and 24, which of the following maps show the correct sequence and recombination distances for genes C, D, and E?
    a. D --------------- 40 ---------------- C --------------- 30 ---------------E.
    b. E -------- 10 ----------- D -------------------- 40 -------------------- C.
    c. C ----------------------- 30 -------------------- E ------- 10 -------- D.
    d. D --------------- 20 ---------------- C --------------- 15 -------------- E.

26. A man is red-green color blind, but his sisters' sons all have normal vision. Which of these statements is most likely?
    a. His mother was red-green color blind.
    b. His father was red-green color blind.
    c. His mother's father was red-green color blind.
    d. His father's mother was red-green color blind.

27. Both husband and wife have normal vision, although both the wife's father and the husband's father were color blind. What is the chance that their first son will be color blind?
    a. 0%.
    b. 25%
    c. 50%
    d. 100%

28. What is the chance that a boy inherits his Y chromosome from his maternal grandfather?
    a. 0%.
    b. 25%
    c. 50%
    d. 100%

29. The cheek lining cells of a man are examined, and one Barr body is observed. Which statement best explains this observation?
    a. This is normal for a male, since males have one X chromosome.
    b. This is really a woman, since females have one Barr body.
    c. This is an individual with Turner syndrome and male phenotype.
    d. This is an individual with Klinefelter syndrome.

30. In the fruit fly *Drosophila melanogaster*, an XO individual would be
    a. phenotypically female, but sterile.
    b. phenotypically male, but sterile.
    c. phenotypically female, and fully fertile.
    d. phenotypically male, and fully fertile.

B. True or False Questions

Mark the following statements either T (True) or F (False).

_____ 31. Some Mendelian genes do not demonstrate the principle of dominance.

_____ 32. A chicken that is only heterozygous for a dominant color inhibitor will be colored.

_____ 33. When pink and red snapdragons are crossed, they yield a true-breeding pink F_1.

_____ 34. An MN blood group heterozygote will have both the M phenotype and the N phenotype.

_____ 35. Corn plants heterozygous (Ww) for mutant colorless leaves result from the cross WW x ww.

_____ 36. A child whose father dies from Huntington's chorea has a 50% chance of developing the disease.

_____ 37. In each of the traits Mendel studied in peas, he considered only two alleles per locus.

_____ 38. For many genetic loci in *Drosophila* there are a number of different alleles known.

_____ 39. The three major alleles in the ABO blood group system give four genotypes and four phenotypes.

_____ 40. In the ABO blood group system, I^A is dominant to I^B, and I^B is dominant to i^O.

_____ 41, Persons with type A blood have the A antigen and produce anti-B antibodies.

_____ 42. Type AB blood tranfused into a type O individual could lead to harmful agglutination of donor cells.

_____ 43. Rh positive mothers have greater risk of erythroblastosis fetalis than do Rh negative mothers.

_____ 44. Genes responsible for albinism in mammals are often epistatic to hair color genes.

_____ 45. The phenotype resulting from some genotypes varies with environmental conditions.

_____ 46. John comes from a long line of tongue-rollers, but no one in his wife's family has ever been a tongue roller. Although John can't roll his tongue, all of his children can. His wife must have provided the dominant allele responsible for tongue-rolling.

_____ 47. Men with pattern baldness can inherit the trait from either their mothers or their fathers.

_____ 48. Boys get their Y chromosomes only from their fathers, and girls get their X chromosomes only from their mothers.

_____ 49. If two genes assort independently at meiosis, then they must be on separate chromosomes.

_____ 50. Two genes that are assorting independently have a recombination frequency of 50%.

C. Fill in the Blanks

Answer the question or complete the statement by filling in the blanks with the correct word or words.

51. Aa is intermediate in phenotype to AA and aa; this is termed _________________.

52. I^A and I^B are both expressed in heterozygotes. This is an example of _________________ of alleles.

53. In mice Tt x Tt always gives 2/3 Tt and 1/3 TT. The t allele must be a _________________.

54. Persons with blood type _________________ produce neither anti-A nor anti-B antibodies.

55. The Rh blood group system gets its designation from _________________.

56. The 9:3:3:1 ratio is modified to a 9:3:4 ratio by recessive _________________.

57. The rare human trait _________________ refers to extra fingers or toes.

58. A trait expressed differently in males and females is a _________________ trait.

59. _________________ results in one mutant gene affecting many different traits.

60. Inherited traits that show continuous variation are often due to _________________ inheritance.

61. Genes that tend to be inherited together form a _________________.

62. If two genes are linked, after meiosis we find fewer _________________ than if the genes were assorting independently

63. _________________ is the physical exchange that is responsible for genetic recombination.

64. If we observe 12% recombination between two linked traits, their genes are 12 _________________ apart.

65. When Morgan's white-eyed male *Drosophila* was mated to his red-eyed sisters, the F_1 progeny had ________________ eyes.

66. Females can be homozygous for X-linked genes, but males are ________________.

67. The Y chromosomal gene believed responsible for maleness in humans is the ________________ gene.

68. One X chromosome may be inactive in human female cells; this is termed the ________________.

69. A man who is color blind expects his daughters to all be ________________.

70. The X linked blood disorder that results is failure of clotting is ________________.

Questions for Discussion

1. Mendel was fortunate to select seven pairs of independently assorting traits to study. He was also fortunate that in each case there was complete dominance rather than partial dominance or codominance. Imagine how his results might have differed! Outline the modifications we now make to Mendelian inheritance. Do these modifications mean that Mendel was incorrect in his conclusions?

2. Why was it much more likely that the X-linked mutant fly that Morgan discovered was male rather than female? Suppose that the first known white-eyed fly had been female. What crosses might Morgan have done to study inheritance of the new mutant? What results would he have obtained? Outline your proposed crosses and your predicted results.

3. There are serious X-linked recessive diseases in humans, such as hemophilia and one type of muscular dystrophy, as well as less serious disorders like red-green color blindness. These are much more common in males than in females. Explain the reason for this difference.

4. Three babies were mixed up in the hospital nursery. Wanting to avoid a lawsuit, the hospital staff compared the babies' blood types with the parents' blood types. With luck, the babies might be matched with the correct parents. Can it be done, or should the hospital seek legal advice?
   Baby 1: type A; Baby 2: type O; Baby 3: type AB.
   Parents X: type AB and type A; Parents Y: type B and type A; Parents Z: type O and type AB.

5. Although skin color in humans is an inherited trait, there are many variations in skin color. Inheritance is not due to a single gene with simple Mendelian inheritance. Outline some of the modifications to Mendelian inheritance that are helpful in explaining the observed variation in skin color.

6. Plant species A occurs over a large range, and plants from different places look very different. A botanist transplants plants from five different sites to a common garden, and after a year, the plants still look strikingly different. Plant species B also occurs over a large range, and plants from different habitats also look very different from one another. However, when the botanist transplants plants of species B from five different sites to the common garden, the plants are indistinguishable after a year. How can you explain the different results with the two species?

Answers to Self-Exam

Multiple Choice Questions

1. c
2. a
3. d
4. d
5. b
6. c
7. c
8. c
9. b
10. c
11. d
12. a
13. b
14. c
15. c
16. a
17. b
18. c
19. c
20. b
21. b
22. d
23. c
24. b
25. c
26. c
27. c
28. a
29. d
30. b

True or False Questions

31. T
32. F
33. F
34. T
35. F
36. T
37. T
38. T
39. F
40. F
41. T
42. T
43. F
44. T
45. T
46. F
47. T
48. F
49. F
50. T

Fill in the Blank Questions

51. partial or incomplete dominance
52. codominance
53. recessive lethal
54. AB
55. the *Rhesus* monkey
56. epistasis
57. polydactyly
58. sex influenced
59. Pleiotropy
60. polygenic
61. linkage group
62. recombinants
63. Crossing over
64. map units
65. red
66. hemizygous
67. testis determination factor (TDF)
68. Lyon effect
69. carriers
70. hemophilia

CHAPTER 12 DNA as the Genetic Material

Learning Objectives

After mastering the material covered in chapter 12, you should be able to confidently do the following tasks:

- Outline the development of circumstantial evidence and experimental evidence leading to the conclusion that DNA is the molecule responsible for hereditary information.

- Identify the chemical constituents of DNA and show how the parts are combined to form nucleotides and complex DNA molecules.

- List the sequence of events that occurs during replication of DNA and specify the enzymes necessary for accurate replication.

- Explain the relationship between genes and proteins that leads to phrases like "one gene, one enzyme."

Chapter Outline

I. Highlights from the Discovery of DNA.

A. The early efforts.
B. Transformation.
C. Hershey and Chase.

II. DNA Structure.

A. DNA nucleotides.
1. Chargaff's rule
2. X-ray diffraction and more puzzles.
B. The Watson and Crick model of DNA.

III. DNA Replication.

A. The chemistry of DNA bonding.
B. Origins of replication.
C. DNA polymerase.
D. Replication in eukaryotes and prokaryotes.

IV. DNA and Information Flow.

A. Genes, enzymes, and inborn errors of metabolism.

Key Words

chromosome
Friedrich Miescher
Robert Feulgen
nucleotide
transformation
Frederick Griffith
Pneumococcus
virulent
smooth colony
rough colony
Avery, MacLeod, and McCarty
Hershey and Chase
bacteriophage
phage
host cell
Escherichia coli
radioactive sulfur (^{35}S)
radioactive phosphorus (^{32}P)
radioactive label
bacteriophage "ghosts"
phosphoric acid groups
deoxyribose
nitrogen base
ATP
DNA polymer
adenine
guanine
thymine
cytosine
pyrimidine
purine
Erwin Chargaff
Chargaff's rule
X-ray crystallography
Maurice Wilkins
Rosalind Franklin
James Watson
Francis Crick
helix
complementary
base pairing
major groove
synthesis
replication
antiparallel
3' -OH
3' end
5' -phosphate
5' end
origins of replication
helicase
replication complex
replication fork
gyrase
single-strand nick
single-strand binding protein
DNA polymerase
primase
primer
leading strand
lagging strand
continuous replication
discontinuous replication
Okazaki fragments
ligase
semiconservative replication
conservative replication
dispersive replication
Meselson and Stahl
cesium chloride gradient
replication "bubble"
A. E. Garrod
inborn errors of metabolism
alkaptonuria
albinism
phenylketonuria
phenylalanine
tyrosine
Beadle and Tatum
Neurospora crassa
"one gene, one enzyme"
central dogma of molecular biology

Exercises

1. In the correct spaces below, draw stick diagrams of the four nucleotides adenine, guanine, cytosine, and thymine. Be sure to show the nitrogen bases, the 5-carbon sugars, and indicate the 3' -OH and the 5' phosphate groups. Can you show how they form complementary base pairs? Check your work with the textbook.

<u>Purines</u> <u>Pyrimidines</u>

When paired,
these form three
hydrogen bonds:

When paired,
these form two
hydrogen bonds:

2. Below is the framework for a diagram of a DNA replication fork. Complete the diagram by showing the newly synthesized DNA on the leading and lagging strands. Be sure to indicate the 3' and 5' ends of the new DNA, and show the direction of synthesis for each strand. Show which strand has Okazaki fragments. Indicate where DNA polymerase acts. Show where ligase joins DNA fragments. Provide clear labels for your diagram. Check your work with figure 12.9 in the text.

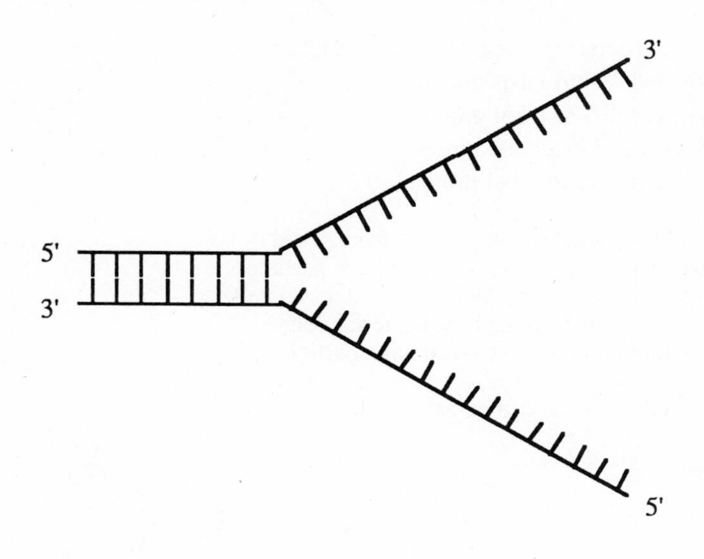

Self-Exam

You should be able to easily answer the following questions after learning the material in chapter 12. If you have difficulty with any question, study the appropriate section in the text and try again.

A. Multiple Choice Questions

Circle one alternative that best completes the statement or answers the question.

1. The scientist(s) given credit for first isolating DNA is (are)
   a. Miescher.
   b. Feulgen.
   c. Avery, MacLeod, and McCarty.
   d. Watson and Crick.

2. In the first half of the 20th century, scientists generally believed that the chemical basis of heredity
   a. was in molecules of nucleic acids.
   b. was in molecules of polypeptide proteins.
   c. was in complex polysaccharides.
   d. was contained in sequences of nucleotides.

3. In his early experiments, Frederick Griffith demonstrated transformation in
   a. genetically different strains of mice.
   b. genetically different bacteriophage strains.
   c. genetically different strains of *Pneumococcus*.
   d. genetically different *Escherichia coli* strains.

4. Virulent strains of *Pneumococcus* have
   a. smooth-looking colonies, and they synthesize a polysaccharide coat.
   b. smooth-looking colonies, but cells lack a polysaccharide coat.
   c. rough-looking colonies, and they synthesize a polysaccharide coat.
   d. rough-looking colonies, but cells lack a polysaccharide coat.

5. Avery, MacLeod, and McCarty showed that nonvirulent strains of bacteria could be transformed by
   a. growing nonvirulent cells together with living virulent strains.
   b. growing nonvirulent cells together with heat-killed virulent cells.
   c. growing nonvirulent cells together with purified protein from heat-killed virulent strains.
   d. growing nonvirulent cells together with purified DNA from heat-killed virulent cells.

6. In their experiments with bacteriophages, Hershey and Chase used
   a. ^{35}S to label DNA and ^{32}P to label protein.
   b. ^{32}P to label DNA and ^{35}S to label protein.
   c. ^{14}N to label DNA and ^{35}S to label protein.
   d. ^{32}P to label DNA and ^{14}N to label protein.

7. In the results of their experiment, Hershey and Chase found
   a. radioactive phosphorus remained in bacteriophage particles.
   b. radioactive phosphorus was found inside host bacterial cells.
   c. radioactive sulfur was found inside host bacterial cells.
   d. radioactive nitrogen remained in bacteriophage particles.

8. The structure to the right is best described as
   a. a nucleotide.
   b. deoxyribose sugar.
   c. a pyrimidine.
   d. a purine.

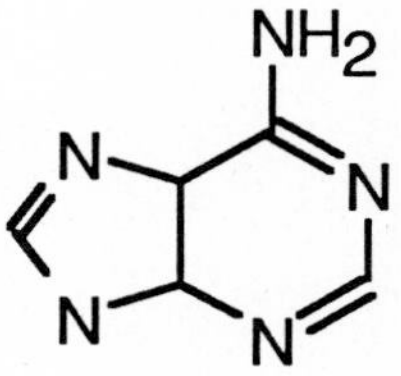

9. In the complete molecule deoxyadenosine triphosphate, to which of the carbons shown at the right would the adenine base be covalently bonded?
   a. 1' carbon.
   b. 2' carbon.
   c. 3' carbon.
   d. 5' carbon.

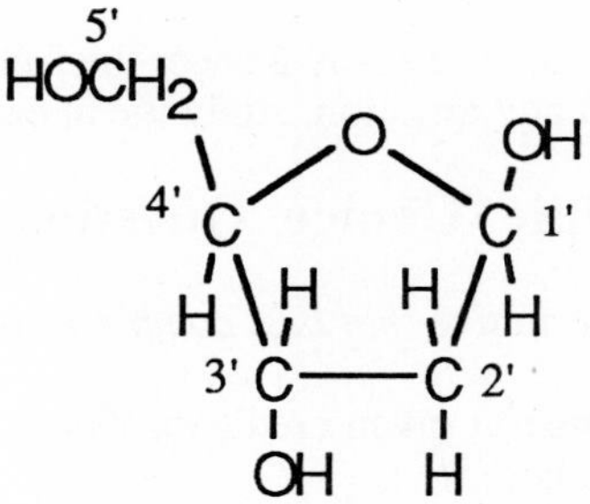

10. Which of the following statements is consistent with Chargaff's rule?
   a. A + T = G + C.
   b. A = G and T = C.
   c. A + G = T + C.
   d. A = C and T = G.

11. Which of the following statements is *incorrect*, with respect to DNA nucleotide bases?
   a. Exactly half of the nucleotide bases are purines.
   b. Exactly half of the nucleotide bases are pyrimidines.
   c. Purines pair with purines, and pyrimidines pair with pyrimidines.
   d. Purines pair with pyrimidines, and pyrimidines pair with purines.

12. Where might we be most likely to find an exception to Chargaff's rule?
   a. In the DNA of a plant cell.
   b. In the DNA of a haploid fungus cell.
   c. In an *E. coli* bacterial cell's DNA.
   d. In the single-stranded DNA of the bacteriophage ΦX 174.

13. Which of the following features of DNA was a major new contribution from Watson and Crick?

a. DNA is a polymer of the four nucleotides G, C, A, and T.
b. X-ray diffraction shows intramolecular distances of 2.6 nm, 0.34 nm, and 3.4 nm.
c. Adenine will specifically pair with thymine and guanine with cytosine, forming hydrogen bonds.
d. The numbers of adenines and thymines are equal, and the numbers of guanines and cytosines are equal.

14. Meselson and Stahl demonstrated that DNA replication was
 a. conservative.
 b. semi-conservative.
 c. liberal.
 d. dispersive.

15. During DNA synthesis, new nucleotides are bonded to
 a. the 3' -OH of the growing DNA strand.
 b. the 5' phosphate of the growing DNA strand.
 c. the 1' carbon in deoxyribose of the growing DNA strand.
 d. the terminal nitrogen base of the growing DNA strand.

16. The enzyme that actually adds new nucleotides to a growing DNA strand is
 a. DNA helicase.
 b. DNA gyrase.
 c. DNA polymerase.
 d. DNA ligase.

17. In DNA replication, primase is an enzyme that copies the template strand and
 a. forms a short sequence of a few bases of DNA to start the new strand.
 b. forms a short sequence of a few bases of RNA to start the new strand.
 c. forms a short sequence of amino acids to start the new strand.
 d. forms a short sequence of enzymes to start the new strand.

18. DNA replication is discontinuous, with the formation of Okazaki fragments,
 a. on the lagging strand at a replication fork.
 b. on the leading strand at a replication fork.
 c. in eukaryotes, but not in prokaryotes.
 d. in prokaryotes, but not in eukaryotes.

19. Which is the best generalization?
 a. Eukaryotic chromosomes and bacterial chromosomes usually contain one long linear DNA molecule.
 b. A eukaryotic chromosome contains one long linear DNA molecule, but in bacteria it is circular.
 c. Eukaryotic chromosomes and bacterial chromosomes typically contain a single circular DNA molecule.
 d. A bacterial chromosome contains one long linear DNA molecule, but in eukaryotes it is circular.

20. An inherited disease characterized by urine that turns black is
 a. hemophilia.
 b. alkaptonuria.
 c. albinism.
 d. phenylketonuria.

21. Beadle and Tatum's "one gene, one enzyme" experimental work was carried out using
 a. *Pneumococcus.*
 b. *Escherichia coli.*
 c. *Neurospora crassa.*
 d. *Drosophila melanogaster.*

B. True or False Questions

Mark the following statements either T (True) or F (False).

_____ 22. James Watson and Francis Crick discovered DNA.

_____ 23. DNA is naturally a dark molecule, which led to the name "chromosome," for a "colored body."

_____ 24. DNA belongs to the class of molecule known as polypeptides.

_____ 25. In the early part of the 20th century, many scientists believed that protein was the hereditary material.

_____ 26. Mice injected with heat-killed *Pneumococcus* of a virulent strain will not becomes diseased.

_____ 27. Griffith transformed *Pneumococcus* by injecting mice with purified DNA along with bacteria.

_____ 28. In the Hershey-Chase studies, ^{32}P was used to radioactively label bacteriophage DNA.

_____ 29. Hershey and Chase were the first to suggest that DNA was the genetic material.

_____ 30. Adenine and guanine are purines, molecules with double rings.

_____ 31. According to Chargaff's rule, cells with 22% adenine should have 28% guanine.

_____ 32. Adenine and thymine form three hydrogen bonds, while guanine and cytosine form two.

_____ 33. The two strands of double-helical DNA have 3' to 5' directions opposite to one another.

_____ 34. In DNA replication, only one of the two original strands serves as a template for new synthesis.

_____ 35. In DNA, adjacent nucleotides are joined by bonds between 3' -OH and 5' phosphate groups.

_____ 36. Unwinding of the DNA double helix for replication is facilitated by the enzyme helicase.

_____ 37. Growing DNA strands elongate in a 3' to 5' direction.

_____ 38. Beadle and Tatum demonstrated conclusively that DNA replication is semi-conservative.

_____ 39. In each eukaryotic chromosome, DNA replication begins at one specific point.

_____ 40. Alkaptonuria is a disorder that is probably due to a defect at a single genetic locus.

C. Fill in the Blanks

Answer the question or complete the statement by filling in the blanks with the correct word or words.

41. In his transformation studies, Griffith used virulent bacteria that can cause ________________.

42. Hershey and Chase used ________________ to selectively label DNA and ________________ to selectively label protein in bacteriophages.

43. The parts of a DNA nucleotide are ________________, ________________, and ________________.

44. The nitrogen bases with a single six-cornered ring of nitrogen and carbon are the ________________.

45. If a human cell DNA has about 58% A and T bases, it has about ________________% C and G bases.

46. Wilkins and Franklin provided information on DNA structure by using ________________ techniques.

47. Watson and Crick concluded that there should be ________________ nucleotide pairs per helical turn.

48. The primary enzyme responsible for DNA replication is ________________.

49. DNA replication is continuous on the ________________ strand.

50. Okazaki fragments are characteristic of the ________________ strand during replication.

51. The enzyme ________________ joins adjacent segments of DNA.

52. Phenylketonuria and alkaptonuria result from defects in enzymes that act in metabolism of the amino acids ________________ and ________________.

53. Meselson and Stahl demonstrated that DNA replication is ________________.

54. Microorganisms are often good for genetic research, since they produce enormous numbers in a short time, and unlike most higher plants and animals, their chromosome number is usually ________________.

Questions for Discussion

1. Trace the development of the idea that molecules of DNA are the vehicles of hereditary information. Begin with the work of Miescher, and continue with the work of Griffith, the work of Avery, MacLeod, and McCarty, and finally that of Hershey and Chase.

2. As Watson and Crick worked on solving the structure of DNA, they were not alone. Outline what was known about the structure of DNA prior to Watson and Crick's remarkable insight.

3. John is an albino because he lacks an enzyme necessary for the production of melanin. Mary is also an albino, and she too lacks an enzyme essential to melanin production. Their children, however, are all normally pigmented. Explain this in terms of Garrod's concept of inborn errors of metabolism and Beadle and Tatum's ideas of one gene, one enzyme.

Answers to Self-Exam

Multiple Choice Questions

1. a
2. b
3. c
4. a
5. d
6. b
7. b
8. d
9. a
10. c
11. c
12. d
13. c
14. b
15. a
16. c
17. b
18. a
19. b
20. b
21. c

True or False Questions

22. F
23. F
24. F
25. T
26. T
27. F
28. T
29. F
30. T
31. T
32. F
33. T
34. F
35. T
36. T
37. T
38. F
39. F
40. T

Fill in the Blank Questions

41. pneumonia
42. ^{32}P; ^{35}S
43. deoxyribose, nitrogen bases, and phosphate groups
44. pyrimidines
45. 42%
46. x-ray crystallography
47. ten
48. DNA polymerase
49. leading
50. lagging
51. ligase
52. phenylalanine; tyrosine
53. semi-conservative
54. haploid

CHAPTER 13 Genes in Action

Learning Objectives

After mastering the material covered in chapter 13, you should be able to confidently do the following tasks:

- Explain the essential features of the central dogma of molecular biology.
- Point out the important differences between DNA and RNA.
- Distinguish the structures and functions of the three varieties of RNA — rRNA, mRNA, and tRNA.
- Show how rRNA and ribosomal proteins make up mature ribosomes.
- Describe how mRNA is synthesized and how the DNA information is transcribed into RNA information.
- Describe the posttranscriptional modifications that occur with eukaryotic mRNA.
- Outline the steps that occur in translation of the information in an mRNA nucleotide sequence into the amino acid sequence of a polypeptide.

Chapter Outline

I. RNA Structure and Transcription.

  A. Comparing RNA to DNA.
  B. Transcription: RNA synthesis.

II. Varieties of RNA.

  A. Ribosomal RNA and the ribosome.
  B. Messenger RNA.
    1. messenger RNA and the genetic code.
    2. eukaryotic mRNA: exons and introns.
    3. messenger RNA in prokaryotes.
  C. Transfer RNA.

III. Translation: How Polypeptides Are Assembled.

  A. Initiation.
  B. Elongation.
  C. Termination.
  D. Polyribosomes.
  E. Free and bound ribosomes.

Key Words

central dogma
DNA
RNA
polypeptide
ribose
deoxyribose
adenine (A)
guanine (G)
cytosine (C)
thymine (T)
uracil (U)
transcription
gene
coding region
RNA polymerase
promoter
template DNA strand
nontranscribed DNA strand
termination signal
ribosomal RNA (rRNA)
ribosome
nucleolus
nucleolar organizing region
primary transcript
18S rRNA
5.8S rRNA
28S rRNA
5S rRNA
ribosomal proteins
large ribosomal subunit
small ribosomal subunit
messenger RNA (mRNA)
codon
colinear
stop codons (UAA, UAG, and UGA)
synonymous codons
degenerate code
initiation codon (AUG)
5' leading region
leader
cistron
3' trailing region
trailer
posttranscriptional
modification
capping
methylated cap
poly-A tail
intervening sequences
intron
splicing
spliceosome
polycistronic
translation
transfer RNA (tRNA)
charging enzymes
anticodon
amino acid attachment site
initiation
P site
A site
methionine
N-formyl methionine
initiation proteins
initiation complex
amino-terminal end
N-terminus
C-terminus
reading frame
elongation
peptide bond
translocation
termination
polyribosomes
polysomes
free ribosomes
bound ribosomes
rough endoplasmic reticulum (rough ER)
signal sequence
receptor site
signal peptidase

Exercises

1. Here is a short DNA sequence, taken from a much longer DNA molecule; it has all the information necessary to produce a short polypeptide:

```
5'  ATGGCAGAGAAGTGTTCTTACTAG  3'
3'  TACCGTCTCTTCACAAGAATGATC  5'
```

In the spaces below, write the nucleotide sequences of the template strand of DNA, the codons of the transcribed mRNA, the anticodons of the appropriate tRNA molecules, and, finally, the sequence of amino acids specified by this nucleotide sequence. To decipher the genetic code, use table 13.1. Be sure to indicate the 5' and 3' ends of the nucleic acids.

DNA template strand sequence
___ ___ ___ ___ ___ ___ ___ ___

mRNA codon sequence
___ ___ ___ ___ ___ ___ ___ ___

tRNA anticodons
___ ___ ___ ___ ___ ___ ___ ___

amino acid sequence
___ ___ ___ ___ ___ ___ ___ ___

2. Below is a diagram of a pre-mRNA copy of a segment of eukaryotic DNA. It will eventually become mRNA. Your task is to label this diagram and then to draw and label a diagram of the mature mRNA resulting from this primary transcript. Label such features as introns, exons, leaders, trailers, initiation and stop codons, methylated caps, poly-A tails, 5' ends, and 3' ends. Indicate which sequences are transcribed, and show which sequences are translated. Be sure to indicate all the modifications that take place to produce mature eukaryotic mRNA.

AUG UGA

Self-Exam

You should be able to easily answer the following questions after learning the material in chapter 13. If you have difficulty with any question, study the appropriate section in the text and try again.

A. Multiple Choice Questions

Circle one alternative that best completes the statement or answers the question.

1. Which of the following direct information transfers is *not* consistent with the so-called central dogma?
   a. DNA → DNA.
   b. DNA → RNA.
   c. RNA → protein.
   d. DNA → protein.

2. Which of the following is *not* a general difference between DNA and RNA?
   a. The sugar in RNA is ribose, not deoxyribose.
   b. RNA contains the base uracil rather than thymine.
   c. DNA is usually double-stranded, while RNA is usually single-stranded.
   d. RNA molecules are generally much longer than DNA molecules.

3. The transcription of information from DNA to RNA is accomplished by the enzyme
   a. DNA polymerase.
   b. RNA polymerase.
   c. DNA ligase.
   d. RNA kinase.

4. The enzyme RNA polymerase identifies and binds to
   a. the promoter region of the DNA sequence for a gene.
   b. the leader region of the DNA sequence for a gene.
   c. the promoter region of the RNA sequence for a gene.
   d. the leader region of the RNA sequence for a gene.

5. Synthesis of a new RNA strand from a DNA template occurs
   a. from the 3' end to the 5' end of the new RNA strand.
   b. from the 5' end to the 3' end of the new RNA strand.
   c. from the 5' end to the 3' end of the template DNA strand.
   d. from both strands of the original DNA.

6. The RNA transcript copied from the DNA strand sequence 3' TACAGACGCTGT 5' would be
   a. 3' TACAGACGCTGT 5'.
   b. 3' UACAGACGCUGU 5'.
   c. 5' AUGUCUGCGACA 3'.
   d. 5' ATGTCTGCGACA 3'.

7. Which of the following is *not* a common feature of mRNA molecules?
   a. promoter sequence.
   b. leader sequence.
   c. start codon.
   d. trailer sequence.

8. A complete ribosome is made up of
   a. a large (60S) subunit, a small (40S) subunit, and about twenty individual amino acids.
   b. mRNA, tRNA, rRNA, and protein.
   c. nucleolar organizing regions.
   d. several different rRNA molecules and dozens of specific ribosomal proteins.

9. In cells, proteins can be synthesized
   a. in the nucleus.
   b. at ribosomes.
   c. in the nucleus and at ribosomes.
   d. in the nucleus and in the endoplasmic reticulum.

10. Peptide bonds are formed between amino acids by the enzyme
    a. RNA polymerase.
    b. peptidyl ligase.
    c. peptidyl transferase.
    d. tRNA charging enzyme.

11. A single codon most often specifies
    a. a single amino acid.
    b. a few amino acids.
    c. a single polypeptide.
    d. a few polypeptides.

12. In eukaryotes, the so-called start codon or initiator is
    a. UAG, and it specifies methionine.
    b. AUG, and it specifies methionine.
    c. AUG, and it specifies arginine.
    d. either AUG or UAG.

13. How many different codons are possible in the genetic code?
    a. 4.
    b. 20.

c. 60.
d. 64.

14. Which of the following is *not* a posttranscriptional modification to eukaryotic mRNA?
 a. introns removed.
 b. 5' end receives a methylated "cap."
 c. 3' end receives a poly-A "tail."
 d. molecule is charged by charging enzyme.

15. In eukaryotes, which classes of RNA are modified after transcription before becoming functional?
 a. mRNA.
 b. mRNA and tRNA.
 c. mRNA and rRNA.
 d. mRNA, tRNA, and rRNA.

16. The 3' end of all mature tRNA molecules ends with the sequence
 a. AUG.
 b. UAA, UAG, or UGA.
 c. CCA.
 d. GGU.

17. In polypeptide assembly, what occurs during translocation?
 a. a tRNA is moved from the P site to the A site on the ribosome.
 b. a tRNA is moved from the A site to the P site on the ribosome.
 c. the mRNA molecule is moved from the P site to the A site.
 d. the mRNA molecule is moved from the A site to the P site.

18. After initiation of protein synthesis, a peptide bond is formed between
 a. the amino group of the methionine and the carboxyl group of the next amino acid.
 b. the carboxyl group of the methionine and the amino group of the next amino acid.
 c. the amino group of the methionine and the amino group of the next amino acid.
 d. the carboxyl group of the methionine and the carboxyl group of the next amino acid.

19. How many amino acids would occur in a polypeptide coded by the sequence
 AUG UAC GCU UGC UAG UGA GUA UAG ?
 a. 8.
 b. 7.
 c. 5.
 d. 4.

20. Which of the following is a feature of a polypeptide destined for the endoplasmic reticulum?
 a. polysomes.
 b. signal sequence.
 c. receptor site.
 d. signal peptidase.

B. True or False Questions

Mark the following statements either T (True) or F (False).

_____ 21. Mature mRNA contains only nucleotide sequences that code for amino acids .

_____ 22. The five-carbon sugar in tRNA is ribose.

_____ 23. In general, DNA molecules are much longer than RNA molecules.

_____ 24. All of the RNA that is transcribed from DNA encodes information specifying amino acids.

_____ 25. A new strand of RNA is synthesized in the 3' to 5' direction, as nucleotides are added to the 5' end.

_____ 26. Transcription of a DNA strand continues until the RNA polymerase reaches a termination signal.

_____ 27. Only a single RNA molecule can be transcribed from a particular gene at one time.

_____ 28. The are many different type of tRNA but only a single type of rRNA.

_____ 29. The smaller ribosomal subunit has about 30 different proteins, and the larger subunit has 45 - 50.

_____ 30. Ribosomes are the only places where proteins are synthesized.

_____ 31. DNA can make proteins directly as well as direct synthesis of RNA copies of DNA.

_____ 32. CCU, CCC, CCA, and CCG all code for proline; they are synonymous codons.

_____ 33. In eukaryotes, all proteins begin with methionine as the first amino acid.

_____ 34. The leader portion of an mRNA molecule occurs at the 3' end.

_____ 35. The removal of introns and splicing of RNA apparently occurs in the nucleus of the cell.

_____ 36. Mature tRNA molecules contain only the nucleotides A, U, G, and C.

_____ 37. The mRNA codon 5' GCU 3' would pair with the tRNA anticodon 3' CGA 5'.

_____ 38. After polypeptide synthesis is complete, the initial methionine will still have a free carboxyl group.

_____ 39. In general, polypeptide synthesis is more rapid in humans than in a bacterium like *E. coli*.

_____ 40. In bacteria, bound ribosomes are seen bound to the surface of the endoplasmic reticulum.

C. Fill in the Blanks

Answer the question or complete the statement by filling in the blanks with the correct word or words.

41. The concept that biological information is transferred from DNA to RNA to protein is termed by biologists the ________________.

42. The enzyme that binds to a promoter sequence to copy genetic information is ________________.

43. DNA transcription proceeds until the enzyme reaches a ________________.

44. The chromosomal regions containing the genes for rRNA are the ________________.

45. Since there is a one-for-one correspondence between mRNA and protein sequence, we say the information they contain is ________________.

46. A genetic code containing synonymous codons is said to be a ________________ code.

47. Noncoding sequences found within the coding portion of mRNA's are termed ________________.

48. The RNA-protein complexes where noncoding sequences are excised are called ________________.

49. Some prokaryotic mRNA's contain information for several polypeptides; they are ________________.

50. The portion of a tRNA molecule that recognizes the mRNA codon is the ________________.

51. Together, the initiation proteins, the smaller subunit, a methionine tRNA, and the initiation codon make up the ribosomal ________________.

52. When a tRNA is move from the A site to the P site of the ribosome, it is termed ________________.

53. The nucleotide triplets UAA, UAG, and UGA are ________________.

54. Several ribosomes bound together by a single mRNA molecule form ________________.

55. Ribosomes attached to one side of endoplasmic reticulum are termed ________________.

56. Polypeptides that are destined for the endoplasmic reticulum identify receptor sites with a segment of amino acids termed a ________________.

Questions for Discussion

1. Today we know the genetic code specifies particular amino acids with triplets of four different nucleotides. But why does each codon have three nucleotides rather than two or four? And why does the code utilize four different nucleotides, rather than some other number? Explain how the genetic code would have differed if there had been a greater variety of nucleotides involved (say, 5 or 10 or 26). What if there were many fewer different amino acids in protein (maybe 10 or 12 rather than 20)? What if there were more than 20 amino acids?

2. The central dogma states, in its simplest terms, that information flows from DNA to RNA to protein. Could protein code for DNA? That is, can you envision a mechanism for reversing information flow, opposite to that of the central dogma? Would additional cellular machinery be necessary? Is discovery of information flow from protein to DNA likely?

3. AUG seems to be universal as an initiation codon. What would happen if the first nucleotide beyond AUG were omitted by a mutation? What if the first two nucleotides beyond AUG were lost? What if the three nucleotides that follow AUG were lost? Is there a way to known whether . . . GAUGAUGAUGAU . . . would specify ...Asp Asp Asp... or ...Met Met Met... or ...stop stop stop...?

Answers to Self-Exam

Multiple Choice Questions

1. d
2. d
3. b
4. a
5. b
6. c
7. a
8. d
9. b
10. c
11. a
12. b
13. d
14. d
15. d
16. c
17. b
18. b
19. d
20. b

True or False Questions

21. F
22. T
23. T
24. F
25. F
26. T
27. F
28. F
29. T
30. T
31. F
32. T
33. T
34. F
35. T
36. F
37. T
38. F
39. F
40. F

Fill in the Blank Questions

41. central dogma
42. RNA polymerase
43. termination signal
44. nucleolar organizing regions
45. colinear
46. degenerate
47. introns or intervening sequences
48. spliceosomes
49. polycistronic
50. anticodon
51. initiation complex
52. translocation
53. stop codons
54. polyribosomes or polysomes
55. bound ribosomes
56. signal sequence

CHAPTER 14 Gene Regulation

Learning Objectives

After mastering the material covered in chapter 14, you should be able to confidently do the following tasks:

- Contrast the approaches to regulation of gene expression taken by prokaryotes and eukaryotes.
- Distinguish between inducible control and repressible control of operons, and give examples of each.
- Outline the places in the pathway of gene expression in eukaryotes where regulation may occur.
- Explain inducible regulation in eukaryotes, and give examples of genes with inducible regulation.
- Show how eukaryotic DNA packaging may affect transcription of genes.
- Describe posttranscriptional events that may be involved in eukaryotic regulation.

Chapter Outline

I. Prokaryotic Gene Regulation: The Operon.

A. Inducible operons.
1. the *lac* operon.

B. Repressible operons.
2. the *trp* operon.

II. Eukaryotic Gene Regulation.

A. Opportunities for regulation.

B. Inducible genes.
1. heat shock genes.
2. steroid hormone induced genes.

C. DNA packaging: a role in regulation.
1. chromatin: heterochromatin and euchromatin.
2. methylation.

D. Posttranscriptional regulation.

Key Words

gene
coding region
regulatory region
5' flanking region
promoter
downstream
upstream
constitutive transcription
operon

repressor
polycistronic
F. Jacob
J. Monod
operator
lac operon
lactose
β galactosidase
permease
trans-acetylase
inducible operon
inducer
lac repressor
i gene
glucose
catabolite activator protein (CAP)
cyclic AMP (cAMP)
positive control
negative control
trp operon
repressible operon
tryptophan
biosynthetic pathway
trp repressor
co-repressor
eukaryotic regulation
heat shock protein (*Hsp*)
thermal induction
TATA box
TATA box binding (TAB) proteins
heat shock factor (HSF)
heat shock element
steroid hormone
endocrine glands
target cells
chicken oviducts
egg albumin
hormone receptor
estrogen receptor protein
adaptor
DNA packaging
chromatin
nucleosome
histone proteins
30nm chromatin fiber
heterochromatin
constitutive heterochromatin
facultative heterochromatin
euchromatin
Barr body
Drosophila larval salivary glands
polytene chromosomes
chromosome puffs
H1 histone
DNA methylation
cytosine residues
posttranscriptional regulation
introns
exons
mouse α amylase gene
homeotic genes
antennapedia
homeo box

Exercises

1. Below is a diagrammatic sketch of the *lac* operon in *E. coli*. Please provide labels to show the relative locations of the following parts of the *lac* operon: the coding sequences for the trans-acetylase (the a gene), β-galactoside permease (the y gene), β-galactosidase (the z gene), and the repressor protein (the i gene). Show the locations of the promoter and the operator, and show where CAP binds. Indicate the "upstream" and "downstream" conventions for this operon. (To help orient you, the coding sequences are shaded.) Check your work with figure 14.2 in the text.

2. The other players in *lac* operon function include the inducer (lactose), the repressor protein, and RNA polymerase. Use the symbols at the right to show where these molecules interact in allowing or restricting transcription if the *lac* operon in (a) the presence of lactose and (b) the absence of lactose.

RNA polymerase   repressor   inducer

a. Lactose present:

b. Lactose absent:

Self-Exam

You should be able to easily answer the following questions after learning the material in chapter 14. If you have difficulty with any question, study the appropriate section in the text and try again.

A. Multiple Choice Questions

Circle one alternative that best completes the statement or answers the question.

1. The place in a gene that RNA polymerase binds to begin transcription is
   a. the coding region.
   b. downstream from the coding region.
   c. upstream from the coding region.
   d. in the mRNA transcript.

2. Which of the following statements describes a feature of prokaryotic gene regulation?
   a. Steroid hormone receptors are often involved in regulation.
   b. Chromosomal proteins help determine which genes are transcribed.
   c. The transcripts are often polycistronic and governed by one operator.
   d. Alternative splicing of exons and introns can be involved in regulating expression.

3. A repressor can block transcription by binding to
   a. the operator.
   b. the inducer.
   c. the RNA polymerase.
   d. the protein products.

4. Which of the following is *not* a product resulting from the *lac* operon?
   a. lactose.
   b. β-galactosidase.
   c. permease.
   d. trans-acetylase.

5. What is the inducer for the *lac* operon?
   a. lactose.
   b. β-galactosidase.
   c. promoter.
   d. repressor protein.

6. The number of *lac* repressor molecules in a typical *E. coli* is approximately
   a. 10 or so.
   b. 100 - 200.
   c. 1,000 - 2,000.
   d. 1,000,000 or more.

7. With the *lac* operon, no transcription will occur if
   a. RNA polymerase can bind to promoter.
   b. inducer is bound to repressor.
   c. repressor is bound to operator.
   d. inducer is bound to operator.

8. Which of the following typically happens when the level of glucose in an *E. coli* cell falls to a low level?
   a. The level of lactose goes up.
   b. The level of cyclic AMP (cAMP) increases.
   c. The level of cyclic AMP (cAMP) decreases.
   d. Catabolite activator protein (CAP) will be unable to bind to the *lac* promoter.

9. Which conditions will result in the greatest rate of transcription of the *lac* operon in *E. coli*?
   a. High levels of both glucose and lactose in the cells.
   b. High level of glucose and absence of lactose in the cells.
   c. High level of lactose and absence of glucose in the cells.
   d. Low levels or absence of both glucose and lactose in the cells.

10. In the operation of the *trp* operon in *E. coli*, molecules of the tryptophan act as
   a. inducer molecules.
   b. repressor molecules.
   c. co-repressor molecules.
   d. promoters.

11. The *trp* operon of *E coli* is likely to be transcribed at the greatest rate
   a. when there is plenty of tryptophan in the cell.
   b. when there is a shortage of tryptophan in the cell.
   c. when the *lac* operon is being transcribed at the maximum rate.
   d. when it is actively repressed.

12. Evidence suggests that regulation of heat shock protein genes in *Drosophila* is accomplished by
   a. an operon much like the *lac* operon of *E. coli.*
   b. steroid hormone receptors in the cell membrane.
   c. selective splicing of transcripts.
   d. protein interaction with the DNA of the promoter upstream from an *Hsp* gene.

13. Efficient transcription of *Hsp* genes in *Drosophila* requires the presence of
   a. RNA polymerase.
   b. RNA polymerase and TAB proteins.
   c. RNA polymerase and heat shock factor, HSF.
   d. RNA polymerase, TAB proteins, and HSF.

14. Albumin in chicken eggs is a protein produced by oviduct cells; synthesis of this protein is regulated by
   a. a repressor - co-repressor feedback system.
   b. a hormone - receptor complex that interacts with the gene's promoter.
   c. an intron that is removed from mRNA to increase translation.
   d. the presence or absence of chromosomal histone H1.

15. Eukaryotic chromosomes that are being transcribed are usually termed
   a. euchromatin.
   b. constitutive heterochromatin.
   c. facultative heterochromatin.
   d. highly condensed chromatin.

16. The chromatin found near centromeres is usually
   a. euchromatin.
   b. constitutive heterochromatin.
   c. facultative heterochromatin.
   d. noncondensed chromatin.

17. A Barr body is a well-known example of
   a. euchromatin.
   b. constitutive heterochromatin.
   c. facultative heterochromatin.
   d. noncondensed chromatin.

18. Which of the nucleotides in some eukaryotic DNA is frequently modified by methylation?
   a. guanine.
   b. cytosine.

c. adenine.
d. thymine.

19. An example of different promoters for a gene functioning in different tissues is seen with
    a. the *lac* operon in *E. coli.*
    b. the albumin gene in chicken oviduct cells.
    c. chromosome puffs in *Drosophila* salivary glands.
    d. the α-amylase gene in mouse.

20. Which of the following is an example of a homeotic gene?
    a. the *trp* operon in *E. coli.*
    b. heat shock protein (*Hsp*) genes in *Drosophila.*
    c. antennapedia gene in *Drosophila.*
    d. the α-amylase gene in mouse.

B. True or False Questions

Mark the following statements either T (True) or F (False).

_____ 21. Most "genes" include both coding sequences and regulatory regions.

_____ 22. The promoter is typically found in the first part of the sequence that codes for amino acids.

_____ 23. Jacob and Monod are the scientists who developed the concept of the operon.

_____ 24. Polycistronic genes are commonly encountered in eukaryotic chromosomes.

_____ 25. Although the *lac* operon is inducible, it can be repressed by the presence of lactose.

_____ 26. In *E. coli*, β-galactosidase, permease, and trans-acetylase are transcribed in the presence of lactose.

_____ 27. If both glucose and lactose are available, *E. coli* will utilize glucose first.

_____ 28. When the level of glucose falls in an *E. coli* cell, the level of cyclic AMP decreases, too.

_____ 29. Like the *lac* operon, the *trp* operon is transcribed only when tryptophan is abundant.

_____ 30. The tryptophan molecule itself is the co-repressor for the *trp* operon.

_____ 31 Regulatory mechanisms are known to be pretty much the same in prokaryotes and eukaryotes.

_____ 32. In *Drosophila*, *Hsp* genes are not usually transcribed but are very rapidly induced if temperature rises.

_____ 33. Eukaryotes can activate a particular gene in certain cells, leaving it inactive in others.

_____ 34. With a hormone-induced gene, the hormone usually affects all cells indiscriminately.

_____ 35. Nucleosomes are present with DNA at all times in eukaryotic chromosomes.

_____ 36. All heterochromatin is considered to be transcriptionally inert.

_____ 37. A Barr body is a heterochromatic chromosome.

_____ 38. The giant larval salivary gland chromosomes in *Drosophila* are totally inactive.

_____ 39. Methylation of cytosines is one way to increase transcription of genes.

_____ 40. In some cases, two different proteins can be produced from the same transcript of a gene.

C. Fill in the Blanks

Answer the question or complete the statement by filling in the blanks with the correct word or words.

41. The portion of a gene that controls its expression is the ________________ region.

42. A gene that is transcribed continuously is said to be ________________.

43. Jacob and Monod called the grouping of regulatory sequences, together with the coding sequences they control, an ________________.

44. The enzymes resulting from the *lac* operon enable *E. coli* to digest ________________.

45. The inducer molecule for the *lac* operon is ________________.

46. The accessory protein that allows RNA polymerase to bind efficiently to the *lac* operon promoter is ________________.

47. Because it contains coding sequences for several enzymes, the *trp* operon is ________________.

48. In regulating transcription of the *trp* operon, tryptophan is the ________________.

49. In *Drosophila*, a sudden shift to elevated temperatures induces production of ________________.

50. In eukaryotic genes, the area in the promoter where RNA polymerase begins transcription is called the ________________.

51. The specific cells that a hormone may influence are termed ________________.

52. Transcription of the genes coding for albumin in chicken oviduct cells is stimulated by ________________.

53. To respond to a hormone, a target cell must have the appropriate ________________ protein.

54. The proteins responsible for nucleosome formation in eukaryotic chromosomes are ________________.

55. The chromosomal material found near the centromeres of eukaryotic chromosomes is termed ________________ heterochromatin.

56. The salivary glands of *Drosophila* have giant "endoreplicated" ________________ chromosomes.

57. In eukaryotic DNA, cytosine residues sometimes are modified by the addition of a ________________ group to the pyrimidine ring.

58. The gene for a-amylase in mouse has two ________________; one functions in salivary gland cells and the other in liver cells.

59. Major regulatory genes that control the expression of other genes are sometimes termed ________________ genes.

60. In *Drosophila*, one type of homeotic mutant has legs rather than antennae on the head; this condition is called ________________.

Questions for Discussion

1. In general, all the cells of a higher organism have all of the genetic information present in the original zygote. For example, we know that a single cell from a mature carrot can be induced to produce a complete carrot plant. However, only a fraction of the genes present in most mature cells are actually expressed. Explain the major levels at which regulation of expression can take place in eukaryotic cells and give examples of mechanisms at each level.

2. We know of pathways in *E. coli* where several proteins are produced at the same time, such as lactose utilization and tryptophan synthesis. We also know of systems in *Drosophila* where several proteins are induced at the same time, such as the heat shock proteins. Compare the regulatory mechanisms used in these examples, and show how different approaches achieve similar results: simultaneous transcription of several protein-coding sequences.

3. Discuss how the following variables can affect the concentration of a particular protein in a cell: rate of transcription, rate of translation, metabolic lifespan of mRNA molecules, metabolic lifespan of the protein molecules.

Answers to Self-Exam

Multiple Choice Questions

1. c
2. c
3. a
4. a
5. a
6. a
7. c
8. b
9. c
10. c
11. b
12. d
13. d
14. b
15. a
16. b
17. c
18. b
19. d
20. c

True or False Questions

21. T
22. F
23. T
24. F
25. F
26. T
27. T
28. F
29. F
30. T
31. F
32. T
33. T
34. F
35. F
36. T
37. T
38. F
39. F
40. T

Fill in the Blank Questions

41. regulatory
42. constitutive
43. operon
44. lactose
45. lactose
46. catabolite activator protein (CAP)
47. polycistronic
48. co-repressor
49. heat shock proteins
50. TATA box
51. target cells
52. estrogen
53. receptor
54. histones
55. constitutive
56. polytene
57. methyl
58. promoters
59. homeotic
60. antennapedia

CHAPTER 15 Prokaryotic Genetics

Learning Objectives

After mastering the material covered in chapter 15, you should be able to confidently do the following tasks:

- Explain why viruses and bacteria are so useful in studying basic genetic phenomena.
- Distinguish DNA viruses from RNA viruses, and describe how their genomes are replicated.
- Compare the lytic cycle and the lysogenic cycle for bacteriophages.
- Describe the ways that genetic recombination can occur between bacterial cells.
- Explain what plasmids are and what role they play in bacterial genetics.
- Show how a genetic map can be made from conjugation studies of *E. coli*.

Chapter Outline

I. Viruses.

A. The life and times of the bacteriophage.
B. Lysogeny, an alternate cycle.
C. Counting viruses.
D. RNA viruses.

II. Bacteria.

A. Locating mutant and recombinant bacteria.
B. Recombination in bacteria.
1. transformation.
2. transduction.
3. plasmids and conjugation.
4. mapping the *E. coli* chromosome.
5. sexduction.
C. Plasmid genes.

Key Words

Escherichia coli
bacteriophage
phage virus
phage T2
polyhedral protein head
tail
tail fibers
restriction enzymes
bacterial host

bacteriophage parasite
lyse
lytic cycle
temperate phage
lysogenize
lysogeny
bacterial lawn
plaque
RNA viruses
reverse transcriptase
RNA replicase
minimal medium (media)
penicillin
wild-type bacteria
nutritional mutant
agar
Lederberg
replica plating
bacterial recombination
transformation
transduction
plasmid
episome
conjugation
pilus (-i)
conjugation tube
sex pilus (-i)
plasmid-plus bacterium
plasmid-minus bacterium
F episome
Lederberg and Tatum
F+ strain
F- strain
markers
unselected loci
fertility factor
F plasmid
Hfr strains
Jacob and Wollman
sexduction
partial diploid
R6 plasmid
antibiotic resistance
rheumatic heart disease
Streptococcus
salmonellosis
Salmonella typhimurium
vectors

Exercises

1. When working with nutritional mutants of bacteria, different nutrient media may be used to select cells with different nutritional requirements. Let's consider an *E. coli* transformation study where transforming DNA comes from a donor strain with the genotype arg^+ thi^+ met^+. .The recipient strain is arg^- thi^- met^-, that is, it requires arginine, thiamine, and methionine be provided in the medium for it to reproduce and grow. After transformation, the recipient strain may have incorporated one or more of the wild-type genes from the recipient, or it may have incorporated none at all. A cell that has become wild type at all three loci will grow on minimal medium, but any other recipient will require one or more of the supplemental nutrients. The following table lists the possible genotypes of recipient cells after transformation. For each of the media listed, place a "+" if a genotype will reproduce and grow, or place a "-" if it will not grow.

Genotype	minimal	minimal +arg	minimal +thi	minimal +met	minimal +arg +thi	minimal +arg +met	minimal +met +thi	minimal +met +arg +thi
arg^- thi^- met^-	______	______	______	______	______	______	______	______
arg^- thi^- met^+	______	______	______	______	______	______	______	______
arg^- thi^+ met^-	______	______	______	______	______	______	______	______
arg^+ thi^- met^-	______	______	______	______	______	______	______	______
arg^- thi^+ met^+	______	______	______	______	______	______	______	______
arg^+ thi^+ met^-	______	______	______	______	______	______	______	______
arg^+ thi^- met^+	______	______	______	______	______	______	______	______
arg^+ thi^+ met^+	______	______	______	______	______	______	______	______

2. In conjugation experiments using Hfr strains of *E. coli* and a blender to interrupt conjugation after various lengths of time, a number of different Hfr strains can be used. These strains differ in the place that the F plasmid is found in the bacterial chromosome, and, as a consequence, they differ in which genetic loci will be the first transferred during conjugation. Except for the location of the F factor, all Hfr strains have the same sequence of genes on the bacterial chromosome. Below are the results that might be obtained from studies using five different Hfr strain. The data given are the minimum transfer times for particular genes in particular strains. Not every gene gets transferred from each Hfr strain. Using these data, complete the map of the *E. coli* chromosome, and indicate the relative time "distances" between each gene. Compare your map with figure 15.11 in the text.

Hfr strain "A" x F^-:
gal, 2 min.; bio, 2.5 min.; try, 10 min.;
his, 24.5 min; tyr, 34.5 min.

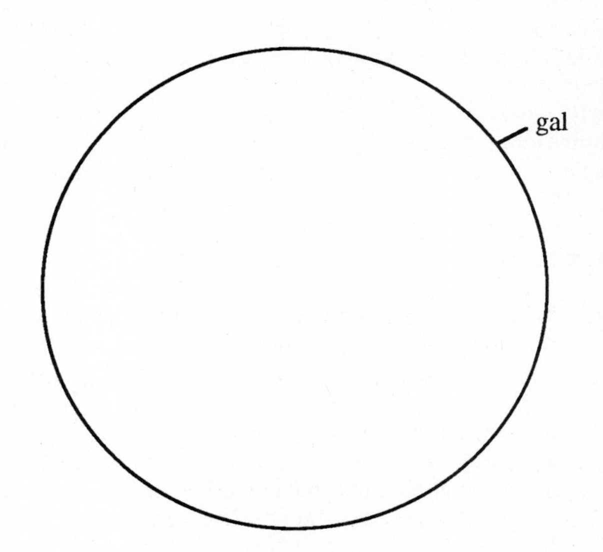

Hfr strain "B" x F^-:
thy, 6 min.; str, 9 min.; cys, 15 min.;
thi, 23 min.; met A, 25 min.

Hfr strain "C" x F^-:
bio, 3.5 min.; gal, 4 min.; lac, 10 min.;
leu, 19.5 min.; met A, 29 min.

Hfr strain "D" x F^-:
his, 1.5 min.; tyr, 11.5 min.; lys, 17.5 min.;
thy, 24 min.; str, 27 min.

Hfr strain "E" x F^-: thi, 3 min.; met A, 5 min.;
leu, 14.5 min.; lac, 24 min.

Self-Exam

You should be able to easily answer the following questions after learning the material in chapter 15. If you have difficulty with any question, study the appropriate section in the text and try again.

A. Multiple Choice Questions

Circle one alternative that best completes the statement or answers the question.

1. How would you best describe the way *E. coli* cells store their genetic information?
   a. They are haploid, with a single linear chromosome.
   b. They are haploid, with a single circular chromosome.
   c. They are diploid, with a pair of linear chromosomes.
   d. They are diploid, with a pair of circular chromosomes.

2. Which of these is a good estimate of how long it might take a culture with 400 *E. coli* cells to reach 800 cells?
   a. 20 seconds.
   b. 20 minutes.
   c. 200 minutes.
   d. 20 hours.

3. What might be a typical host for phage T2?
   a. a virus.
   b. an *E. coli* cell.
   c. a susceptible *Drosophila* cell.
   d. a susceptible human cell.

4. Restriction enzymes are a class of enzymes
   a. produced by eukaryotic cells to restrict bacterial invasion.
   b. produced by bacteriophages to resist host cell defenses.
   c. produced by bacterial cell to degrade certain DNA sequences.
   d. produced by RNA viruses to copy the RNA into DNA.

5. The cycle of bacteriophage invasion of a host, reproduction, and release of progeny phages is termed
   a. a lytic cycle.
   b. a lysogenic cycle.
   c. a transduction cycle.
   d. a serial dilution cycle.

6. In which of the following might Chargaff's rule concerning base composition *not* hold?
   a. a *Drosophila* reproductive cell.
   b. an *E. coli* cell.
   c. a DNA virus.
   d. an RNA virus.

7. How might a minimal medium be best described?
   a. It is a medium with no nutrients whatsoever.
   b. It is a medium that provides only the nutrients required to support wild-type cells and no more.
   c. It is a medium that will support nutritional mutants, at a minimum.
   d. It is a medium augmented with nutrients to support nutritional mutants.

8. An *E. coli* cell line that is trp^+, arg^-
   a. will grow on minimal medium.
   b. will grow on minimal medium supplemented with tryptophan.
   c. will grow on minimal medium supplemented with arginine.
   d. will grow on minimal medium supplemented with tryptophan and arginine.

9. Which of the following is *not* known to lead to bacterial recombination?
   a. transformation.
   b. transduction.
   c. reverse transcription.
   d. conjugation.

10. Which of the following processes involves a virus as a vector?
   a. transformation.
   b. conjugation.
   c. transduction.
   d. sexduction.

11. Which of the following is actually a plasmid?
   a. bacteriophage T2.
   b. the sex pilus.
   c. the *E. coli* chromosome.
   d. the fertility factor, F.

12. Hfr *E. coli* cells that are met^+ bio^- are allowed to conjugate with F^- met^- bio^+ cells. Which resulting cells will grow on a minimal medium supplemented with methionine?
   a. met^- bio^+.
   b. met^+ bio^-.
   c. met^+ bio^+.
   d. met^- bio^+ and met^+ bio^+.

13. *E. coli* cells that are F^+ are allowed to conjugate with F^- cells. What is the most likely result?
   a. The F^- cells are converted to F^+ cells.

b. The F^+ cells are converted to F^- cells.
c. The cells will no longer grow on minimal medium.
d. The F^+ become Hfr cells.

14, Which is the best description of the fertility factor F?
a. It is part of the typical *E. coli* chromosome.
b. It is normally a closed circle of DNA, independent of the bacterial chromosome.
c. It is normally a small linear piece of DNA, a plasmid.
d. It is responsible for conferring antibiotic resistance to many kinds of bacteria.

15. About how long does it take for an Hfr chromosome of *E. coli* to be completely transferred in conjugation?
a. about 20 minutes.
b. about one and one-half hours.
c. about 20 hours.
d. about one day.

16. In the process of sexduction, genes are transferred from one cell to another via
a. a bacteriophage or virus particle.
b. a piece of extracellular DNA.
c. bacterial DNA incorporated into the F factor.
d. another bacterial cell.

17. The R plasmid is best known for carrying genes that can
a. determine the fertility of the bacterial host cell.
b. confer resistance to some antibiotic to the bacterial host cell.
c. allow transformation to occur.
d. produce antibiotics.

18. Rheumatic heart disease is due to infections by
a. phage T2.
b. the R plasmid of many bacteria.
c. *Salmonella typhimurium.*
d. *Streptococcus.*

B. True or False Questions

Mark the following statements either T (True) or F (False).

_____ 19. A bacteriophage is a type of bacterium.

_____ 20. .Restriction enzymes will cleave DNA molecules but only at specific nucleotide sequences.

_____ 21. The enzyme RNA polymerase will readily transcribe bacterial DNA but is unable to transcribe viral DNA.

_____ 22. When a virus enters a lysogenic cycle, it causes the host bacterial cell to rupture.

_____ 23. A temperate bacteriophage does not necessarily kill its host cell.

_____ 24. Bacteriophages are normally grown in the laboratory on sterile nutrient agar plates free of all bacteria.

_____ 25. Some RNA viruses make RNA copies directly from RNA templates using RNA replicase.

_____ 26. Penicillin kills actively growing bacteria by interfering with cell-wall synthesis.

_____ 27. The minimal medium that wild-type bacterial cells grow on has no nutrients.

_____ 28. A strain of *E. coli* labeled trp^+, arg^- would require a medium supplemented with arginine to grow and reproduce.

_____ 29. Donor and recipient cells must be in direct contact for transformation to pass genes from one bacterium to another.

_____ 30. In transduction, a bacteriophage carries DNA from one bacterium to another.

_____ 31. Plasmids carry some of the genes necessary for conjugation in *E. coli.*

_____ 32. With *E. coli*, conjugation can occur between F^+ and F^- cells and between Hfr and F^- cells, but not between F^+ cells and Hfr cells.

_____ 33. The *E. coli* chromosome was discovered to be a circular DNA molecule.

_____ 34. In conjugations between Hfr and F^- cells, the entire *E. coli* chromosome is usually transferred.

_____ 35. Some *E. coli* cells can be completely diploid.

_____ 36. Most microbiologists believe there is little risk to widespread use of antibiotics in agriculture, even when there is no need for the antibiotics.

C. Fill in the Blanks

Answer the question or complete the statement by filling in the blanks with the correct word or words.

37. The most commonly studied organism in bacterial genetics is the colon bacterium _________________.

38. A virus that has a bacterial host is termed a _________________.

39. Many bacteria produce _________________ enzymes that cut foreign DNA at specific sequences.

40. _________________ phages are able to lysogenize host cells.

41. The hole or clearing in a bacterial "lawn" resulting from phage infection and lysis of cells is called a _________________.

42. Some RNA viruses direct synthesis of DNA copies of viral RNA using the enzyme _________________.

43. A food source that contains only those nutrients needed for wild-type bacterial cells to grow and reproduce is termed _________________.

44. An *E. coli* nutritional mutant that is trp^+ arg^- will grow on minimal medium supplemented with _________________.

45. Bacterial recombination can result from _________________, where bacterial DNA is carried from one bacterial cell to another by a bacteriophage.

46. The protein tube through which DNA passes during bacterial conjugation is a _________________.

47. For two *E. coli* cells to conjugate, one of them must have _________________.

48. A strain of *E. coli* cells with the F plasmid inserted in the main bacterial chromosome is today called an _________________ strain.

49. Genetically determined resistance to antibiotics can be transferred from one bacterium to another via ____________________.

50. We can refer to the carriers of DNA sequences between organisms, such as viruses, phages, and plasmids, as ____________________.

Questions for Discussion

1. Bacteriophages and other viruses are not placed in one of the kingdoms of living organisms. Why not? Outline the typical infectious cycles for phages, both lytic and lysogenic. Are phages themselves cells, or are they just chemical particles? How can phages be explained in defining "living" versus "nonliving"? What properties of living things are absent in phages?

2. Suppose you discover a new nutritional mutation in a strain of *E. coli*, mutant "X". Cells of this strain are unable to synthesize "X", which they must have to grow and reproduce. Outline how you might go about determining the location of the "X" gene on the *E. coli* chromosome. Remember, you have all the tools of the bacterial geneticist to use, such as marker strains with nutritional mutations, different minimal media and media with nutritional supplements, and various Hfr strains.

3. We have recently heard discussion about the health risks to humans from *Salmonella* bacteria that are often found in commercially available chicken that is not properly stored and cooked. There is little risk if the chicken is properly processed, stored at cold temperatures, and cooked completely, and if preparation areas and utensils are thoroughly cleaned. Why is such caution a better protection against *Salmonella* poisoning than would be widespread use of antibiotics? Why not just treat processed chicken with antibiotics before shipping it to market?

Answers to Self-Exam

Multiple Choice Questions

1. a
2. b
3. b
4. c
5. a
6. d
7. b
8. c
9. c
10. c
11. d
12. d
13. a
14. b
15. b
16. c
17. b
18. d

True or False Questions

19. F
20. T
21. F
22. F
23. T
24. F
25. T
26. T
27. F
28. T
29. F
30. T
31. T
32. T
33. T
34. F
35. F
36. F

Fill in the Blank Questions

37. *Escherichia coli*
38. bacteriophage
39. restriction
40. Temperate
41. plaque
42. reverse transcriptase
43. minimal medium
44. arginine
45. transduction
46. sex pilus or conjugation tube
47. an F plasmid
48. Hfr
49. an R plasmid
50. vectors

CHAPTER 16 Genetic Engineering, the Frontier

Learning Objectives

After mastering the material covered in chapter 16, you should be able to confidently do the following tasks:

- Explain the scope of activities that scientists call genetic engineering.

- Describe the molecular biology tools that are used in genetic engineering.

- Explain how a gene can be isolated, placed in a vector, and cloned.

- List the current practical applications of genetic engineering techniques.

- Discuss the societal and ethical implications of potential application of genetic engineering.

Chapter Outline

I. Just What Is Genetic Engineering?

II. DNA Technology.

A. Restriction enzymes.
B. Splicing and cloning DNA molecules.
1. libraries.
2. creating a library.
3. screening a library.

III. Applications of the Technology.

A. Manufacture of biomolecules.
B. Agriculture.
C. Medicine.
1. diagnosis.
2. gene replacement therapy.

IV. A Social Perspective.

Key Words

biotechnology
genetic engineering
"foreign" DNA
clone
vector
recombinant DNA
restriction endonuclease
restriction enzyme
sticky ends

restriction site
recognition sequence
Eco RI
restriction fragment
vector
plasmid
DNA ligase
transformation
positive selection
ampicillin
ampicillin resistance
pUC 19
tetracycline
tetracycline resistance
DNA library
genomic library
cDNA
cDNA library
reverse transcriptase
library screening
DNA hybridization probe
autoradiography
gene machine
kinase
insulin
signal peptide
essential amino acids
Agrobacterium tumefaciens
Ti plasmid
crown gall disease
dicotyledons
Tay-Sachs disease
sickle cell anemia
Huntington disease
β-globin gene
electrophoresis
amniocentesis
DNA fingerprint
hypogonadism
gonadotropin-releasing hormone
gene replacement therapy
human growth hormone
oncogenes
Southern blot
DNA sequencing
phage M13
dideoxy nucleotide triphosphate

Exercises

1. Draw a schematic diagram of how a foreign piece of DNA could be cloned. Begin with the diagram below of a bacterial cell with its main chromosome and a plasmid that can be used as a cloning vector. Use the restriction enzyme Eco RI. You have all the other molecular biology tools you need. Indicate at a minimum the roles of restriction endonuclease, sticky ends, ligase, host cell, vectors, etc. Check you work with figure 16.2 in the text.

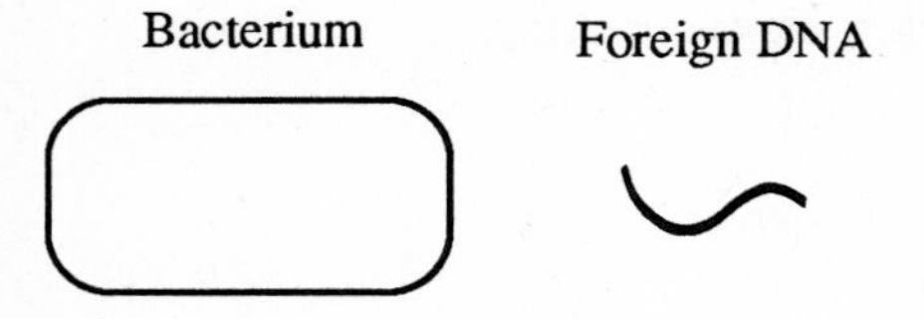

2. DNA sequencing has become a very important tool to the genetic engineer. Once the technique is mastered, interpreting the results is straightforward. Let's look at the results from sequencing a short piece of DNA using the Sanger technique discussed in chapter 16. Remember, a single nucleotide strand is synthesized complementary to one strand from the DNA whose sequence we are interested in. Four separate reactions terminate replication with either dideoxy A, dideoxy G, dideoxy C, or dideoxy T. Interpret the following diagram of a sequencing autoradiogram, and write the sequence of bases pairs (that is, both strands of the original DNA). Be sure to indicate 3' and 5' ends.

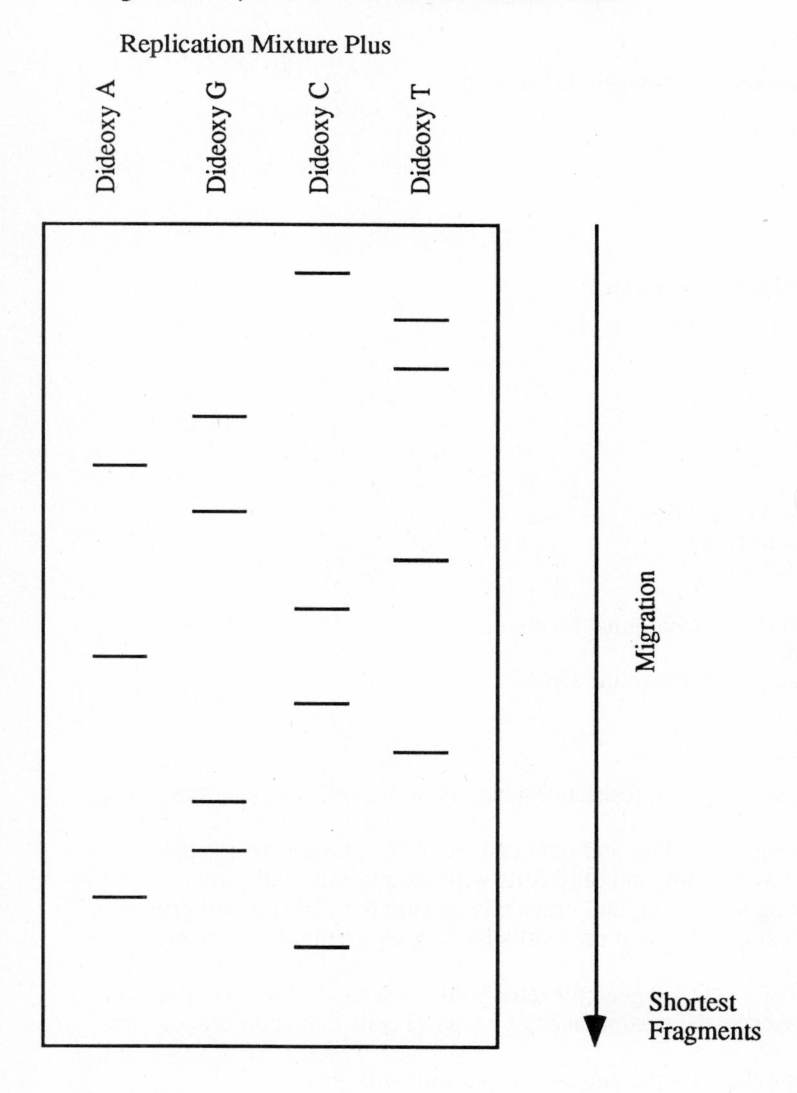

Self-Exam

You should be able to easily answer the following questions after learning the material in chapter 16. If you have difficulty with any question, study the appropriate section in the text and try again.

A. Multiple Choice Questions

Circle one alternative that best completes the statement or answers the question.

1. Which statement best describe what today's genetic engineers do?
   a. Genetic engineers are animal and plant breeders.
   b. Genetic engineers are trying to create new species of plants and animals.

c. Genetic engineers use biotechnology, for example, to produce wine by fermentation.
d. Genetic engineers manipulate DNA and introduce foreign DNA into cells.

2. An enzyme that cuts DNA at a specific nucleotide sequence is
   a. a restriction site.
   b. a restriction endonuclease.
   c. a restriction fragment.
   d. a DNA polymerase.

3. Which enzyme would be most likely to recognize the nucleotide sequence 5'...GAATTC...3' / 3' CTTAAG...5' ?
   a. DNA polymerase.
   b. reverse transcriptase.
   c. Eco RI.
   d. DNA ligase.

4. The most common vectors for cloning DNA fragments are
   a. bacterial plasmids and viruses.
   b. restriction endonucleases.
   c. *Drosophila* cells.
   d. genetic engineers.

5. What is HindIII most likely to do?
   a. It can cause an infection.
   b. It can degrade DNA to its constituent nucleotides.
   c. It can cut DNA at a certain nucleotide sequence.
   d. It can clone a DNA sequence in a bacterium.

6. To transform bacterial cells with plasmid DNA, what must be done?
   a. The bacterial cells must be disrupted.
   b. The bacterial cell walls must be made permeable to the DNA.
   c. The bacteria must be dormant.
   d. The bacteria must be conjugating.

7. Assume a plasmid cloning vector carries an ampicillin resistance gene. How are cells carrying the vector selected from those without the vector?
   a. Cells are cultured on a medium lacking ampicillin, and only cells with the plasmid will grow.
   b. Cells are cultured on a medium with ampicillin, and only cells with the plasmid will grow.
   c. Cells are cultured on a medium lacking ampicillin, and only cells lacking the plasmid will grow.
   d. Cells are cultured on a medium with ampicillin, and only cells lacking the plasmid will grow.

8. A foreign gene is spliced into the middle of a plasmid gene for tetracycline resistance, but most plasmids remain intact. Which of the following describe the performance of bacterial cells that carry one of these plasmids?
   a. On a medium with tetracycline, only cells with the engineered plasmid will grow.
   b. On a medium with tetracycline, only cells with the intact original plasmid will grow.
   c. On a medium lacking tetracycline, only cells with the engineered plasmid will grow.
   d. On a medium lacking tetracycline, only cells with the intact original plasmid will grow.

9. Which of the following is sometimes used as a cloning vector?
   a. Eco RI.
   b. *E. coli*.
   c. phage T2.
   d. pUC19.

10. If all of the DNA from an individual is cloned, what is the resulting collection?
    a. a genomic library.
    b. a cDNA library.

c. a restriction library.
d. a complementary library.

11. cDNA is obtained by using
 a. genomic DNA and a good cloning vector.
 b. mRNA, rRNA, tRNA, and RNA polymerase.
 c. mRNA and reverse transcriptase.
 d. electrophoresis and Southern blots.

12. A DNA library can be screened for the DNA of interest by using
 a. an appropriate cloning vector, like pUC19.
 b. a radioactive DNA probe, complementary to the DNA being sought.
 c. restriction endonucleases and electrophoresis.
 d. tissue culture and cell regeneration.

13. DNA fragments of different sizes can be separated by
 a. electrophoresis.
 b. a Southern blot.
 c. autoradiography.
 d. cloning.

14. If you wish to synthesize a radioactive DNA probe, how is it usually made radioactive?
 a. By adding radioactive nucleotide precursors during synthesis.
 b. By adding a radioactive phosphate group to the 3' end.
 c. By adding a radioactive phosphate group to the 5' end.
 d. By adding to the probe with an autoradiogram.

15. Which of the following was *not* part of the procedure used to produce human insulin commercially?
 a. mRNA for human insulin was isolated.
 b. A cDNA clone for human insulin was isolated.
 c. The cDNA for human insulin was spliced downstream from a promoter.
 d. Insulin was isolated from pig and cow pancreases and prepared for use.

16. A commonly used vector for introducing foreign DNA into plant cells is
 a. Eco RI.
 b. pUC19 plasmid.
 c. Ti plasmid.
 d. cDNA.

17. The disease caused by *Agrobacterium tumefaciens* is
 a. diabetes.
 b. crown gall.
 c. sickle cell anemia.
 d. Tay-Sachs.

18. Which kind of traits in crop plants might be most successfully altered by genetic engineering techniques?
 a. growth rate.
 b. size of fruit.
 c. time of harvest.
 d. herbicide resistance.

19. Genetic engineering techniques have seen the *least* application in which area of human concern?
 a. production of medically useful drugs and molecules.
 b. diagnosis of genetic disease.
 c. forensic techniques to identify criminals.
 d. gene replacement therapy to treat genetic disease.

20. Humans fail to achieve normal adult height if they are deficient in
    a. insulin.
    b. gonadotropin-releasing hormone.
    c. human growth hormone.
    d. introns and exons.

B. True or False Questions

Mark the following statements either T (True) or F (False).

_____ 21. Restriction enzymes are naturally occurring nucleases.

_____ 22. Restriction endonucleases will cut only invading viral DNA.

_____ 23. The enzyme Eco RI will cut double-stranded DNA every sixth base.

_____ 24. Restriction fragments generated by Eco RI have sticky ends.

_____ 25. To clone a particular DNA sequence, the naked DNA is first placed in bacterial cells.

_____ 26. The cloning vector pUC19 has both an ampicillin resistance gene and the lacZ gene.

_____ 27. A complete human genomic library could probably be maintained in a few thousand different clones.

_____ 28. An *E. coli* genomic library would be much smaller than a human genomic library.

_____ 29. The cDNA library of a fruit fly would include all expressed genes, including introns and exons.

_____ 30. A good genetic probe will hybridize only with DNA sequences that are complementary.

_____ 31. Radioactively labeled mRNA can sometimes be used as a probe for specific DNA sequences.

_____ 32. Natural human insulin is produced by human pancreatic tissue.

_____ 33. Genetic engineering techniques now offer the first opportunity to selectively improve crop plants.

_____ 34. Complete carrot plants can be grown from single carrot cells that have altered genes.

_____ 35. The pUC plasmid of *Agrobacterium tumefaciens* causes crown gall disease in dicotyledons.

_____ 36. Many quantitative plant traits such as growth rate and yield are polygenic traits.

_____ 37. The sickle cell genotype of a fetus at risk can be determined prenatally, using DNA techniques.

_____ 38. Many forensic scientists believe that an individual's DNA "fingerprint" is essentially unique.

_____ 39. Many human genetic diseases are effectively treated today with gene replacement therapy.

_____ 40. We now know that some cancers are caused by oncogenes in an individuals DNA.

C. Fill in the Blanks

Answer the question or complete the statement by filling in the blanks with the correct word or words.

41. A group of cells derived from a single cell and genetically identical are called a _________________.

42. The particular DNA sequence recognized by an enzyme like Eco RI is a ________________.

43. When Eco RI cleaves DNA, it leaves unpaired nucleotides that are called ________________.

44. The enzyme DNA ________________ can covalently join restriction fragments to vector DNA.

45. The enzyme Eco RI was isolated from the bacterium ________________.

46. A growth medium used for positive selection of cells will often contain an ________________.

47. A collection of thousands of different DNA clones derived from mRNA molecules is a ________________.

48. A particular DNA sequence may be detected among many other sequences by using a ________________.

49. A radioactive phosphate group can be added to a DNA sequence with the enzyme ________________.

50. A secretory protein has a ________________ at its N-terminal end.

51. The amino acids not manufactured by humans are termed ________________.

52. Plants that can be transformed by the Ti plasmid are all ________________.

53. Traits that are controlled by many genes are called ________________ traits.

54. In individuals with sickle cell anemia, the defective gene is the ________________ gene.

55. DNA from human fetuses can be obtained by means of ________________.

56. Individuals who fail to produce gonadotropin-releasing hormone have the condition ________________.

57. Prior to recombinant DNA techniques, the only source of human growth hormone was the ________________ of human cadavers.

58. Genes that cause cancer are termed ________________.

59. DNA fragments of different sizes can be easily separated using an ________________ process.

60. A nucleotide lacking a 3'-OH group is a ________________ nucleotide.

Questions for Discussion

1. Now that scientists have the potential to use the techniques of genetic engineering to alter some genetic characteristics of organisms, there has been some concern about humans "tinkering with nature" or "playing God." Is this really a totally new ability, or is it just application of new techniques to the long-time practice of crop improvement, selective breeding of domestic animals, and trying to minimize the incidence of genetic disease? Outline some of the ways that humans have tried to interfere with the heredity of plants, animals, and even humans, prior to the genetic engineering revolution. What kinds of new risks are there with the new techniques?

2. Outline the sequence of events that a team of scientists might follow in trying to begin commercial production of insulin in large-scale bacterial cultures. Be sure to include such steps as isolation and identification of the responsible gene, splicing the gene into a vector, and introducing it into bacterial cells in a way that the gene is expressed and insulin is produced.

3. Why would a large agricultural products corporation that manufactures herbicides be interested in using genetic engineering to develop plants that are genetically resistant to their herbicides?

4. Sickle cell disease is a genetic disease. It is considered to be lethal when homozygous. There is no way to "cure" the responsible allele. Simple tests can tell individuals if they are carriers of the disease and thus at risk of conceiving a homozygous recessive child. But until recently, the child's genotype could not be determined until after birth. The techniques of molecular biology now allow determining the fetal genotype early in pregnancy, via amniocentesis. What are the advantages of premarital testing of couples that are at risk? What are the advantages of prenatal testing of fetuses? Are there disadvantages to such testing?

Answers to Self-Exam

Multiple Choice Questions

1. d
2. b
3. c
4. a
5. c
6. b
7. b
8. b
9. d
10. a
11. c
12. b
13. a
14. c
15. d
16. c
17. b
18. d
19. d
20. c

True or False Questions

21. T
22. F
23. F
24. T
25. F
26. T
27. F
28. T
29. F
30. T
31. T
32. T
33. F
34. T
35. F
36. T
37. T
38. T
39. F
40. T

Fill in the Blank Questions

41. clone
42. restriction site
43. sticky ends
44. ligase
45. *Escherichia coli*
46. antibiotic
47. cDNA library
48. DNA hybridization probe
49. kinase
50. signal peptide
51. essential amino acids
52. dicotyledons
53. polygenic
54. β-globin
55. amniocentesis
56. hypogonadism
57. pituitary gland
58. oncogenes
59. electrophoresis
60. dideoxy

CHAPTER 17 DNA Change

Learning Objectives

After mastering the material covered in chapter 17, you should be able to confidently do the following tasks:

- Explain why mutation leading to change in DNA is generally rare.
- Describe the common DNA repair systems that occur in cells.
- List the consequences of the different types of micromutations: base substitutions, additions, and deletions.
- Show how chromosomal macromutations occur, and describe their effects.
- Explain the roles of transposons and transposable elements in genetic change.

Chapter Outline

I. Stability of DNA.

II. The Nature of Mutation.

III. DNA Repair.
- A. Photoreactivation repair.
- B. Excision repair.

IV. Micromutations.
- A. Substitutions.
- B. Chain termination substitutions.
- C. Additions, deletions, and frameshift mutations.
- D. Mutations in noncoding regions.

V. Macromutations.
- A. Breakage and repair.
- B. Deletions, inversions, and duplications.
- C. Translocations.

VI. Jumping Genes.
- A. Transposons.
- B. Transposons as mutagenic agents.

Key Words

mutation
somatic mutation
point mutation
chromosomal mutation
transposition
mutagen
electromagnetic radiations
gamma ray
X-ray
alpha particle
beta particle
ultraviolet (UV) radiation
sodium nitrite
primary lesion
pyrimidine dimer
chemical mutagen
intercalating agent
acridine orange
base analog
DNA modifying agent
DNA adduct
mustard gas
ionizing radiation
free radicals
DNA repair processes
photo-reactivation repair
excision repair
xeroderma pigmentosum
micromutation
base substitution
β-globin gene
silent mutation
neutral mutation
chain-termination mutation
lethal mutation
base addition
base deletion
frameshift mutation
thalessemia
macromutation
centric fragment
acentric fragment
chromosomal deletion
cri-du-chat syndrome
ring chromosome
chromosomal inversion
chromosomal duplication
translocation
translocation Down syndrome
transposons
Barbara McClintock
controlling element
transposable element
transposase
inverted repeat
complex transposon
protooncogene
oncogene
ras gene

Exercises

1. Mutations are changes that occur in DNA, but they can be seen also in the mRNA transcripts of the DNA. Below are given portions of the sequences for a "wild type" mRNA and for several additional mRNA's from DNA's that have been mutated. Using table 13.1 in your text, give the amino acid sequence specified by the wild type mRNA. Then use an arrow to mark the location of the alteration to the other mRNA's, and give the sequence of amino acids specified by each. In case, identify the type of mutation responsible for the change, if there is a change (e.g., base substitution, silent, frameshift, chain termination).

"wild type" ...A U G C U C U C A A A G G U C G C C U G A G A G . . .

mutant "A" ...A U G C U C C C A A A G G U C G C C U G A G A G . . .

mutant "B" ...A U G C U C U C C A A G G U C G C C U G A G A G . . .

mutant "C" ...A U G C U C U C A U A G G U C G C C U G A G A G . . .

mutant "D" ...A U G C U C U C A A A G G U C G C C U G C G A G . . .

mutant "E" ...A U G C U C C A A A G G U C G C C U G A G A G . . .

2. Chromosomal damage can lead to several types of macromutations or chromosomal rearrangements. Below are diagrams for two different intact chromosomes. Sketch similar diagrams to show chromosomes like these that have the following changes:

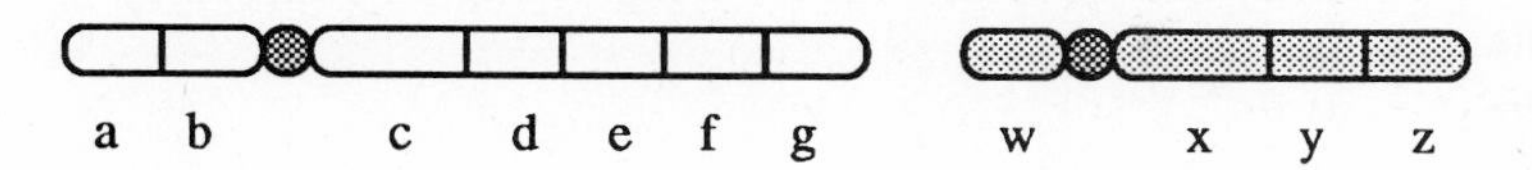

A. Deletion of segment d.

B. Duplication of segment f.

C. Inversion of segment d e f.

D. Terminal deletion of segment f g.

E. Breakage between segments x and y giving centric and acentric fragments (label).

F. Translocation of segment y z to the remainder of the larger chromosome in D above.

Self-Exam

You should be able to easily answer the following questions after learning the material in chapter 17. If you have difficulty with any question, study the appropriate section in the text and try again.

A. Multiple Choice Questions

Circle one alternative that best completes the statement or answers the question.

1. Which is the best description of mutation?
   a. Mutation is a change in DNA.
   b. Mutation is a change in an organism's phenotype.
   c. Mutation is a change in the polypeptide specified by a gene.
   d. Mutation is usually harmful but usually reversible.

2. Which of the following types of change is *not* considered mutational change?
   a. A single base substitution in the DNA.
   b. The inversion of a chromosomal segment.
   c. Recombination between chromosomes during meiosis.
   d. Transposition of elements that disrupt gene function.

3. Gamma and X-rays are examples of potentially mutagenic
   a. base analogs.

b. DNA-modifying agents.
c. electromagnetic radiations.
d. intercalating agents.

4. Pyrimidine dimers are primary lesions to DNA that we expect might be caused by
   a. sodium nitrite.
   b. acridine orange.
   c. free radicals.
   d. ultraviolet radiation.

5. Mustard gas is an example of which type of potential mutagen?
   a. base analog.
   b. DNA modifying agent.
   c. electromagnetic radiation.
   d. intercalating agent.

6. Mutagens that act by squeezing between adjacent nucleotides in DNA are called
   a. base analogs.
   b. DNA modifying agents.
   c. electromagnetic radiations.
   d. intercalating agents.

7. Which of the following activates a repair enzyme in photo-reactivation repair of pyrimidine dimers?
   a. UV wavelengths.
   b. infrared wavelengths.
   c. blue wavelengths.
   d. ionizing wavelengths.

8. The hereditary disease xeroderma pigmentosum is a result of
   a. dry skin.
   b. too much sunlight.
   c. failure of excision repair enzymes.
   d. skin cancer.

9. A single nucleotide is substituted in an coding sequence, but none of the amino acids is changed. This is
   a. a frameshift mutation.
   b. a silent mutation.
   c. a chain-termination mutation.
   d. a lethal mutation.

10. Which of the following mutations is likely to have the greatest effect on the resulting polypeptide?
    a. deletion of a single nucleotide.
    b. base substitution mutation.
    c. silent mutation.
    d. neutral mutation.

11. Considering chromosomal breakage, studies with *Drosophila* sperm show
    a. few, if any, breaks are corrected by repair mechanisms.
    b. as much as one fourth of the breaks are corrected by repair mechanisms.
    c. about half of the breaks are corrected by repair mechanisms.
    d. almost all of the breaks are corrected by repair mechanisms.

12. The congenital condition cri-du-chat syndrome is apparently due to
    a. a small chromosomal deletion.
    b. a small chromosomal duplication.
    c. a small chromosomal inversion.
    d. a small chromosomal translocation.

13. Translocation of a part of human chromosome 21 to chromosome 14 can sometimes lead to symptoms of
    a. thalassemia.
    b. sickle cell anemia.
    c. cri-du-chat syndrome.
    d. Down syndrome.

14. How many breaks must there be in a chromosome for the central part to become inverted with respect to the ends?
    a. none.
    b. one.
    c. two.
    d. three.

15. If we compare the number of chromosomes in the great ape species, which do we observe?
    a. Humans, gorillas, chimps, and orangutans all have the same number of chromosomes.
    b. Humans have more chromosomes than do gorillas, chimps, and orangutans.
    c. Humans have fewer chromosomes than do gorillas, chimps, and orangutans.
    d. Humans and chimps have fewer chromosomes than do gorillas and orangutans.

16. Which of the following discoveries is today credited to Barbara McClintock?
    a. the discovery of Indian corn.
    b. the discovery of "jumping genes."
    c. the discovery of the enzyme transposase.
    d. the discovery of complex transposons.

17. In general, a simple transposable element consists of
    a. transposase.
    b. the gene for transposase.
    c. inverted repeats.
    d. the transposase gene flanked by inverted repeats.

18. The term "inverted repeat" refers to
    a. complementary DNA sequences in the two strands of DNA.
    b. an upside down DNA sequence at the beginning of a gene.
    c. identical DNA sequences that have opposite orientations.
    d. complementary DNA sequences that have opposite orientations.

19. We have discovered that normal human cells contain
    a. oncogenic viruses.
    b. protooncogenes.
    c. oncogenes.
    d. abnormal *ras* genes.

20. Which processes are believed responsible for the large number of genes in many present day organisms?
    a. transposition and gene duplication.
    b. mutation and migration.
    c. inversion and translocation.
    d. base substitution and addition.

B. True or False Questions

Mark the following statements either T (True) or F (False).

_____ 21. Somatic mutations are not passed on to subsequent generations.

_____ 22. Packaging of DNA in eukaryotic chromosomes provides additional protection from damage.

_____ 23. Single nucleotide changes are termed chromosomal mutations.

_____ 24. Because it is less powerful than gamma or X-rays, UV radiation is not a significant mutagen.

_____ 25. Exposure to X-rays can lead to single strand breaks in DNA.

_____ 26. When acridine orange intercalates in DNA, it distorts the molecule.

_____ 27. DNA adducts can interfere with correct transcription and replication of DNA.

_____ 28. Blue wavelengths of light are an important cause of thymine dimers.

_____ 29. In humans, many skin mutations are due to exposure to sunlight.

_____ 30. Point mutations include deletions, inversions, translocations, and duplications.

_____ 31. The presence of silent mutations can be easily detected in an organism's phenotype.

_____ 32. Sickle cell anemia is due to a neutral mutation of the β-globin gene.

_____ 33. Mutation of the codon UAC to UAG is a chain termination mutation.

_____ 34. Many chain termination mutations are lethal mutations in microorganisms.

_____ 35. Frameshift mutations are generally rather benign in their effects on polypeptide structure.

_____ 36. Thalassemia-causing mutations occur in exons, introns, and promoter regions of DNA.

_____ 37. During meiosis, acentric fragments often lag behind and are lost.

_____ 38. There exist healthy humans with ring-shaped chromosomes in their cells.

_____ 39. The most common cause of Down syndrome is an inherited translocation.

_____ 40. Transposons are mobile genetic elements, but much of their movement appears to be random.

C. Fill in the Blanks

Answer the question or complete the statement by filling in the blanks with the correct word or words.

41. Mutations occurring in cells that are not in the germ-cell line are _________________ mutations.

42. Changes involving one or a few nucleotides in DNA are termed _________________ mutations.

43. _________________ are insertions of copies of DNA in new positions in the genome.

44. A common type of primary lesion to DNA exposed to UV light is formation of a _________________.

45. A chemical mutagen molecule that can squeeze between adjacent DNA base pairs and distort the DNA molecule is an _________________.

46. A _________________ is a molecule that is quite similar to one of the normal nucleotides and can be incorrectly incorporated into DNA during the synthesis process.

47. Ionizing radiation can create highly reactive _________________, which can alter the bases of DNA.

48. In the light, thymine dimers can be repaired by the _________________ repair mechanism.

49. In the dark, thymine dimers can be repaired by the ________________ repair mechanism.

50. ________________ is an inherited disorder in humans where DNA excision repair is deficient.

51. A base substitution resulting in a codon coding for the same amino acid is a ________________ mutation.

52. The mutant sickle cell allele of the β-globin gene is a ________________ mutation.

53. In microorganisms, many chain-termination mutations are ________________ mutations, resulting in nonfunctional polypeptides.

54. The addition or deletion of a single nucleotide in a coding sequence is a ________________ mutation, resulting in a polypeptide with incorrect amino acids beyond the mutation.

55. A piece of chromosome that has no centromere is termed an ________________ fragment.

56. In humans, deletion of a small piece of one end of chromosome 5 is seen in ________________.

57. The inherited form of Down syndrome results from a translocation of part of chromosome number ________________ to chromosome number ________________.

58. Sequences of DNA that seem to move from place to place in the genome have been termed ________________.

59. ________________ is the enzyme that allows transposable elements to insert copies of themselves elsewhere.

60. The ________________ gene codes for a regulatory protein; its mutation may activate an oncogene.

Questions for Discussion

1. DNA is the genetic material. Why is it important for DNA to be a molecule that is resistant to being changed by chemicals, energy, and other factors? What are some of the properties of DNA that lend to its stability and help resist change? If there were absolutely no mutation to DNA, what would be the consequences for evolutionary processes?

2. Much attention has been directed recently to the increasing risks of potent mutagens in the environment as a result of human activities. Human activity has apparently diminished the upper atmosphere ozone layer. Ozone in the upper atmosphere absorbs UV radiation from the sun and greatly reduces the UV radiation reaching the earth's surface. Discuss the effects, human and otherwise, that may result from increased UV radiation.

3. Consider the point mutations that can occur in a gene: base substitutions, silent or otherwise, base additions, base deletions. Rank the various point mutations from potentially least serious to potentially most serious in terms of the effects on the gene, the gene product, and the organism. Which is least likely to be detected? Which is most likely to be lethal?

4. Theoretically, transposons could jump around repeatedly in an individual genome without any noticeable effect. However, transposon movement can lead to heritable genetic change. How do transposons act as mutagenic agents?

5. Some DNA sequences in higher organisms known as "pseudogenes" have sequences quite similar to other functional genes. However, the "pseudogenes" themselves appear to have no function in the genome, as there is no corresponding gene product produced. How might these "pseudogenes" have arisen? It has been suggested that these sequences are old genes that no longer function or new potential genes that have never yet functioned. Is either idea more reasonable?

Answers to Self-Exam

Multiple Choice Questions

1. a
2. c
3. c
4. d
5. b
6. d
7. c
8. c
9. b
10. a
11. d
12. a
13. d
14. c
15. c
16. b
17. d
18. c
19. b
20. a

True or False Questions

21. T
22. T
23. F
24. F
25. T
26. T
27. T
28. F
29. T
30. F
31. F
32. F
33. T
34. T
35. F
36. T
37. T
38. T
39. F
40. T

Fill in the Blank Questions

41. somatic
42. point
43. Transpositions
44. pyrimidine dimer
45. intercalating agent
46. base analog
47. free radicals
48. photo-reactivation
49. excision
50. Xeroderma pigmentosum
51. silent
52. base substitution, point
53. lethal
54. frameshift
55. acentric
56. cri-du-chat syndrome
57. 21; 14
58. transposons
59. transposase
60. *ras*

CHAPTER 18 Evolution: Mechanisms and Evidence

Learning Objectives

After mastering the material covered in chapter 18, you should be able to confidently do the following tasks:

- Outline highlights of the development of ideas concerning the origin of species, from the ancient Greek philosophers through the appearance of Darwin's theory to present-day understanding of the origin of species.
- State the basic rationale of Darwin's theory of natural selection.
- Explain the importance of genetic variation, identify its sources, and list mechanisms that can maintain genetic variation in populations.
- Show how studies with the peppered moth illustrate the action of natural selection in actual populations.
- Outline the multiple lines of evidence, from fossils and biogeography to anatomy and biochemistry, that can be cited to support our present-day understanding of evolution.

Chapter Outline

I. History of Evolutionary Ideas.

A. Pre-Darwinian ideas.
B. Enter Charles Darwin.

II. Darwin's Theory.

A. Gathering evidence.
B. The origin of species.
C. The new synthesis.

III. Genetic Variation.

A. Sources of variation.
B. Maintaining genetic diversity.
1. balancing selection.
2. heterozygote advantage.
3. frequency-dependent selection.

IV. Natural Selection.

A. The peppered moth, natural selection in a natural population.

B. How the leopard got its spots.
C. Fitness.

V. Evidence of Evolution.

A. The fossil record.
B. Biogeography.
C. Comparative anatomy.
D. Comparative biochemistry of polypeptides.
E. Comparative biochemistry of nucleic acids.

Key Words

Plato
Aristotle
natural scale
immutability of species
Linnaeus
Buffon
Lamarck
"law of use and disuse"
Charles Darwin
Galapagos islands
Galapagos finches
"descent with modification"
natural selection
artificial selection
Malthus
geometric increase
dispersal
Wallace
the New Synthesis
variation
physical variation
behavioral variation
biochemical variation
mutation
genetic recombination
genetic diversity
balancing selection
balanced polymorphism
morph
heterozygote advantage
overdominance
sickle cell anemia
malaria
frequency-dependent selection
equilibrium frequency
Biston betularia
peppered moth
Haldane
Kettlewell
fitness
reproductive success
relative fitness
selection coefficient
Marsh
Equus
fossils
species selection
biogeography
marsupial
monotreme
comparative anatomy
homologous structures
analogous structures
vestigial structures
comparative biochemistry
cytochrome c
molecular clock
DNA-DNA hybridization
genetic similarity
neutral selection
artificial selection

Exercises

1. Most species have considerable genetic variation for many traits, and selection can act on that variation to change the average expression of the selected trait. In one well-known study, researchers applied artificial selection to corn and were able to strikingly alter the oil content of the grain. Two lines were selected: one for high oil content and one for low oil content. Each season, seed was planted from plants with the very highest oil content or the very lowest oil content, and this process was repeated year after year. On the below graph, plot the changes in oil content you might expect over 20 years, plotting the high and low lines separately.

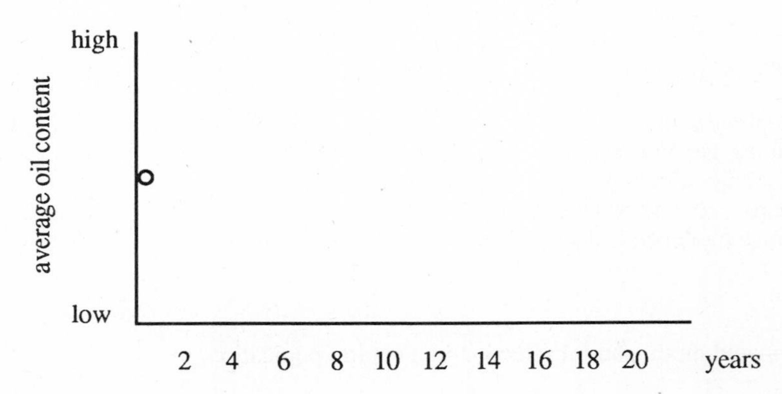

2. You have learned that evolutionary changes in the amino acid sequence of proteins lead to differences between species. The slow rate of amino acid substitution allows us to use such differences to estimate the times of divergence of evolutionary lines. The table below lists the number of amino acid differences in cytochrome c molecules between pairs of five present-day plant species, labeled **A - E**. Use these data to estimate which lineages diverged earliest and which diverged more recently. Then label the ends of the dendrogram emerging from the hypothetical ancestral species.

Number of Amino Acid Differences

	A	**B**	**C**	**D**	**E**
A	0	4	3	1	4
B	4	0	4	4	2
C	3	4	0	3	4
D	1	4	3	0	4
E	4	2	4	4	0

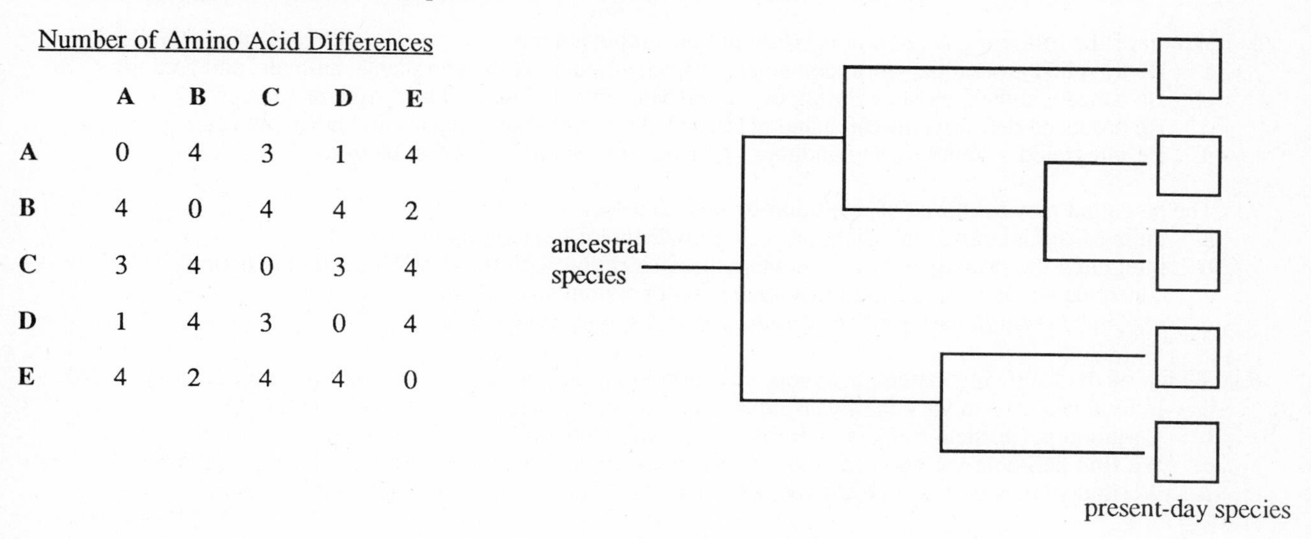

Self-Exam

You should be able to easily answer the following questions after learning the material in chapter 18. If you have difficulty with any question, study the appropriate section in the text and try again.

A. Multiple Choice Questions

Circle one alternative that best completes the statement or answers the question.

1. The concept of the immutability of species — species do not change — was held earliest by
   a. Plato
   b. Linnaeus
   c. Lamarck
   d. Darwin

2. The botanist who devised a system for classifying all living organisms was
   a. Plato
   b. Linnaeus
   c. Lamarck
   d. Darwin

3. Inheritance of acquired characteristics and the "law of use and disuse" were espoused by
   a. Aristotle
   b. Buffon
   c. Lamarck
   d. Darwin

4. The idea that is associated with Thomas Malthus today is
   a. that one species can change into another, through natural selection.
   b. that many related but different species may evolve in isolation on different islands.
   c. the number of individuals in populations increases geometrically, until limited by starvation or disease.
   d. that organs can become larger through use, and such change can be passed on to progeny.

5. Which of the following statements is consistent with Darwin's idea of natural selection?
   a. Larger individuals tend to have more offspring.
   b. Individuals better adapted to the environment tend to leave more offspring.
   c. Individual with variation tend to survive longer.
   d. Survival of individuals is more important in natural selection than is reproduction.

6. Which of the following is *not* among Darwin's accomplishments?
   a. He travelled around the world and amassed great amounts of data on plants, animals, and geology.
   b. He conceived the idea of evolution by natural selection, and wrote *The Origin of Species.*
   c. He produced definitive research and published extensively on barnacles and other subjects.
   d. He integrated a sound understanding of genetics into natural selection theory.

7. The so-called new synthesis of evolution by natural selection
   a. united Darwinian natural selection with knowledge of biogeography.
   b. integrated the principles of Darwinian natural selection with those of Mendelian genetics.
   c. consolidated Darwinian and Lamarckian ideas of evolutionary change.
   d. explained evolutionary processes at the level of synthesis of DNA.

8. Which of the following statements about variation in natural populations is most accurate?
   a. Individuals tend to vary in physical characteristic but generally not in biochemical traits.
   b. Variation in physical traits is often heritable, but behavioral variation is not inherited.
   c. We find heritable variation in physical, biochemical, and behavioral traits within populations.
   d. We find some variation in most species but seldom find variation in single populations.

9. The genetic differences between individuals in populations result
   a. only from mutations giving rise to different alleles for genetic loci.
   b. from the action of natural selection.
   c. from mutations producing novel alleles and from recombination producing new combinations of traits.
   d. only from recombination shuffling existing alleles during sexual reproduction.

10. New genetic alleles can arise in populations
    a. only from mutation of existing alleles.
    b. only from natural selection.
    c. only from recombination.
    d. from either mutation or recombination.

11. The high frequency of the sickle cell allele for β-hemoglobin in malarial regions is attributed to
    a. frequency-dependent selection.
    b. heterozygote advantage.
    c. polymorphism in a patchy environment.
    d. lower availability of health care services.

12. When frequency-dependent selection maintains a polymorphism, a genotype has a selective advantage
    a. in a patchy environment.
    b. when it is rare.
    c. when it is heterozygous.
    d. when it is in high frequency.

13. Considering the peppered moth, *Biston betularia*, collections from the early 18th century
    a. showed about the same frequency of peppered and black morphs.
    b. were mostly peppered moths, the black morphs being rare.
    c. were mostly black moths, the peppered morphs being rare.
    d. had peppered and black morphs at about equal frequencies.

14. The agent of selection that has been shown to alter the frequency of peppered moth morphs is
    a. soot from factories.
    b. lichens growing on trees.

c. birds that eat moths.
d. lepidopterists that collect moths.

15. In Kettlewell's mark-recapture experiment with peppered moths, in a soot-covered woodland
    a. all marked black morphs were recaptured.
    b. a greater fraction of marked black morphs were recaptured than marked peppered morphs.
    c. a greater fraction of marked peppered morphs were recaptured than marked black morphs.
    d. all peppered morphs were recaptured.

16. In measuring the effects of selection, an individual with higher fitness
    a. survives longer than those with lower fitness.
    b. has greater reproductive success than those with lower fitness.
    c. is more vigorous and healthy than less fit individuals.
    d. has a higher selection coefficient than a less fit individual.

17. The reproductively most successful genotype in a population
    a. has a relative fitness of 1.0.
    b. has a selection coefficient of 1.0.
    c. has a relative fitness of zero.
    d. has a relative fitness value equal to its selection coefficient.

18. Which is the most accurate statement about evolution's fossil evidence?
    a. There is a complete record of fossils for the ancestors of most plants and animals.
    b. Paleontologists believe that most extinct forms of life formed fossils.
    c. We have complete fossil records for bones, shells, wood, as well as for soft-bodied organisms
    d. There are few fairly complete fossil lineages, such as that for the ancestors of *Equus*.

19. Which is a valid observation concerning the world's mammals?
    a. Until recent times, all nonflying mammals in Australia were marsupials and monotremes.
    b. Marsupials tend to be the dominant mammal on most oceanic islands.
    c. Marsupial mammals are absent in North America, except in zoos.
    d. When different species are compared, it is unlikely that all marsupials are closely related.

20. Homologous structures in different species
    a. have common embryonic origin.
    b. appear similar but form differently in the embryo.
    c. are presumed to have different evolutionary origins.
    d. are considered to be analogous.

21. With respect to the rate of evolutionary change of the amino acid sequence of cytochrome c,
    a. it changes quite rapidly, and there is little similarity between mammals and reptiles.
    b. it changes rather slowly, but there are usually differences between individuals of one species.
    c. change is very slow, perhaps one amino acid change per 17 million years, on the average.
    d. generalizations cannot be made, since the rate seems to vary so widely in different species.

B. True or False Questions

Mark the following statements either T (True) or F (False).

_____ 22. Some scientists before Darwin, such as Lamarck, believed that one species could give rise to another.

_____ 23. Lamarck believed that giraffes came to have long necks through stretching their necks.

_____ 24. Darwin was little influenced by his travels to South America, the Galapagos, and elsewhere.

_____ 25. Artificial selection of domestic animals became common only after *Origin of Species* was published.

_____ 26. Malthus suggested animal number increased arithmetically (2, 4, 6, 8, and so on).

_____ 27. Natural selection should act only on variation that can be inherited.

_____ 28. The naturalist Wallace, like Darwin, also developed the concept of evolution by natural selection.

_____ 29. Darwin had published his conclusions on natural selection long before he heard Wallace's ideas.

_____ 30. Darwin clearly took Mendel's work on heredity into account in his own work on natural selection.

_____ 31. Genotypic differences between individuals can arise via mutation or genetic recombination.

_____ 32. Biologists think that many mutations are deleterious and are selected against.

_____ 33. A polymorphism can be maintained in a patchy environment by balancing selection.

_____ 34. Evolution can be defined as the changing of the ratio of alleles in a population over time.

_____ 35. In the peppered moth *Biston betularia*, the black morph results from a single recessive allele.

_____ 36. In industrial areas of England, as industrial pollution increased, the frequency of black morph of the peppered moth decreased.

_____ 37. In general, reproductive success can be used as a measure of fitness.

_____ 38. Although there are a few fairly complete lines of fossil evidence, such as for the ancestors of the horse, the fossil record is generally fragmentary and incomplete.

_____ 39. Darwin observed that most oceanic islands have a mammalian fauna like the continents.

_____ 40. The relatedness of all vertebrates is supported by the presence of gill arches in the embryonic development of each species, even those which lack gills when mature.

_____ 41. Some scientists base an evolutionary molecular clock on the rate of amino acid substitution in very conservative, slowly changing polypeptides.

_____ 42. Neutral selectionists maintain that most mutations are either beneficial or harmful and subject to the action of natural selection.

C. Fill in the Blanks

Answer the question or complete the statement by filling in the blanks with the correct word or words.

43. Some biologists, such as Aristotle, believed there was a ________________ of life, from the very simple to the most complex forms, with mankind at the top.

44. The French proponent of the "law of use and disuse" was ________________.

45. In his journey on the *Beagle*, Darwin visited the ________________ islands, which many believe provided important insights that led to his conclusions about natural selection.

46. Darwin was impressed with an essay on geometric population growth written by ________________.

47. The integration of evolution and genetics in the 20th century has been called ________________.

48. In explaining variation, we know that new alleles arise only from ________________.

49. Genetic recombination is more important in producing genotypic variation in species that reproduce __________________.

50. When selection maintains two or more variants in a population, the result is a __________________.

51. In frequency-dependent selection, the frequencies at which two morphs remain stable are termed the __________________ frequencies.

52. In peppered moth studies it has been found that industrial pollution has killed __________________ on trees in industrial areas.

53. The value (1 - relative fitness) is called the __________________.

54. Darwin found fossils of extinct giant armadillos in __________________.

55. The primitive monotreme mammals from Australia are the __________________ and __________________.

56. The wings of insects and the wings of bats are __________________ structures.

57. There are 15 differences in the amino acid sequences of __________________ from reptiles and mammals.

58. The differences between Great Danes and toy poodles result from __________________.

Questions for Discussion

1. Charles Darwin played a central role in the development of 20th century biology; most biologists consider his ideas pivotal in our present-day understanding of the diversity of life. But, what if Darwin had *not* signed on as naturalist for the *Beagle*? What if he had not visited South America, the Galapagos, and other places? What if Darwin had failed to write *The Origin of Species*? Based on what you know about the other researchers and the available evidence, do you think that we would today have essentially the same understanding of evolutionary processes? Or do you think that we would still adhere to the ideas of Aristotle, Linnaeus, and Lamarck?

2. Is change through natural selection merely an untested theory, or is there evidence that supports change in populations through natural selection? Like Rudyard Kipling, we can create Just-So Stories, but scientists rely on data. Cite evidence for selection.

3. Compare the concepts of homologous structure and analogous structures. Give examples of each. Which types of structures provide more information about evolutionary relatedness? Why?

4. Explain how the amino acid sequences of a particular polypeptide from different taxa can be compared and used as a "molecular clock" to estimated the time of divergence of evolutionary lines. What types of polypeptides would be most useful in comparing distantly related groups, such as major phyla? What types would be most useful in comparing more closely related groups, such as related species or genera?

Answers to Self-Exam

Multiple Choice Questions

1. a
2. b
3. c
4. c
5. b
6. d
7. b
8. c
9. c
10. a
11. b
12. b
13. b
14. c
15. b
16. b
17. a
18. d
19. a
20. a
21. c

True or False Questions

22. T
23. T
24. F
25. F
26. F
27. T
28. T
29. F
30. F
31. T
32. T
33. T
34. T
35. F
36. F
37. T
38. T
39. F
40. T
41. T
42. F

Fill in the Blank Questions

43. natural scale
44. Lamarck
45. Galapagos
46. Malthus
47. the New Synthesis
48. mutation
49. sexually
50. balanced polymorphism
51. equilibrium
52. lichens
53. selection coefficient
54. South America
55. duck-billed platypus; spiny anteater
56. analogous
57. cytochrome *c*
58. artificial selection

CHAPTER 19 Microevolution: Changing Alleles

Learning Objectives

After mastering the material covered in chapter 19, you should be able to confidently do the following tasks:

- Calculate allele frequencies and genotype frequencies for genes in populations

- Explain the Hardy-Weinberg Law and how it works.

- Show the role of mutation in providing the variation required for evolutionary change.

- Explain the ways that natural selection can act to change allele frequencies or to stabilize them.

- Demonstrate the effects of random change in such events as genetic drift, bottlenecks of population size, and the founder effect.

- Explain what is meant by a selectively neutral mutation and describe the controversy between so-called selectionists and neutralists.

Chapter Outline

I. Alleles and Allele Frequencies.

II. The Hardy-Weinberg Law.

A. How the Hardy-Weinberg law works.
B. An algebraic equivalent.
C. The restrictions of the Hardy-Weinberg law.

III. Mutation: The Raw Material of Evolution.

A. Balancing mutation and selection.

IV. Natural Selection.

A. Stabilizing selection.
B. Directional selection.
C. Disruptive selection.

V. Random Change: Genetic Drift, Population Bottlenecks, and the Founder Effect.

VI. The Genetic Future of *Homo sapiens.*

Key Words

allele
allele frequency
population
gene pool
Hardy-Weinberg law
genetic equilibrium
reciprocal matings
genotypic ratios
p
q
$p + q = 1$
p^2
$2pq$
q^2
$p^2 + 2pq + q^2 = 1$
random mating
achondroplasia
hemophilia
albinism
sickle-cell anemia
phenylketonuria
schizophrenia
cystic fibrosis
Tay-Sachs disease
normal distribution
stabilizing selection
directional selection
disruptive selection
bimodal distribution
genetic drift
population bottleneck
founder effect
porphyria variegata
selectionist
neutralist
selectively neutral mutation
fixation
pyloric stenosis
myopia

Exercises

1. Here are some data from which you can calculate allele frequencies for the alleles responsible for the MN blood group phenotypes. MM individuals have 2 "M" alleles, NN individuals have 2 "N" alleles, and MN individuals are heterozygotes. In a study of 5,000 new army recruits, the following phenotypes were observed:
   1,805 MM   2,390 MN   805 NN   (5,000 total)

   A. Calculate p, the frequency of the M allele, and q, the frequency of the N allele.

   $$p = \frac{2\,(\text{no. of MM}) + \text{no. of MN}}{2\ (\text{total})} =$$

   $$q = \frac{2\,(\text{no. of NN}) + \text{no. of MN}}{2\ (\text{total})} =$$

   B. Now calculate the number of individuals you would predict to have the MM, MN, and NN genotypes, if the army recruits were drawn from a population in Hardy-Weinberg equilibrium. Remember, there must be 5,000 recruits in all.

   No. of MM = p^2 x 5,000 =

   No. of MN = 2pq x 5,000 =

   No. of NN = q^2 x 5,000 =

   C. Do you think the population from which the recruits were drawn was in Hardy-Weinberg equilibrium for the MN genotypes? Why do you draw this conclusion? How might you test your conclusion?

2. There exists a particular type of wild bean that has been the victim of bean weevils for years. These weevils are small beetles which place their eggs on beans, and the developing larvae eat the beans. In different bean populations, however, the effects of the bean weevils have not always been the same. In some populations of beans, beans of all sizes are equally infested with weevils. In other populations, the weevils seem to select beans of certain sizes to serve as hosts for their larvae. In bean population A, weevils select only the largest beans for their larvae to feed on. In bean population B, large beans and small beans are rejected, and only beans of intermediate size are chosen. And in population C, weevils place their eggs on only the largest or smallest beans, rejecting the beans that have sizes near the mean value. Let us assume that the size of beans is an inherited trait. In uninfested bean populations, the distribution of bean sizes can be described with the curve shown below. On the axes given below, sketch the distributions of bean sizes you might expect in populations A, B, and C after a number of generations of weevils. If the mean bean size changes, show the new mean value.

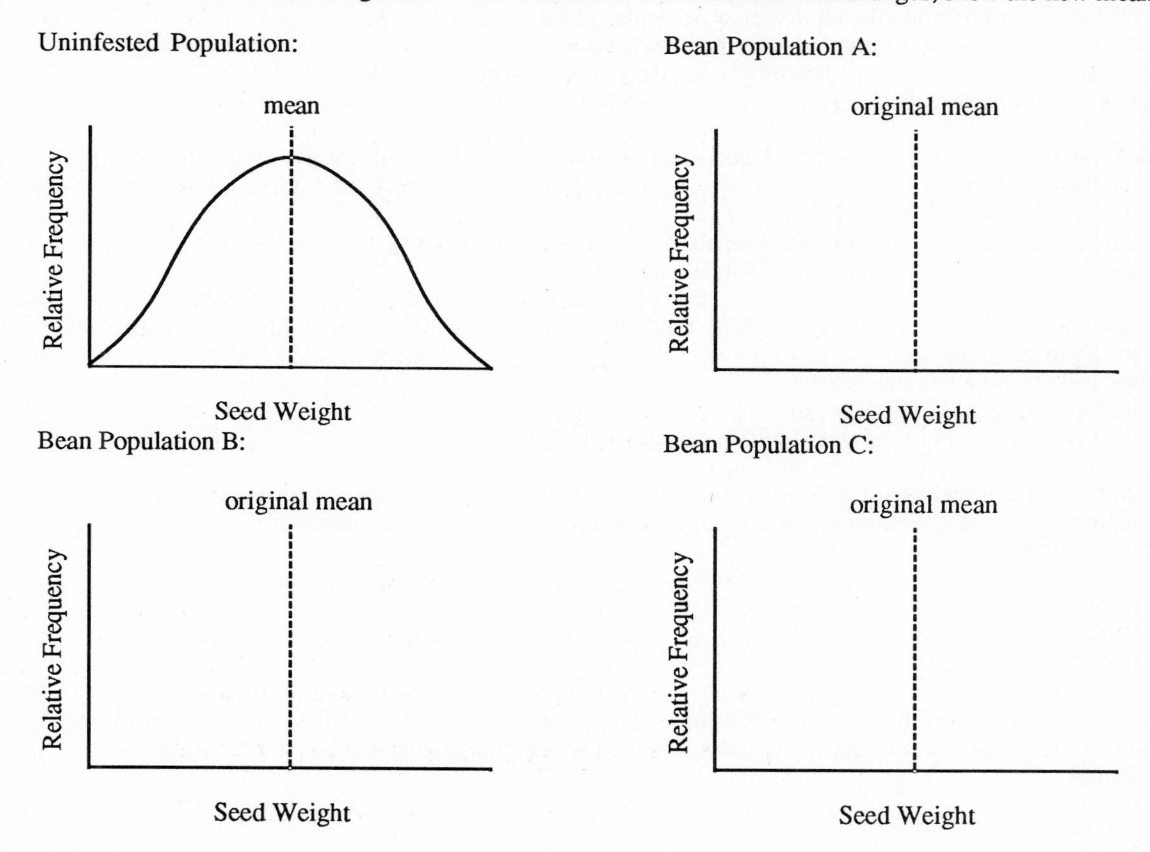

Self-Exam

You should be able to easily answer the following questions after learning the material in chapter 19. If you have difficulty with any question, study the appropriate section in the text and try again.

A. Multiple Choice Questions

Circle one alternative that best completes the statement or answers the question.

1. Which is an accurate statement regarding allele frequencies?
   a. The values of allele frequencies are always between 0 and 1.
   b. Allele frequencies are always values greater than 1.
   c. In considering evolution, allele frequencies do not change.
   d. Allele frequencies are the same as genotypic proportions.

2. Which is the best statement regarding the progress of evolution?
   a. Evolution is always a result of natural selection.

b. Evolution is a change in the number of individuals from generation to generation.
c. Evolution involves a change in allele frequencies in a population.
d. Evolution involves a change in an individual's genotype.

3. If a population is in genetic equilibrium, in the absence of natural selection,
   a. the number of individuals will not change.
   b. the relative proportions of alternative phenotypes will not change.
   c. there will be one genotype and one phenotype.
   d. mutation cannot occur.

4. A biological population can best be described as
   a. all the plants and animals that live at a particular place.
   b. a group of individuals of one species that live together.
   c. a group of interbreeding or potentially interbreeding individuals.
   d. a lineage of parents, F_1's, F_2's, F_3's, F_4's, and so forth.

5. A plant population is 50% BB and 50% bb. After random mating among all these individuals, we expect
   a. the F_1 to be 100% Bb.
   b. the F_1 to be 50% BB and 50% bb.
   c. the F_1 to be 25% BB, 50% Bb, and 25% bb.
   d. the F_1 to be randomly composed of BB, Bb, and bb individuals.

6. Another plant population is 100% Bb. After random mating among all these individuals, we expect
   a. the F_1 to be 100% Bb.
   b. the F_1 to be 50% BB and 50% bb.
   c. the F_1 to be 25% BB, 50% Bb, and 25% bb.
   d. the F_1 to be randomly composed of BB, Bb, and bb individuals.

7. A butterfly population has a rare homozygous recessive form ww that occurs at a frequency of 0.0004. The population is in Hardy-Weinberg equilibrium for the trait. What is the frequency of the homozygous dominant form?
   a. 0.02.
   b. 0.9604.
   c. 0.98.
   d. 0.9996.

8. Assume the frequency of human X chromosomes carrying the sex-linked recessive red-green color blindness allele is 0.13. What is the frequency of color blind females in the population?
   a. 0.0169.
   b. 0.2262.
   c. 0.7569.
   d. 0.13.

9. Hardy-Weinberg equilibrium assumes there is no mutation. However, for a typical locus, mutations occur
   a. in about 1 of every 10 gametes.
   b. in about 1 of every 1,000 gametes.
   c. in about 1 of every 100,000 gametes.
   d. in about 1 of every 10 million gametes.

10. For which type of mutation would it be easiest to measure the rate of occurrence?
    a. Mutations that are lethal to early embryos.
    b. Dominant mutations with mild but noticeable effects.
    c. Recessive mutations with mild but noticeable effects.
    d. Biochemical mutations that cause no disease.

11. The type of natural selection that would favor individuals near the mean and select against extremes is
    a. stabilizing selection.
    b. directional selection.

c. disruptive selection.
d. artificial selection.

12. The type of natural selection that would lead to a bimodal distribution curve is
 a. stabilizing selection.
 b. directional selection.
 c. disruptive selection.
 d. artificial selection.

13. A random change in allele frequency from generation to generation due to chance events is termed
 a. artificial selection.
 b. mutation pressure.
 c. genetic equilibrium.
 d. genetic drift.

14. Genetic drift is most likely to be an important evolutionary force in
 a. artificial populations.
 b. natural populations.
 c. very small populations.
 d. very large populations.

15. A small, isolated religious group has a relatively high frequency of polydactyly, an ordinarily rare condition where individuals may have extra fingers or toes. The best explanation for the high frequency in this group is
 a. natural selection.
 b. artificial selection.
 c. repeated mutation.
 d. founder effect.

16. Elephant seals were common in the early 19th century but were hunted almost to extinction. Today their numbers have increased again, but the species has little genetic variation. We can say the species has experienced
 a. natural selection.
 b. mutation pressure.
 c. a bottleneck.
 d. founder effect.

17. Which of the following would argue that much variation at the molecular level is due to genetic drift?
 a. selectionists.
 b. neutralists.
 c. Afrikaaners.
 d. Darwinists.

18. If we compare the amino acid sequences of human beta globin with horse beta globin, we find
 a. they are identical.
 b. there are many more similarities than differences.
 c. there are many more differences than similarities.
 d. there is no similarity whatsoever.

19. Traditionally, human males with the genetic disease hemophilia often died before they reached reproductive age. If modern medical treatment allows most hemophiliacs to lead long and fruitful lives, what will happen to the frequency of the allele responsible for hemophilia?
 a. It should increase in frequency.
 b. It should decrease in frequency.
 c. It should not change in frequency.
 d. Change in frequency will be unpredictable due to random events.

20. Which of the following statements is most accurate at present?
 a. Genetic disease cannot be treated.
 b. The alleles that are responsible for genetic disease can be replaced with "normal" alleles.

c. The occurrence of many genetic diseases can be predicted in families with a history of a disease.
d. Many genetic diseases can be prevented in families with a history of a disease.

B. True or False Questions

Mark the following statements either T (True) or F (False).

_____ 21. Individuals evolve, not populations.

_____ 22. Evolution is not always the result of natural selection.

_____ 23. Godfrey Hardy, one of the "authors" of the Hardy-Weinberg law, was a biologist by training.

_____ 24. A biological population includes all the members of a single species.

_____ 25. According to the Hardy-Weinberg law, in the absence of forces acting to change allele frequencies, the proportions of genotypes in a population will remain constant from generation to generation.

_____ 26. If the frequencies of the A allele is p and the frequency of the a allele is q, then the Hardy-Weinberg frequency of Aa heterozygotes will be 2pq.

_____ 27. According to Hardy-Weinberg equilibrium, $p + q = p^2 + 2pq + q^2$.

_____ 28. If the frequency of a rare recessive allele is 0.003, we expect the frequency of homozygous recessives to be 0.009.

_____ 29. Many natural populations meet the conditions required for Hardy-Weinberg equilibrium.

_____ 30. Mutation is an important source of genetic variation, and it affects equilibrium allele frequencies.

_____ 31. Typically, about 1 in 100 gametes carries a mutation for a particular allele.

_____ 32. Most human individuals probably carry at least one new gene mutation.

_____ 33. All mutations are deleterious, and selection acts against them.

_____ 34. If we measure some trait from a large number of individuals, the variation observed often shows a normal distribution.

_____ 35. Birth weight in humans provides a good example of a trait affected by disruptive selection.

_____ 36. If a plant breeder wished to increase the yield of a crop plant like corn, she might use directional selection.

_____ 37. Genetic drift is usually an important factor in populations with a large number of individuals.

_____ 38. One type of porphyria is relatively more common among Afrikaaners than Europeans. This is probably due to founder effect.

_____ 39. Selectionists and neutralists agree on the relative importance of chance events in leading to evolutionary change.

_____ 40. Treating the symptoms of genetic disease in organisms like humans can actually lead to an increase in the frequency of the responsible alleles.

C. Fill in the Blanks

Answer the question or complete the statement by filling in the blanks with the correct word or words.

41. The relative proportion of a given allele in a population is given as its _________________.

42. The Hardy-Weinberg law states that allele and genotype frequencies remain in _________________ unless altered by outside forces like mutation or selection.

43. A biological population is a group of _________________ individuals.

44. If a population is to remain at Hardy-Weinberg equilibrium, the mating pattern must be _________________.

45. The source of new variation in populations is _________________.

46. Variation in nature is often distributed as in a bell-shaped curve, a _________________ distribution.

47. When selection favors individuals near the mean and opposes individuals at the extremes it is termed _________________ selection.

48. Disruptive selection can lead to a _________________ distribution of a trait.

49. _________________ is the term for random changes in allele frequencies due to chance events.

50. When a population is severely reduced in numbers for a brief period, it is termed a _________________.

51. An unusually high frequency of a rare allele in an isolated population might be due to _________________.

52. A new mutation that has the same fitness as the old allele is termed a _________________ mutation.

53. Some geneticists might argue that modern medicine has reduced the strength of _________________ against deleterious alleles for some human traits.

Questions for Discussion

1. How does a biological population differ from other populations such as local populations of villages or student body populations of a school? What are the important features of biological populations?

2. The conditions for the Hardy-Weinberg law do not occur in most populations. What are the restrictions that apply when we predict Hardy-Weinberg equilibrium frequencies? Although these conditions are often not met, why is the Hardy-Weinberg principle still useful?

3. A farmer wishes to increase the average size of pumpkins he grows in his field, but he does not wish to use additional fertilizer, pesticides, or labor. Outline how he might accomplish his goal using his knowledge of natural selection.

4. There is a hereditary trait in cats that causes them to have six toes on each foot. The trait is usually rare, but there are many cats like this on Galveston Island. This has been the case since at least the early 19th century, when Galveston became an important port. Why might there be such a high frequency of six-toed cats there?

5. Many human genetic diseases reduce fitness, but modern medicine has reduced the strength of selection against the responsible alleles. Does this mean the frequencies of such deleterious alleles will increase? Could an increase in allele frequency be countered in any way? What are the pros and cons of trying to influence the genetic composition of human populations?

Answers to Self-Exam

Multiple Choice Questions

1. a
2. c
3. b
4. c
5. c
6. c
7. b
8. a
9. c
10. b
11. a
12. c
13. d
14. c
15. d
16. c
17. b
18. b
19. a
20. c

True or False Questions

21. F
22. T
23. F
24. F
25. T
26. T
27. T
28. F
29. F
30. T
31. F
32. T
33. F
34. T
35. F
36. T
37. F
38. T
39. F
40. T

Fill in the Blank Questions

41. allele frequency
42. equilibrium
43. interbreeding
44. random
45. mutation
46. normal
47. stabilizing
48. bimodal
49. Genetic drift
50. population bottleneck
51. founder effect
52. neutral
53. natural selection

CHAPTER 20 Macroevolution: The Origin of Species

Learning Objectives

After mastering the material covered in chapter 20, you should be able to confidently do the following tasks:

- Give a working definition of what a species is.

- Explain the practical difficulties with the species concept.

- Show how a species is named and placed in categories at levels from kingdom through species.

- Contrast the traditional and present day methods that are used in taxonomic classification.

- Outline the known mechanisms of speciation, including allopatric, parapatric, and sympatric mechanisms.

- List prezygotic and postzygotic reproductive isolating mechanisms.

- Describe the common patterns of evolution observed in many groups of organisms.

- Compare the concepts of evolutionary gradualism and punctuated equilibrium.

Chapter Outline

I. What Is a Species?

II. Taxonomy and Systematics.

A. How species are named.
B. Taxonomic organization: kingdom to species.
C. The five kingdoms.

III. Methods in Taxonomy.

A. The classical approach.
B. The phenetic approach.
C. Cladistics.

IV. The Mechanisms of Speciation.

A. Allopatry and speciation.
B. Parapatry and speciation.
C. Sympatry and speciation.
1. sympatric speciation in plants.

2. sympatric speciation in animals.

V. Reproductive Isolating Mechanisms.

A. Prezygotic barriers.
1. geographical barriers.
2. ecological barriers.
3. behavioral barriers.
4. mechanical barriers.
5. gametic barriers.

B. Postzygotic barriers.
1. hybrid failure.
2. hybrid inviability.
3. hybrid sterility.
4. hybrid breakdown.

VI. Patterns of Evolution.

A. Divergent evolution.
B. Convergent evolution.
C. Coevolution.
D. Gradualism versus punctuated equilibrium.
E. Extinction and evolution.

Key Words

macroevolution
species
Linneaus
taxonomy
grackle
jackal
timber wolf
domestic dog
golden jackal
side-striped jackal
black-backed jackal
red wolf
coyote
Canis
salamander
subspecies
asexually reproducing species
binomial system
genus (genera)
Latin names
systematics
family
order
class
phylum (phyla)
division
kingdom
taxon (taxa)
Monera
Protista
Fungi
Plantae
Animalia
monophyletic
polyphyletic
primitive traits
derived traits
platypus
lumpers
splitters
Felis
Panthera
classical taxonomy
homology
comparative anatomy
paleontology
comparative embryology
echinoderm
chordata
phenetics
shared characters
meadowlarks
ocotillo
allauidia
convergent evolution
cladistics
cladogram
shared primitive characters
shared derived characters
speciation
allopatry
allopatric speciation
continental drift
parapatry
parapatric speciation
sympatry
sympatric speciation
hybridization
hybrid swarms
polyploidy
tetraploid
autotetraploid
allotetraploid
triploid
hexaploid
allohexaploid
Triticale
reproductive isolating mechanism
prezygotic barrier
geographical barrier
tiglon
liger
ecological barrier
behavioral barrier
temporal isolation
mechanical barrier
gametic barrier
postzygotic barrier
hybrid failure
hybrid inviability
hybrid sterility
hybrid breakdown

divergent evolution
adaptive radiation
Galapagos finches
convergent evolution
South American mara
placental mammals
marsupial mammals
coevolution
predator-prey relationship
host-parasite relationship
flowering plant–pollinator relationship
passion-flower
Heliconius butterfly
gradualism
fossil record
punctuated equilibrium
extinction
continental drift
Pangaea
Laurasia
Gondwanaland

Exercises

1. A plant taxonomist is studying a large group of species in a certain genus, and for a number of characters she has determined what seems to be the primitive trait versus the derived trait. These are:

   Primitive: woody trees or shrubs with smooth, simple, undivided leaves, white flowers, and elongate fruit.

   Derived: herbaceous plants, compound leaves, hairy leaves, red flowers, and round fruits.

   Help classify these six species (A - F) by assuming that plants that share derived traits are more closely related than those that do not share derived traits. Locate these species at the ends of the branches in the evolutionary diagram below, and indicate at the branch points which characters separate the lineages.

	Growth Form	Leaf Type	Leaf Surface	Flower Color	Fruit Shape
Species A	tree	compound	smooth	white	elongate
Species B	tree	simple	smooth	red	elongate
Species C	tree	simple	hairy	red	elongate
Species D	herb	simple	smooth	white	round
Species E	herb	simple	smooth	white	elongate
Species F	tree	simple	smooth	white	elongate

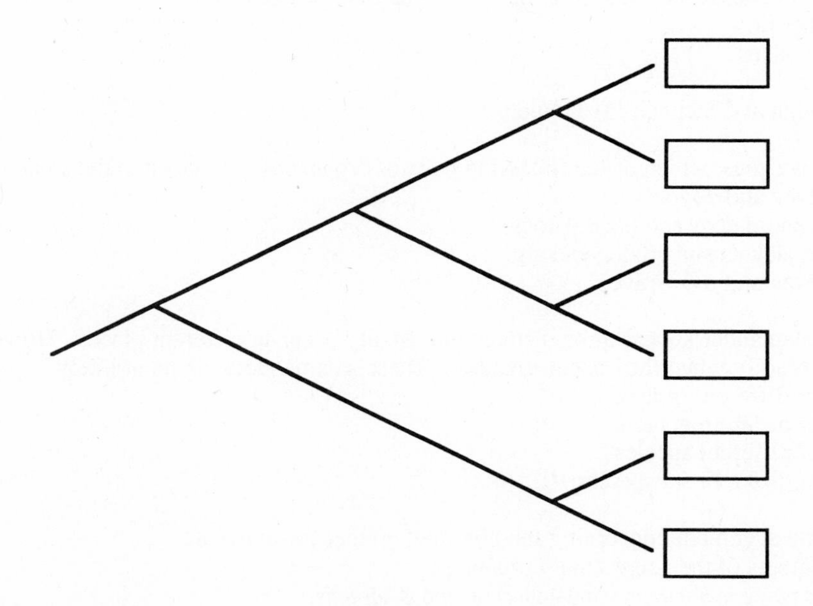

2. Fill in the following table as completely as you can. You may need to consult additional sources to complete all the blanks.

	Human	Domestic Dog	Oak Tree	Corn	Brewer's Yeast
Kingdom	___________	___________	___________	___________	___________
Phylum or Division	___________	___________	___________	___________	___________
Class	___________	___________	___________	___________	___________
Order	___________	___________	___________	___________	___________
Family	___________	___________	___________	___________	___________
Genus	___________	___________	___________	___________	___________
Species	___________	___________	___________	___________	___________

Self-Exam

You should be able to easily answer the following questions after learning the material in chapter 20. If you have difficulty with any question, study the appropriate section in the text and try again.

A. Multiple Choice Questions

Circle one alternative that best completes the statement or answers the question.

1. The person(s) credited as the founder(s) of modern taxonomy is(are)
   a. Charles Darwin.
   b. Carolus Linnaeus.
   c. Ernst Mayr.
   d. Niles Eldridge and Stephen Jay Gould.

2. Within the genus *Canis*, which of the following pairs of organisms are placed in the *same* species?
   a. timber wolves and coyotes.
   b. german shepherd dogs and poodle dogs.
   c. side-striped jackals and golden jackals.
   d. timber wolves and red wolves.

3. Two types of salamander appear quite distinct and usually occur in different places. However, the two types will interbreed readily when they occur together. These salamanders are most likely
   a. members of different orders.
   b. members of different genera.
   c. members of different species.
   d. members of different subspecies.

4. Which is the best statement concerning the binomial epithet for humans?
   a. We are members of the genus *Homo sapiens*.
   b. Our species name is *homo* and and generic name is *sapiens*.

c. Our generic name is *Homo* and our specific name is *Homo sapiens*.
d. We are members of the species *sapiens*.

5. From most inclusive to most specific, the taxonomic categories for animals are
   a. kingdom, division, class, order, family, genus, species.
   b. kingdom, phylum, class, order, family, genus, species.
   c. kingdom, division, order, class, family, genus, species.
   d. kingdom, phylum, order, class, family, genus, species.

6. Which statement concerning the taxonomy of viruses is currently accepted?
   a. Viruses are placed in the kingdom Monera.
   b. Viruses are placed in the kingdom Protista.
   c. Viruses are placed in the kingdom Fungi.
   d. Viruses are not placed in any of the five kingdoms at present.

7. We call characteristics that are believed to have evolved very early in the evolutionary history of a group
   a. primitive traits.
   b. derived traits.
   c. advanced traits.
   d. specific traits.

8. The classical approach to classifying organisms is based on
   a. homologies of structures.
   b. the relative number of shared characteristics, whether primitive or derived.
   c. shared derived traits and time of divergence.
   d. whether two organisms look alike or not.

9. The phenetic approach to classifying organisms is based on
   a. homologies of structures.
   b. the relative number of shared characteristics, whether primitive or derived.
   c. shared derived traits and time of divergence.
   d. whether two organisms look alike or not.

10. The cladistic approach to classifying organisms is based on
    a. homologies of structures.
    b. the relative number of shared characteristics, whether primitive or derived.
    c. shared derived traits and time of divergence.
    d. whether two organisms look alike or not.

11. Which group would classify crocodiles and birds as more closely related than crocodiles and lizards?
    a. classical taxonomists.
    b. numerical pheneticists.
    c. cladists.
    d. none of these; all agree that crocodiles and lizards are more closely related.

12. When one evolutionary lineage is separated geographically and gives rise to two distinct species
    a. it is termed allopatric speciation.
    b. it is termed parapatric speciation.
    c. it is termed sympatric speciation.
    d. it is termed instant speciation.

13. Sympatric speciation is believed to be more common
    a. in geographically separated groups.
    b. during periods of continental drift.
    c. in development of animal species.
    d. in development of plants species.

14. Features that lead to sympatric speciation can include
    a. barriers like mountain ranges.
    b. isolation, such as on islands.
    c. hybridization and polyploidy.
    d. mating rituals.

15. Which is most likely to be fertile and produce fertile offspring?
    a. an animal hybrid, such as a mule.
    b. a triploid plant.
    c. an autopolyploid plant.
    d. an allopolyploid plant.

16. Polyploidy is believed to be more common
    a. among plant species in tropical regions.
    b. among plant species in far northern latitudes.
    c. among temperate animal species.
    d. among mammalian hybrids.

17. Which reproductive isolating mechanism operates postzygotically?
    a. geographical barriers.
    b. hybrid sterility.
    c. behavioral differences.
    d. habitat differences.

18. Which of the following properties must two species have in order to be "good" biological species?
    a. They must have different appearances.
    b. They must live in different habitats.
    c. They must be separated by reproductive isolating mechanisms.
    d. They must be descended from different ancestors.

19. There are many species of finch on the Galapagos Islands. The wide diversity is usually described as a result of
    a. convergent evolution.
    b. adaptive radiation.
    c. continental drift.
    d. punctuated equilibrium.

20. Similarities between marsupials in Australia and placental mammals elsewhere are usually attributed to
    a. convergent evolution.
    b. adaptive radiation.
    c. continental drift.
    d. punctuated equilibrium.

21. The very close and specific relationship between many flowering plants and their insect pollinators is often cited as an example of
    a. convergent evolution.
    b. coevolution.
    c. adaptive radiation.
    d. evolutionary gradualism.

22. In relationships between parasites and their hosts that have been "fine-tuned" by coevolution,
    a. the host will always kill the parasite.
    b. the parasite will always kill the host.
    c. the host will tolerate the parasite, and the parasite will do little damage to the host.
    d. the host will be entirely dependent on the parasite.

23. Darwin's conception of the pace of evolution could best be described as
    a. evolutionary gradualism.
    b. punctuated equilibrium.

c. biological revolution.
d. adaptive radiation.

24. The first of the five great extinctions for which there is considerable evidence occurred
    a. at the end of the Cambrian period (about 500 million years ago).
    b. at the end of the Devonian period.
    c. at the close of the Permian period.
    d. at the end of the Mesozoic era.

25. The large land mass which existed some 200 million years ago and gave rise to the continents of today has been given the name.
    a. Panarctica.
    b. Pangaea.
    c. Laurasia.
    d. Gondwanaland.

B. True or False Questions

Mark the following statements either T (True) or F (False).

_____ 26. Macroevolution includes short-term events at the population level, such as allele frequency fluctuations.

_____ 27. Linnaeus, the founder of modern taxonomy, developed the binomial system of nomenclature for naming organisms.

_____ 28. Members of two different species cannot ever successfully form hybrids.

_____ 29. The golden jackal (*Canis aureus*) and the side-striped jackal (*Canis adjustus*) frequently mate with one another and produce hybrids.

_____ 30. There are two California salamanders which are very different in appearance, yet they interbreed routinely in nature. These are different species, since they differ in appearance.

_____ 31. In general, almost anyone could correctly classify a large organism as a plant or an animal.

_____ 32. A taxonomic order will often contain several to many different taxonomic families.

_____ 33. Two species within the same class share more recent ancestry than do two species in different classes.

_____ 34. The kingdom Protista includes the bacteria.

_____ 35. The goal of taxonomy is to describe monophyletic taxa, not polyphyletic taxa.

_____ 36. Derived traits are sometimes called "advanced" since they have undergone more recent change.

_____ 37. Although humans have a number of primitive traits such as bipedalism, our jaw is quite advanced.

_____ 38. Classical taxonomists often look for homologous structures by studying comparative anatomy and embryology.

_____ 39. Fossil evidence suggests that bears are relatively closely related to dogs.

_____ 40. The phenetic approach to classification requires determining whether each character studied is a primitive or a derived trait.

_____ 41. Because of its unbiased numerical approach, phenetic classification can sometimes be confused by evolutionary convergence.

_____ 42. In cladistics, closely related taxa will tend to have more shared derived characters.

_____ 43. A primitive character shared by all of group of taxa is quite useful in differentiating the groups.

_____ 44. If speciation has occurred allopatrically, the two species will hybridize if they come together.

_____ 45. Continental drift has probably led to allopatric speciation in the past.

_____ 46. If parapatric speciation occurs, it probably occurs in a "patchy" environment where selection may be strong in the different patches.

_____ 47. In plants, sympatric speciation may occur through hybridization and stabilization of the hybrid.

_____ 48. Polyploidy is much more important in species formation in animals than it is in plants.

_____ 49. Polyploid plant species are more common in northern latitudes than in tropical regions.

_____ 50. Behavioral isolation is a typical postzygotic reproductive isolating mechanism.

_____ 51. Horses and donkeys remain separate species due to the sterility of their hybrids.

_____ 52. The Galapagos finches provide a good example of adaptive radiation.

_____ 53. Placental mice and marsupial mice are closely related; they should be in the same genus.

_____ 54. An example of evolutionary co-adaptation might be a flowering plant and its insect pollinator.

_____ 55. For the last several hundred million years, extinction has occurred at a very uniform rate, varying very little.

C. Fill in the Blanks

Answer the question or complete the statement by filling in the blanks with the correct word or words.

56. Jackals, wolves and dogs are all members of the same taxonomic _________________.

57. If a species has distinct geographical subunits that are dissimilar in appearance, they may be classified as_________________ .

58. Within the plant kingdom, classes are grouped into _________________.

59. The polyphyletic group that includes mostly single-celled eukaryotes is kingdom _________________.

60. In the mammal platypus, egg-laying and the presence of a cloaca are considered _________________traits.

61. Scientist who study fossils of organisms are called _________________.

62. Humans are members of the phylum _________________.

63. Wings of birds and forelimbs of cats are considered to be _________________ structures, being descended from the same structure of some remote ancestor.

64. Although they are quite unrelated, two species can appear very similar due to _________________ evolution.

65. A _________________ is the diagrammatic evolutionary tree resulting from cladistic analysis.

66. Among chordates, a dorsal nerve cord is a _________________ character, according to cladistics.

67. When evolutionary divergence leads to speciation in geographically separated groups, the process is termed _________________ speciation.

68. The South American and African continents and their plants and animals were separated some 65 million years ago in the continuing process of _________________.

69. Descendents of apple maggots have experienced rapid local speciation in North America involving a change of host plants; this is probably an example of _________________ speciation in animals.

70. In zones of contact between interfertile plant species, there often exists _________________ of plants with ancestry from both species.

71. A tetraploid plant descended from two different diploid ancestors is a(an) _________________.

72. The specific birdsongs that many bird species recognize are examples of _________________ reproductive isolating mechanisms.

73. Hybrid inviability is an example of a _________________ barrier to ensure reproductive isolation.

74. The Darwinian idea that species change slowly through time, accumulating many small changes is today termed evolutionary _________________.

75. We believe that the ancestral continent Pangaea gave rise over 200 million years ago to a northern landmass called _________________ and a southern landmass called _________________.

Questions for Discussion

1. How is a biological species defined? How can a scientist recognize a species? What are some of the problems with the "species concept"?

2. Outline the different approaches used today in classifying organisms. Compare the methods used. Do you think that any one of the methods is clearly the "correct" way to classify organisms?

3. What do biologists mean by "speciation"? Is it a process that can be readily observed, or must scientists deduce its operation from indirect evidence? Give some examples of the ways that speciation can proceed.

4. For species to maintain their distinct identities, they must be reproductively isolated from one another. Outline the categories of mechanisms that can enforce reproductive isolation, and give examples.

5. Contrast evolutionary gradualism and punctuated equilibrium models of the long-term process of macroevolution. How does the fossil record bear on the differences between these two schemes?

Answers to Self-Exam

Multiple Choice Questions

1. b
2. b
3. d
4. c
5. b
6. d
7. a
8. a
9. b
10. c
11. c
12. a
13. d
14. c
15. d
16. b
17. b
18. c
19. b
20. a
21. a
22. c
23. a
24. a
25. b

True or False Questions

26. F
27. T
28. F
29. F
30. F
31. T
32. T
33. T
34. F
35. T
36. T
37. F
38. T
39. T
40. F
41. T
42. T
43. F
44. F
45. T
46. T
47. T
48. F
49. T
50. F
51. T
52. T
53. F
54. T
55. F

Fill in the Blank Questions

56. genus
57. subspecies
58. divisions
59. Protista
60. primitive
61. paleontologists
62. chordata
63. homologous
64. convergent
65. cladogram
66. shared primitive
67. allopatric
68. continental drift
69. sympatric
70. hybrid swarms
71. allotetraploid
72. behavioral
73. postzygotic
74. gradualism
75. Laurasia, Gondwanaland

CHAPTER 21 Origin of Life

Learning Objectives

After mastering the material covered in chapter 21, you should be able to confidently do the following tasks:

- Describe the supposed conditions on the very early earth.
- Outline the Haldane-Oparin hypothesis that describes a mechanism for the spontaneous origin of life on earth.
- Describe the Miller-Urey experiment that tested this hypothesis.
- Evaluate the hypothesis in light of today's scientific knowledge
- Describe such model protocells as coacervates and thermal proteinoids.
- Explain some of the properties that have been proposed for the earliest cells.
- Account for the development of the oxygen-rich atmosphere of today.
- Place on a broad time scale estimates for the origin of the earth, the appearance of the first cells, and the development of the present atmosphere.

Chapter Outline

I. A Hypothesis on the Origin of Life.

A. The Haldane-Oparin hypothesis.
B. The Miller-Urey experiment.

II. The Early Earth

A. The formation of the earth.
B. The constituents of the early atmosphere.
C. Early energy sources on earth.

III. The Hypothesis Today

A. Aspects of the hypothesis
1. physical conditions of primitive earth.
2. presence of essential monomers.
3. spontaneous formation of familiar polymers.
4. spontaneous formation of bounded protocells.
5. production of self-replicating systems with genetic information.

B. The monomers.
C. The polymers.
D. Self-replicating systems.

IV. Model Protocells

A. Oparin and the coacervate.
B. Fox and thermal proteinoids.
C. Further considerations.
   1. Clay and polymerization.
   2. Speculation on self-replication.

V. The Earliest Cells

A. Autotrophic cells.
B. Heterotrophic cells.
C. Atmospheric change.

Key Words

polymerization
coacervates
autocatalytic
protocells
Haldane-Oparin hypothesis
Miller-Urey experiment
ozone
shock energy
dehydration linkages
monomers
polymers
thermal proteinoids
proteinoid microspheres
cyanobacteria
autotroph
chemoautotroph
photoautotoph
reducing atmosphere
oxidizing atmosphere
free oxygen
oxygen sinks
heterotroph
anaerobes
prokaryotes
eukaryotes

Exercises

1. In the space below, draw and label a simple, clear diagram of the apparatus that Miller and Urey used in their experiment to simulate the physical conditions of early earth.

What materials did they place into the apparatus? List these under "reactants".
What molecules were produced? List these under "products".
What was the energy source for the experiment? Show where this was involved.

Reactants:

Products:

2. On the below "time line," which runs from 5 billion years ago to the present, show the correct relative locations of each of the entries in the list on the left. If you have difficulty, refer to figure 21.4, and keep trying until you can get the sequence correct.

a. first land plants

b. age of dinosaurs

c. first eukaryotes

d. earth takes form

e. first humans

f. first prokaryotes

g. era of chemical evolution

h. appearance of primitive animal phyla

- 5 billion years ago
-
- 4 billion years ago
-
- 3 billion years ago
-
- 2 billion years ago
-
- 1 billion years ago
-
- the present

Self-Exam

You should be able to easily answer the following questions after learning the material in chapter 21. If you have difficulty with any question, study the appropriate section in the text and try again.

A. Multiple Choice Questions

Circle one alternative that best completes the statement or answers the question.

1. The first serious proposals concerning the spontaneous origin of life came more than fifty years ago from the Scottish scientist and Russian scientist
   a. Haldane and Fox.
   b. Haldane and Oparin.
   c. Miller and Urey.
   d. Urey and Fox.

2. The simulated primitive atmosphere of the Urey-Miller experiment included all the following *except*
   a. methane.
   b. ammonia.
   c. water vapor.
   d. free oxygen.

3. Oparin's model included the formation of crude membrane-surrounded droplets called
   a. cells.
   b. autotrophs.
   c. coacervates.
   d. thin soup.

4. A coacervate that had taken in autocatalytic molecules would be termed a(an)
   a. cell.
   b. protocell.
   c. phototroph.
   d. autotroph.

5. After the Urey-Miller experiment, all of the following molecules were found *except*
   a. proteins.
   b. amino acids.
   c. aldehydes.
   d. carboxylic acids.

6. Some estimates suggest that the earth took form approximately how many years ago?
   a. 4.6 billion.
   b. 46 million.
   c. 4.6 million.
   d. 460,000.

7. Which of the following was probably *not* present in the primitive atmosphere of earth?
   a. ozone.
   b. ultraviolet light.
   c. lightning.
   d. carbon dioxide.

8. Most scientists agree that the primitive atmosphere had conditions that would have made very likely the formation of
   a. organic polymers.
   b. organic monomers.
   c. nucleic acid polymers.
   d. enzymes.

9. The formation of biological polymers like polypeptides and polynucleotides involves
   a. glycolysis.
   b. hydrolysis.
   c. dehydration linkages.
   d. vitamins.

10. For which of the following aspects of the hypothesis is there the *least* evidence today?
    a. physical conditions of primitive earth.
    b. production of essential monomers.
    c. formation of bounded cell-like bodies.
    d. production of self-replicating systems.

11. Some scientists believe that self-perpetuating, genetic systems arose first and that these were probably in the form of
    a. protein.
    b. amino acids.
    c. DNA or RNA.
    d. enzymes.

12. Primitive autotrophs of today obtain their energy from
    a. sunlight.
    b. hydrogen sulfide.
    c. other organisms.
    d. DNA or RNA.

13. Photosynthesis by the early cyanobacteria led to a substantial increase in the amount of which of the following, in the earth's atmosphere?
    a. hydrogen sulfide.
    b. hydrogen.
    c. carbon dioxide.
    d. oxygen.

14. An increased level of oxygen in the atmosphere would be toxic to
    a. anaerobic organisms.
    b. aerobic organisms.
    c. most photoautotrophs.
    d. eukaryotes.

15. Which of the following probably flourished earliest on the earth?
    a. dinosaurs.
    b. vertebrates.
    c. eukaryotes.
    d. cyanobacteria.

B. True or False Questions

Mark the following statements either T (True) or F (False).

_____ 16. By careful investigation, we should be able to prove how life came to be on this planet.

_____ 17. Monomers which have now been synthesized in the laboratory include sugars, all of the amino acids, and many essential vitamins.

_____ 18. Protein is an example of an autocatalytic molecule.

_____ 19. As the early earth coalesced and took form, much heat was generated, and there was no liquid water on the earth's surface.

_____ 20. Little ultraviolet light reached the surface of the early earth before photosynthesis appeared.

_____ 21. The experiments of Miller and Urey are singular - they have never really been repeated.

_____ 22. Current evidence suggests that nucleic acid polymers can be produced spontaneously.

_____ 23. Coacervates are known to actually "grow".

_____ 24. While coacervates are not living entities, thermal proteinoids are considered to be alive.

_____ 25. The surfaces of clay particles are likely places for spontaneous polymerization to occur.

_____ 26. A virus is a good model of a self-reproducing system, and it can easily replicate itself, all by itself.

_____ 27. Cyanobacteria are known from about 3.5 billion years ago.

_____ 28. Chemoautotrophic organisms derive their energy from light.

_____ 29. Elemental iron was an "oxygen sink" on early earth, and most of it had to be saturated before free oxygen could become a significant atmospheric gas.

_____ 30 Photosynthetic, oxygen-producing phototrophs preceded aerobically respiring, oxygen utilizing heterotrophs.

C. Fill in the Blanks

Answer the question or complete the statement by filling in the blanks with the correct word or words.

31. The crude membrane-surrounded droplets, conglomerates of early enzyme, as envisioned by Oparin, are called ___________________.

32. A coacervate containing autocatalytic molecules could be termed a ___________________.

33. In their experiment, Miller and Urey used ___________________as a source of energy for reactions.

34. It is estimated that the earth took form about ___________________years ago.

35. Sources of energy for chemical reactions on early earth are thought to include UV light from the sun and lighting, heat, and shock energy from ___________________.

36. Biological polymerizations involve ___________________between monomers.

37. Researchers have shown spontaneous formation of polymers of as many as 200 amino acids under ___________________conditions, such as along the edges of volcanos.

38. When thermal proteinoids are placed in water, they form clusters, called ___________________by researcher Sidney Fox.

39. When water is the hydrogen source for photosynthesis, the waste product is ___________________.

40. The present-day organisms believed to be the direct descendants of the earliest forms of life on earth are the ___________________.

Questions for Discussion

1. The statement is made in the text that it is generally agreed by scientists that life arose only once in earth's history - a unique event. Why do you think this is the consensus? What types of evidence suggest that this might be the case? Why is it unlikely that life as we know it would arise again today on the earth?

2. Restate the components of the Haldane-Oparin hypothesis on the spontaneous origin of life on earth. Describe the current evidence that bears on each point and assess how well each is supported. What is the weakest element of the hypothesis at present?

3. Describe and distinguish Oparin's coacervates and Fox's thermal proteinoid microspheres. How do these relate to the origin of cells?

4. Describe the changes in the composition of the earth's atmosphere that are believed to have taken place from prebiotic times until the present day (you may ignore today's ozone holes). What explains the changes that have occurred over the last few billion years?

5. If life arose spontaneously on earth under a unique set of conditions, what do you think the likelihood is that life of a similar sort has also arisen at some other place in the universe? Is there any evidence? What types of characteristics would you expect a likely location to have? Could there be "life" of a very different sort? How would you recognize it?

Answers to Self-Exam

Multiple Choice Questions

1. b
2. d
3. c
4. b
5. a
6. a
7. a
8. b
9. c
10. d
11. c
12. a
13. d
14. a
15. d

True or False Questions

16. F
17. T
18. F
19. T
20. F
21. F
22. F
23. T
24. F
25. T
26. F
27. T
28. F
29. T
30. T

Fill in the Blank Questions

31. coacervates
32. protocell
33. electrical discharges
34. 4.6 billion
35. earthquakes
36. dehydration linkages
37. hot, drying
38. proteinoid microspheres
39. oxygen
40. prokaryotes

CHAPTER 22 Prokaryotes and Viruses

Learning Objectives

After mastering the material covered in chapter 22, you should be able to confidently do the following tasks:

- Give evidence that prokaryotes form the most ancient group of organisms on earth, and explain why eubacteria and archaebacteria are today considered to be separate prokaryotic lineages.
- Outline the basic taxonomy of bacteria and viruses, and give the main characteristics of the majors groups.
- Explain the ways bacteria move and reproduce.
- List the principal modes of bacterial nutrition, distinguish the different forms of heterotrophy and autotrophy, and explain the ecological roles of prokaryotes.
- Describe important types of plant and animal pathogens, the way they are transmitted between hosts, and methods of disease prevention.
- Explain why science today believes that archaebacteria are so different from eubacteria.
- Tell what viruses are, show how they invade cells and reproduce, and give some of the diseases they can cause in plants and animals.
- Describe the biological basis for AIDS and show how the responsible virus, HIV-III, infects individuals and causes the symptoms of AIDS.

Chapter Outline

I. Prokaryote Origins

II. Biology of Prokaryotes

A. Cell Structure.
1. Cell Walls, Capsules, and Antibiotics.
B. Cell Form and Arrangement.
1. Mycoplasma: The Smallest Cells.
C. Bacterial Movement.
D. Bacterial Reproduction.
E. Metabolic Activities of Bacteria.
1. Bacterial Energetics.
F. Bacterial Heterotrophs.
1. Decomposers.
a. Inhibiting Decomposers.

b. Useful Decomposers.
2. Pathogens of Animals.
a. Sexually Transmitted Diseases.
3. Pathogens of Plants.
G. Bacterial Autotrophs.
1. Photoautotrophs.
2. Chemoautotrophs.
H. The Nitrogen Fixation Process.

III. Eubacteria, Archaebacteria, and a New Phylogeny.

A. Archaebacterial Life.

IV. The Viruses

A. The Discovery of Viruses.
B. The Biology of Viruses.
1. Viral Shapes.
2. Biochemical Differences.
C. Viruses and the Host Cell.
1. Bacteriophages.
2. Plant Viruses.
3. Animal Viruses.
4. Herpes simplex.
D. The Viral Genome and its Replication.
E. HIV-III: Portrait of a Killer.
1. The Origin of AIDS.

Key Words

monera
prokaryotes
bacteria
stromatolites
cyanobacteria
eubacteria
archaebacteria
thylakoid
mesosome
pilus (-i)
pilin
sex pilus (-i)
flagellum (-a)
polyribosomes
plasmids
peptidoglycan
"outer membrane"
techoic acid
Gram positive
Gram negative
penicillin
streptomycin
tetracycline
capsule
endospores
bacillus (-i)
coccus (-i)
spirilla
spirochaetes
vibrio
cholera
diplococcus
streptococcus
staphylococcus
sarcina
mycoplasmas
flagellin
myxobacteria
fission
Escherichia coli
conjugation
heterotrophs
decomposers
parasites
pathogens
autotrophs
photoautotrophs
chemoautotrophs
aerobes
anaerobes
obligate anaerobes
obligate aerobes
facultative anaerobes
saprobe
sterilization
pasteurization
actinomycetes
Streptomyces
Salmonella
salmonellosis
botulism
endospores
Clostridium botulinum
Clostridium tetani
tetanus
Clostridium perfringens
gas gangrene
Streptococcus pyrogenes
rickettsias
Rocky Mountain spotted fever
Neisseria gonorrhoeae
gonorrhea
Treponema pallidum
syphilis
chlamydia
blights
wilts
galls
rots
Erwinia amylovora
fire blight

Corynebacterium
crown gall
Agrobacterium tumefaciens
cyanobacteria
Oscillatoria
chlorophyll *a*
chlorophyll *b*
phycobilins
heterocysts
halobacteria
bacteriorhodopsin
nitrogen-fixation
nitrogenase
methanogens
halophiles
thermophiles
thermoacidophiles
virus
oncogenic viruses
vaccine
tobacco mosaic virus
virion
capsid
bacteriophage
T-even bacteriophages
phage lysozyme
lytic phages
lysogeny
temperate phages
phage lambda
prophage
adenoviruses
myxoviruses
Herpes simplex I
Herpes simplex II
genital herpes
retroviruses
reverse transcriptase
reoviruses
Human Immunodeficiency Viruses
HIV-I
HIV-II
HIV-III
leukemia
AIDS
Acquired Immune Deficiency Syndrome
lymphocytes
helper T-cells
pneumocystic pneumonia
Kaposi's sarcoma
epidemic
pandemic

Exercises

1. Below is a diagram of a typical prokaryotic cell. Supply a label for each of the indicated parts.

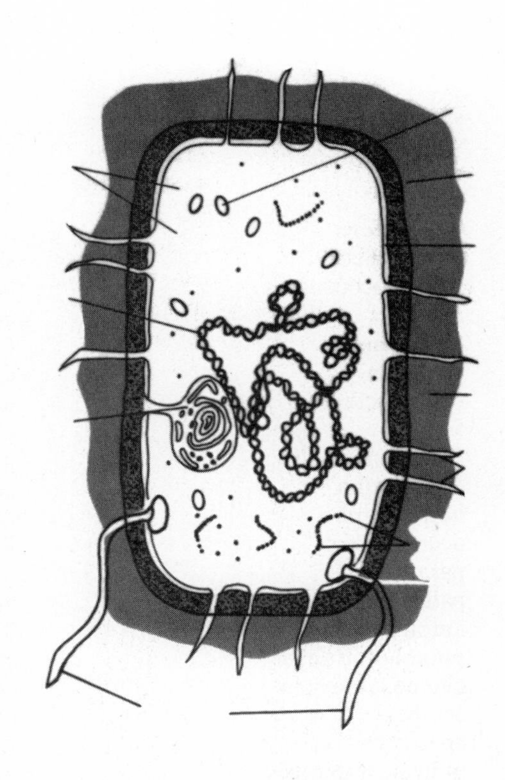

2. In the spaces below, draw diagrams showing the shapes of the indicated bacteria.

a. coccus	b. bacillus	c. spirochaete
d. diplococcus	e. bacillus chain	f. streptococcus
g. staphylococcus	h. sarcina	i. vibrio

Self-Exam

You should be able to easily answer the following questions after learning the material in chapter 22. If you have difficulty with any question, study the appropriate section in the text and try again.

A. Multiple Choice Questions

Circle one alternative that best completes the statement or answers the question.

1. All prokaryotes are members of the Kingdom
   a. Protista.
   b. Monera.
   c. Fungi.
   d. Animalia.

2. Prokaryotes have which of the following?
   a. membrane-surrounded organelles.
   b. an organized nucleus.
   c. mitochondria.
   d. ribosomes.

3. How old are some of the oldest the deposits in which fossils of bacteria have been found?
   a. 2.5 million year old.
   b. 3.5 million years old.

c. 3.5 billion years old.
d. 4.6 billion years old.

4. The unique formations called stromatolites are the results of activity of
   a. plants.
   b. heterotrophs.
   c. autotrophs.
   d. saprobes.

5. Although there is obviously a lot of variation in size among the bacteria, a generalization would characterize them as about _____ the size of a typical eukaryotic cell.
   a. one-hundredth.
   b. one-tenth.
   c. one-half.
   d. twice

6. The bacterial cell membrane may contain features such as
   a. respiratory and phosphorylating enzymes.
   b. photosynthetic pigments and enzymes.
   c. a mesosome that may function in cell division.
   d. all of the above can be found there.

7. A bacterial cell pilus is made up of the protein called
   a. actin.
   b. tubulin.
   c. pilin.
   d. flagellin.

8. Peptidoglycan is a building block of the cell wall in
   a. eubacteria.
   b. archaebacteria.
   c. fungi.
   d. plants.

9. The bacteria with the thicker type of cell wall
   a. retain crystal violet and are Gram negative.
   b. retain crystal violet and are Gram positive.
   b. are sensitive to crystal violet, but positive.
   d. do not retain crystal violet.

10. Many Gram negative bacteria are not susceptible to the antibiotic drug
   a. penicillin.
   b. tetracycline.
   c. streptomycin.
   d. all of these.

11. The antibiotic penicillin acts against some types of bacteria by
   a. inhibiting protein synthesis.
   b. inhibiting nucleic acid synthesis.
   c. inhibiting cell wall synthesis.
   d. inhibiting viruses.

12. Antibiotics like tetracycline and streptomycin are effective against some bacteria by
   a. inhibiting protein synthesis.
   b. inhibiting nucleic acid synthesis.
   c. inhibiting cell wall synthesis.
   d. inhibiting viruses.

13. Some types of bacteria form endospores. In general, these have which of the following properties?
    a. They are thick-walled, but can be killed by brief boiling.
    b. They are thin-walled, but can survive for very long periods.
    c. Although they are dormant, they can be killed by prolonged boiling or very high temperatures.
    d. Although they are dehydrated, they still have considerable metabolic activity.

14. A bacillus is
    a. a spherical cell.
    b. a rod-like cell.
    c. a comma-shaped cell.
    d. a spiral cell.

15. A coccus is
    a. a spherical cell.
    b. a rod-like cell.
    c. a comma-shaped cell.
    d. a spiral cell.

16. A vibrio is
    a. a spherical cell.
    b. a rod-like cell.
    c. a comma-shaped cell.
    d. a spiral cell.

17. As a generalization, the smallest prokaryotes are the
    a. cyanobacteria.
    b. spirochaetes.
    c. mycoplasmas.
    d. myxobacteria.

18. The bacterial flagellum is a stiff structure composed of the protein
    a. actin.
    b. tubulin.
    c. pilin.
    d. flagellin.

19. Bacterial movement can occur via
    a. flagellum action, or sweeping by cilia.
    b. sweeping by cilia, or by gliding.
    c. flagellum action, or by gliding.
    d. flagellum action, or sweeping by cilia, or by gliding.

20. The so-called "gliding bacteria" are the
    a. cyanobacteria.
    b. spirochaetes.
    c. mycoplasmas.
    d. myxobacteria.

21. The most common mode of reproduction among bacteria is
    a. sexual, via fission.
    b. asexual, via fission.
    c. sexual, via conjugation.
    d. asexual, via conjugation.

22. Bacteria that cause disease are best described as
    a. chemoautotrophs.
    b. photoautotrophs.

c. pathogens.
d. decomposers

23. Decomposers are examples of
    a. heterotrophs.
    b. photoautotrophs.
    c. chemoautotrophs.
    d. parasites.

24. Photoautotrophic bacteria synthesize their essential molecules from simple
    a. organic molecules, using light energy.
    b. inorganic molecules, using light energy.
    c. inorganic molecules, using chemical energy.
    d. organic molecules, using chemical energy.

25. Bacteria that synthesize their essential molecules from inorganic substances, using energy derived from chemical reaction they bring about, are termed
    a. chemoautotrophs.
    b. photoautotrophs.
    c. pathogens.
    d. decomposers.

26. Food spoilage is usually caused by
    a. chemoautotrophs.
    b. photoautotrophs.
    c. pathogens.
    d. decomposers.

27. Bacteria of the genus *Salmonella* are known to be the cause of
    a. botulism.
    b. "lockjaw".
    c. food poisoning.
    d. plague.

28. Bacteria of the genus *Clostridium* are known to be the cause of
    a. botulism.
    b. typhoid fever.
    c. food poisoning.
    d. plague.

29. The condition known as "strep throat" is caused by bacteria of the genus
    a. *Salmonella.*
    b. *Clostridium.*
    c. *Streptococcus.*
    d. *Staphylococcus*

30. The bacteria *Neisseria gonorrhoeae* and *Treponema pallidum* are responsible for
    a. food poisoning.
    b. sexually transmitted diseases.
    c. wilts and blights.
    d. herpes and AIDS.

31. Bacteria of the species *Agrobacterium tumefaciens* are the cause of
    a. blights.
    b. wilts.
    c. galls.
    d. rots.

32. The characteristic blue-green color of many cyanobacteria is due to which type of pigment?
    a. chlorophyll *a*.
    b. chlorophyll *b*.
    c. beta carotene.
    d. phycobilins.

33. Nitrogen-fixing bacteria characteristically do which of the following?
    a. oxidize sulfur compounds to nitrogen.
    b. reduce nitrogen compounds to sulfur.
    c. reduce atmospheric nitrogen to ammonia.
    d. reduce ammonia to nitrogen gas.

34. Methanogens form the largest group of
    a. archaebacteria.
    b. eubacteria.
    c. rickettsias.
    d. viruses.

35. Bacteria that thrive in the hot springs of deep-ocean rifts can be characterized as
    a. extreme halophiles.
    b. extreme thermophiles.
    c. extreme photophiles.
    d. acidophiles.

36. Which of the following diseases is not caused by a virus?
    a. tuberculosis.
    b. polio.
    c. smallpox.
    d. chickenpox.

37. Viruses can be characterized as
    a. autotrophs.
    b. saprobes.
    c. parasites.
    d. decomposers.

38. he genetic material of viruses can be
    a. protein.
    b. DNA, but not RNA.
    c. RNA, but not DNA.
    d. DNA or RNA.

39. The infectious agent that leads to AIDS belongs to which group?
    a. reovirus.
    b. retrovirus.
    c. myxovirus.
    d. arbovirus.

40. The type of cell most commonly attacked by HIV-III viruses when a human is infected is the
    a. helper T-cell.
    b. helper B-cell.
    c. nerve cell.
    d. sex cell.

B. True or False Questions

Mark the following statements either T (True) or F (False).

_____ 41. Kingdom Monera, as defined at present, includes both bacteria and protista.

_____ 42. Bacterial cells have a plasma membrane with a typical phospholipid bilayer structure.

_____ 43. The mesosome is a completely internal membrane-surrounded organelle in bacteria.

_____ 44. Archaebacteria possess a cell wall composed mainly of peptidoglycan.

_____ 45. Most Gram negative bacteria are susceptible to penicillin.

_____ 46. Endospores are dehydrated cells in a state of dormancy.

_____ 47. Grapelike clusters of spherical cells are typical of staphylococci.

_____ 48. Cholera is a viral disease.

_____ 49. In nature our colon bacteria, *E. coli*, seldom divide by fission but rather conjugate frequently.

_____ 50. Obligate anaerobes are bacteria that may live without oxygen but will utilize it if available.

_____ 51. The energy to power the bacterial flagellum is derived from a proton gradient, as is ATP production.

_____ 52. Salt and sugar can be used to treat food to inhibit the growth of food-spoiling bacteria.

_____ 53. Tetanus and botulism are caused by related bacteria that are placed in the same genus.

_____ 54. If untreated, syphilis will usually run its course, and complete recovery follows.

_____ 55. Bacteria are not known to cause disease in plants.

_____ 56. Cyanobacteria may be blue-green, black, purple, brown, or red.

_____ 57. Heterocysts are nitrogen-fixing cells of cyanobacteria.

_____ 58. Although they are not considered bacteria, viruses are very small cells.

_____ 59. A temperate phage is called a prophage when it is integrated into its host's chromosome.

_____ 60. Infection by *Herpes simplex I*, which causes cold sores, can lead to sexually transmitted genital herpes.

C. Fill in the Blanks

Answer the question or complete the statement by filling in the blanks with the correct word or words.

61. All prokaryotes are members of the kingdom ____________________.

62. The columnlike deposits providing evidence of ancient cyanobacteria are called ____________________.

63. A membranous infolding in bacteria that apparently functions during cell division is the ____________________.

64. Eubacterial cell walls utilize a molecular building block called ____________________.

65. Many bacteria have, in addition to their cell walls, an outer slimy layer called a ____________________.

66. As it moves a bacterial cell, the bacterial flagellum ____________________ on its axis.

67. During conjugation between bacterial cells, a one-way transfer on genes occurs through the ___________________.

68. Bacterial cells that must have oxygen to survive and reproduce are called ___________________.

69. Rocky Mountain spotted fever is caused by extremely small parasites called ___________________.

70. Silver nitrate is routinely placed in newborn babies' eyes to prevent ___________________.

71. When bacteria invade a plant's vascular system, hindering water movement, the result can be a disease called a ___________________.

72. Halobacteria capture light energy with the pigment ___________________.

73. The first virus observed by scientists was ___________________ virus.

74. Retroviruses use the enzyme ___________________ to make DNA copies of their RNA genomes.

75. The most likely geographic origin of the virus HIV-III that is responsible for AIDS is ___________________.

Questions for Discussion

1. Cyanobacteria were formerly called blue-green algae and were considered to be types of plants. Today we treat them as specialized photosynthetic bacteria. Outline the reasons that these organisms are today considered to be part of the kingdom Monera and not part of the plant kingdom. Explain the important similarities they do have with members of the plant kingdom. What evolutionary connection might there be between cyanobacteria and plants?

2. Although bacteria often reproduce by simple fission, such processes as conjugation allow for the possibility of genetic exchange between different bacterial cells. Why is the potential for genetic exchange important? Considering bacteria that are human pathogens, what are the practical consequences of such processes in terms of antibiotic effectiveness?

3. Today, many biologists think the the kingdom Monera contains two very different groups, the archaebacteria and the eubacteria. Outline the reasons for this conclusion by describing the basic similarities and differences between the two proposed groups. Do you think either of these groups is related to eukaryotes? Why?

4. Assume the general statement can be made that antibiotics that are effective against bacterial pathogens have no effect on pathogenic viruses. On the basis of your knowledge of the way some antibiotics act on bacteria, explain this generalization.

5. Infection of a human with HIV-III leads to AIDS. The infectious agent is a retrovirus. On the basis of your knowledge of the reproductive cycle, suggest some approaches that might be taken to stop proliferation of viruses once infection has occurred. On the basis of your knowledge of how the virus is transmitted between human hosts, suggest some biologically sound ways to prevent infection of additional humans with this virus.

Answers to Self-Exam

Multiple Choice Questions

1. b
2. d
3. c
4. c
5. b
6. d
7. c
8. a
9. b
10. a
11. c
12. a
13. c
14. b
15. a
16. c
17. c
18. d
19. c
20. d
21. b
22. c
23. a
24. b
25. a
26. d
27. c
28. a
29. c
30. b
31. c
32. d
33. c
34. a
35. b
36. a
37. c
38. d
39. b
40. a

True or False Questions

41. F
42. T
43. F
44. F
45. F
46. T
47. T
48. F
49. F
50. F
51. T
52. T
53. T
54. F
55. F
56. T
57. T
58. F
59. T
60. F

Fill in the Blank Questions

61. Monera
62. stromatolites
63. mesosome
64. peptidoglycan
65. capsule
66. rotates
67. sex pilus
68. obligate aerobes
69. rickettsias
70. gonorrhea
71. wilt
72. bacteriorhodopsin
73. tobacco mosaic
74. reverse transcriptase
75. equatorial Africa

CHAPTER 23 Protists

Learning Objectives

After mastering the material covered in chapter 23, you should be able to confidently do the following tasks:

- Present a hypothesis proposing endosymbiosis as an origin of eukaryotic cells, cite evidence that supports this notion, and evaluate the strength of the hypothesis.

- Describe and distinguish the basic life cycles exhibited by present day eukaryotic organisms.

- List the main groups of protists, give their important distinguishing traits, and relate them to their presumed ancestors.

- For the major phyla of Protozoa, describe the structure, life cycle, and reproduction of representative species.

- Outline the complete infectious cycle of the parasite that causes malaria, describing each stage and identifying the hosts.

- Distinguish the major phyla of algae by comparing complexity of organization, type of life cycle, pattern of reproduction, and similarities or differences in structure, pigments, and carbohydrates.

Chapter Outline

I. Origin of the Protists.

A. The serial endosymbiosis hypothesis.

II. Characteristics of Protists.

A. Life cycles in protists and other eukaryotes.
1. the zygotic cycle.
2. the gametic cycle.
B. Phylogeny of today's protists.

III. The Protozoans

A. Sarcomastigophora.
1. the flagellates.
2. the amebas.
B. Apicomplexa: the sporozoans.
C. Ciliophora: the ciliates.

IV. The Myxomycota and Acrasiomycota: Slime Molds.

V. The Oomycetes: the Water Molds.

VI. The Algae.

A. Dinoflagellates.
B. Euglenophyta.
C. Chrysophyta: the yellow-green and golden brown algae.
  1. The Diatoms.
D. Multicellular Algae.
  1. Rhodophyta: the red algae.
  2. Phaeophyta: the brown algae.
E. Chlorophyta: the green algae.
  1. single-celled green algae.
  2. colonial green algae.
  3. multicellular green algae.

Key Words

endosymbionts
symbiosis
mutualism
protoeukaryote
mutualistic symbiosis
tubulin
pseudopods
fragmentation
autogamy
isogametes
heterogametes
zygotic cycle
gametic cycle
sporic cycle
spores
gametophytes
sporophytes
alternation of generations
polyphyletic
protozoans
algae
slime molds
Euglena
phylum Sarcomastigophora
subphylum Mastigophora
Trypanosoma gambiense
African sleeping sickness
subphylum Sarcodina
Entamoeba histolytica
amebic dysentery
ectoplasm
endoplasm
hyalin cap
phagocytosis
exocytosis
food vacuoles
test
heliozoans
silicon dioxide
axopodia
radiolarians
radiolarian ooze
foraminiferans
calcium carbonate
phylum Apicomplexa
sporozoans
Plasmodium vivax
malaria
Anopheles .mosquito
DDT
sporozoites
merozoite
gamonts
phylum Ciliophora
ciliates
Paramecium
cytostome
pellicle
trichocysts
contractile vacuoles
oral groove
cytopyge
Paramecium caudatum
micronucleus
macronucleus
Myxomycota
acellular slime molds
Acrasiomycota
cellular slime molds
Physarum polycephalum
plasmodium
sporangiophores
sporangium (-ia)
myxameba
swarm cell
Oomycetes
water mold
Phytophthora infestans
downy mildews
blights
oogonium (-ia)
antheridium (-ia)
phytoplankton
lichens
dinoflagellates
Pyrrophyta
red tide
algal blooms
Eugenophyta
euglenoids
Euglena
eye-spot
photoreceptor
gullet
paramylon
Chrysophyta
yellow-green algae
golden brown algae
carotenoids
fucoxanthin
leucosin
diatoms
auxospore
diatomaceous earth
Rhodophyta
red algae
phycocyanin
phycoerythrin
seaweed
holdfasts
floridean starch
agar-agar
Irish moss
carrageenan

isomorphic alternation
heteromorphic alternation
Polysiphonia
spermatium (-ia)
spermatangia
carpogonium (-ia)
trichogyne
carposporophyte
carpospores
Phaeophyta
brown algae
laminarin
mannitol
algin
Sargassum
Sargasso Sea
kelp
Macrocystis
Nereocystis
Laminaria
rockweed
Fucus
antheridium (-ia)
oogonium (-ia)
stipes
blades
bladders
Chlorophyta
green algae
Chlamydomonas
zygospore
meiospore
Volvox
siphonous algae
coenocyte
Acetabularia
compound nucleus
sea lettuce
Ulva
Ulothrix

Exercises

1. Below is a diagram of the life cycle of a brown alga, *Laminaria*. Apply labels to the diagram, using the terms in the list below. Indicate whether each structure is haploid (n) or diploid (2n). Show where meiosis occurs and where fertilization occurs. Is this life cycle gametic, zygotic, or sporic?

sporophyte
male gametophyte
female gametophyte
zygote
gametes
spores
eggs
sperms
oogonium
antheridium

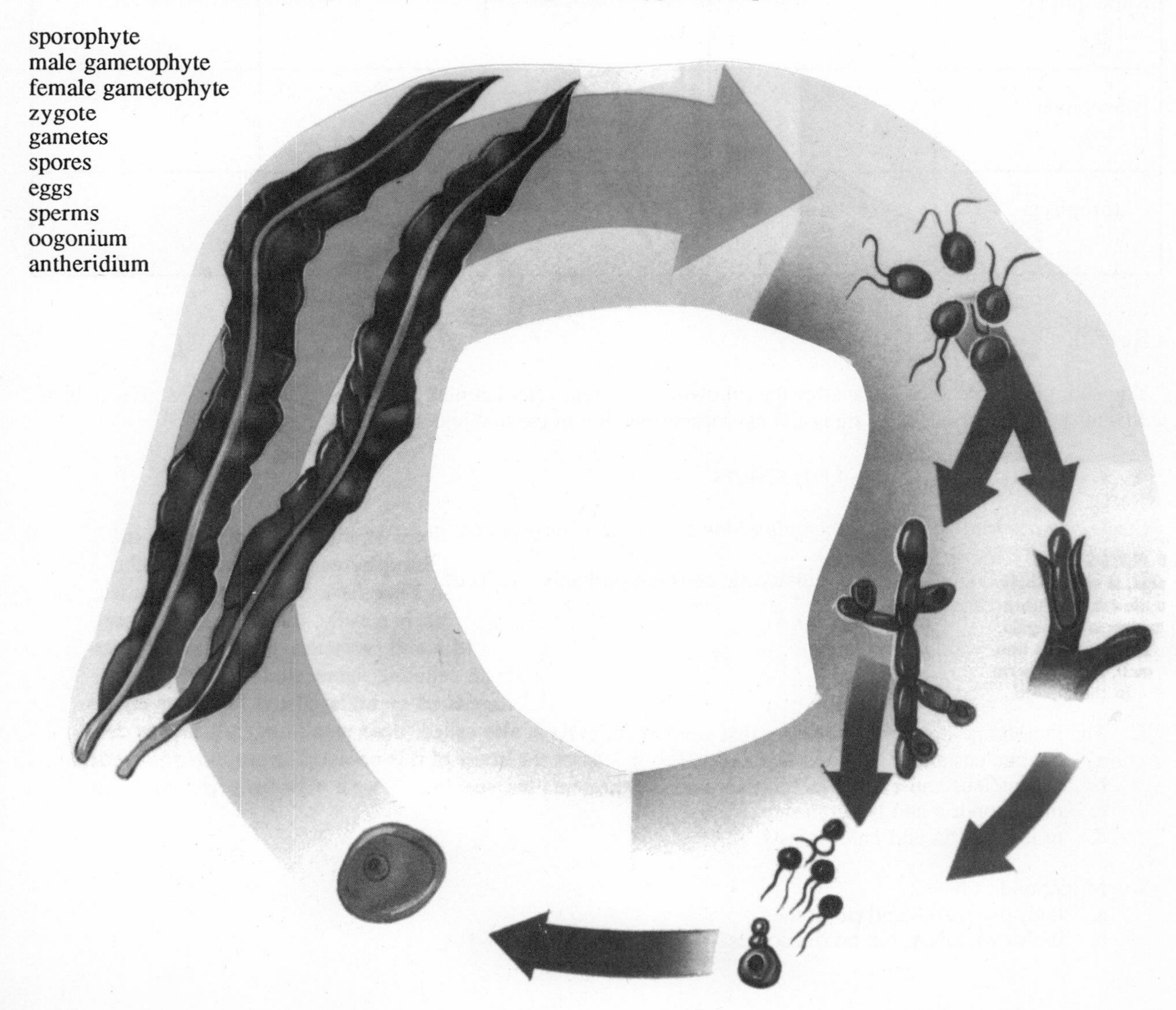

2. Below is an empty table that can be used to outline the characteristics of the different groups of algae. Fill in this table as completely as possible.

Algal Group	Type of Chlorophyll	Other Pigments	Cell Wall	Storage Carbohydrates
Dinoflagellates				
Euglenophyta				
Chrysophyta				
Rhodophyta				
Phaeophyta				
Chlorophyta				

Self-Exam

You should be able to easily answer the following questions after learning the material in chapter 23. If you have difficulty with any question, study the appropriate section in the text and try again.

A. Multiple Choice Questions

Circle one alternative that best completes the statement or answers the question.

1. The oldest known fossils of eukaryotic cells are probably fossils of
   a. blue-green algae.
   b. amebas.
   c. red and green algae.
   d. fungi.

2. The membrane-bounded organelles that are thought to have originated as endosymbionts are
   a. the nucleus and mitochondria.
   b. the nucleus and chloroplasts.
   c. mitochondria and lysosomes.
   d. mitochondria and chloroplasts.

3. Mitochondria have
   a. their own DNA and ribosomes.
   b. their own DNA, but no ribosomes.

c. their own ribosomes, but no DNA.
d. neither DNA nor ribosomes of their own.

4. Common modes of nutrition among protists include
   a. predation, but not autotrophy.
   b. photosynthesis, but not parasitism.
   c. both heterotrophy and autotrophy.
   d. both chemoautotrophy and photoautotrophy.

5. Reproduction of protists where mitosis is followed by cell division is termed __________ and is _________.
   a. fission; sexual.
   b. fission; asexual.
   c. fragmentation, sexual.
   d. fragmentation; asexual.

6. In the type of reproduction called autogamy, we observe
   a. mitosis, but not meiosis.
   b. mitosis, followed by exchange of sex cells.
   c. meiosis, but no transfer of gametes.
   d. neither mitosis nor meiosis.

7. The type of sexual life cycle considered to be primitive is probably the
   a. zygotic cycle.
   b. gametic cycle.
   c. sporic cycle.
   d. genic cycle.

8. Gametophytes are ____________ and produce ___________.
   a. diploid; gametes.
   b. haploid; spores.
   c. diploid; spores.
   d. haploid; gametes.

9..In the sporophyte generation of plants, the process of ____________ leads to production of _____________.
   a. mitosis; spores.
   b. meiosis; spores.
   c. mitosis; gametes.
   d. meiosis; gametes.

10. Flagellates and amebas are members of the phylum
   a. Sarcomastigophora.
   b. Ciliophora.
   c. Dinoflagellates.
   d. Apicomplexa.

11. The disease African sleeping sickness is caused by the flagellate
   a. *Trypanosoma gambiense.*
   b. *Entamoeba histolytica.*
   c. *Plasmodium vivax.*
   d. *Paramecium caudatum.*

12. The disease malaria is caused by the parasite
   a. *Trypanosoma gambiense.*
   b. *Entamoeba histolytica.*
   c. *Plasmodium vivax.*
   d. *Paramecium caudatum.*

13. In cells that exhibit amoeboid movement, the firmer, jelly-like state of the cytoplasm is found along the perimeter of the cell and is called
    a. protoplasm.
    b. ectoplasm.
    c. endoplasm.
    d. sarcoplasm.

14. Calcium carbonate is characteristic of the shells of
    a. diatoms.
    b. sarcodines.
    c. radiolarians.
    d. foraminiferans.

15. The malarial parasite has characteristics of which group?
    a. Sporozoans.
    b. Mastigophora.
    c. Sarcodina.
    d. Ciliophora.

16. Which is the correct sequence of stages in the life cycle of the malarial parasite?
    a. zygote ⇒ sporozoite ⇒ gamont ⇒ merozoite ⇒ gamete ⇒ zygote.
    b. zygote ⇒ sporozoite ⇒ merozoite ⇒ gamont ⇒ gamete ⇒ zygote.
    c. zygote ⇒ merozoite ⇒ sporozoite ⇒ gamont ⇒ gamete ⇒ zygote.
    d. zygote ⇒ gamont ⇒ gamete ⇒ sporozoite ⇒ merozoite ⇒ zygote.

17. In *Paramecium*, the DNA that transmits information to descendent cells is found in the
    a. cytostome.
    b. macronucleus.
    c. micronucleus.
    d. cytopyge.

18. Most of the slime molds could be described as
    a. predators.
    b. parasites.
    c. autotrophs.
    d. decomposers.

19. Unlike the fungi, oomycetes have cells walls that contain
    a. cellulose.
    b. peptidoglycan.
    c. chitin.
    d. calcium.

20. A lichen is a symbiontic association between
    a. myxomycota and oomycetes.
    b. algal and fungal cells.
    c. an alga and a slime mold.
    d. an alga and a water mold.

21. The organisms responsible for "red tides" are found among which group?
    a. dinoflagellates.
    b. red algae.
    c. euglenoids.
    d. diatoms.

22. The photosynthetic system in *Euglena* is similar to that found in green plants because it uses
    a. chlorophylls *a* and *b*.
    b. chlorophylls *a* and *c*.

c. chlorophylls *a* and *d*.
d. fucoxanthin.

23. Single-celled algae that live in glass shell are the
 a. dinoflagellates.
 b. diatoms.
 c. rhodophyta.
 d. uglenoids.

24. The polysaccharides agar and carrageenan are produced by members of which group?
 a. green algae.
 b. golden brown algae.
 c. brown algae.
 d. red algae.

25. When red algae are entering a sexual reproductive phase, egg cells are formed in the
 a. oogonium.
 b. archegonium.
 c. carpogonium.
 d. sporophyte.

26. The algae that dominate the Sargasso Sea are members of which group?
 a. Chrysophyta.
 b. Rhodophyta.
 c. Phaeophyta.
 d. Chlorophyta.

27. The largest of the multicellular algae would probably be found among the
 a. rockweeds.
 b. kelps.
 c. sea lettuce.
 d. *Acetabularia.*

28. In sexual reproduction, cells of *Chlamydomonas* produce
 a. antheridia and oogonia.
 b. sperms and eggs.
 c. isogametes.
 d. heterogametes.

29. A good example of a colonial green alga is
 a. *Volvox* .
 b. *Chlamydomonas.*
 c. *Acetabularia.*
 d. *Ulva.*

30. Sea lettuce exhibits an alternation of generations, although the sporophyte and the gametophyte are quite similar; it is termed
 a. isosporic.
 b. isomorphic.
 c. heteromorphic.
 d. zygotic.

B. True or False Questions

Mark the following statements either T (True) or F (False).

_____ 31. Protoeukaryotes, as described by the Margulis hypothesis, are well represented on earth today.

_____ 32. All of the protists are clearly single-celled organisms.

_____ 33. When both sexual and asexual reproduction occur in a species, asexual reproduction is commonly found in populations that are rapidly increasing in size.

_____ 34. The gametic type of life cycle occurs in animals such as humans.

_____ 35. Organisms with the zygotic cycle alternate between sporophytes and gametophytes.

_____ 36. Today's protists are polyphyletic, indicating that plants, animals, and fungi all originated from the same ancestral group.

_____ 37. Sarcodines will usually feed by phagocytosis.

_____ 38. Radiolarians have coverings of silicon dioxide, while foraminiferans have shells of calcium carbonate.

_____ 39. Amebic dysentery is transmitted by the female *Anopheles* mosquito.

_____ 40. Malarial gamonts mature in human red blood cells.

_____ 41. *Paramecium* can reproduce sexually, at which time macronuclei fuse with one another.

_____ 42 The acellular slime molds, the myxomycota, are often multinucleate.

_____ 43. The organism responsible for the Irish potato famine of the mid-19th century was a type of water mold.

_____ 44. Euglenoids were once considered animals because their eye-spots allow them to see fairly clear images.

_____ 45. Euglenoids store their carbohydrates in a form called leucosin.

_____ 46. Fucoxanthin is an abundant carotenoid of the Chrysophyta.

_____ 47. Diatoms reproduce asexually, with the daughter cells being unequal in size; the larger cell eventually sheds its glass shell and produces gametes.

_____ 48. Marine red algae grow on rocky coasts, where they are attached by specialized roots.

_____ 49. The sperm cells of red algae have typical eukaryotic flagella.

_____ 50. Kelps have specialized tissues and organs, such as stem-like stipes, leaf-like blades, and hollow bladders to keep photosynthetic tissue near the ocean surface.

C. Fill in the Blanks

Answer the question or complete the statement by filling in the blanks with the correct word or words.

51. The first person to report seeing organisms that we today call protists was ____________________.

52. The hypothesis explaining the origin of today's mitochondria in eukaryotic cells proposes that formerly there existed a mutualistic ____________________.

53. A form of sexual reproduction in which the products of meiosis are recombined within the same organism is called ____________________.

54. Organisms that exhibit the zygotic life cycle spend most of their lives in the ____________________ state, while those with the gametic life cycle have cells that are usually ____________________.

55. In species with alternation of generations, there is typically a diploid ____________________ alternating with a haploid ____________________.

56. The group of flagellated protozoans, whose members mostly are propelled by falgella, are placed in the subphylum ____________________.

57. A human acquires the malarial parasite when a mosquito injects ____________________ of *Plasmodium* into the blood.

58. The ciliate *Paramecium* has DNA in both the diploid ____________________ and the polyploid ____________________.

59. When the slime mold *Physarum polycephalum* enters its reproductive phase, spore-forming sporangia are formed atop vertical ____________________.

60. In water molds, egg cells are produced in spherical ____________________.

61. The floating, aquatic algal protists are an important part of the ____________________.

62. Some of the photosynthetic ____________________ are symbiontic with corals, and their numbers can be very great.

63. Through use of their photoreceptors, euglenoids can detect light and move ____________________ it.

64. *Euglena* stores starch in the form of ____________________.

65. The "golden plants," with high concentrations of carotenoids, are the ____________________.

66. The alga-derived material used variously as an abrasive polish, filter, and insulating material is called ____________________.

67. The red alga Irish moss provides the polysaccharide ____________________.

68. The characteristic brown pigment of the brown algae is ____________________.

69. The habitat of most of the species of green algae is ____________________.

70. A multinucleate mass of cytoplasm, such as occurs in some *Acetabularia* species, is termed a ____________________.

Questions for Discussion

1. What evidence can you cite that supports the serial endosymbiosis hypothesis for the origins of eukaryotic organelles? What is the weakest aspect of the hypothesis? Why? Is there evidence today of processes similar to those which would have occurred in the past?

2. It has been suggested that each of the three nonprotist eukaryotic kingdoms has its roots in some group of protists. Identify which protist group has the greatest affinity with each of these kingdoms, and give reasons for selecting these groups. Does this mean that these eukaryotic kingdoms evolved from the Protista? Can you give alternative suggestions?

3. There are some similarities between a Saecodine like *Entamoeba histolytica* and an acellular slime mold like *Physarum polycephalum*, but there are some significant differences as well. Discuss their similarities and differences in structure, behavior, feeding, and reproduction. Why are acellular slime molds and amebas placed in separate phyla? How are acellular slime molds similar to fungi?

4. Malaria has long been a disease with tremendous impact on human populations. The first effective approach toward controlling malaria was to control the mosquito vector with DDT. However, malaria remains a serious problem in many areas, and new control techniques are still needed. Consider the entire life cycle of the malarial parasite: What other ways can you propose to stop the transmission of the parasite or to prevent the disease from developing in infected humans?

5. The statement might be made "All algae are closely related, and they should all be placed in the same phylum." Can this statement be defended, or should it be refuted? In reaching a conclusion, you should cite evidence to support your position.

Answers to Self-Exam

Multiple Choice Questions

1. c
2. d
3. a
4. c
5. b
6. c
7. a
8. d
9. b
10. a
11.
12. a
13. c
14. b
15. d
16. a
17. b
18. c
19. d
20. a
21. a
22. a
23. b
24. d
25. c
26. c
27. b
28. c
29. a
30. b

True or False Questions

31. F
32. F
33. T
34. T
35. F
36. F
37. T
38. T
39. F
40. F
41. F
42. T
43. T
44. F
45. F
46. T
47. F
48. F
49. F
50. T

Fill in the Blank Questions

51. van Leeuwenhoek
52. symbiosis
53. autogamy
54. haploid, diploid
55. sporophyte, gametophyte
56. Mastigophora
57. sporozoites
58. micronucleus, macronucleus
59. sporangiophores
60. oogonia
61. phytoplankton
62. dinoflagellates
63. toward
64. paramylon
65. Chrysophyta
66. diatomaceous earth
67. carrageenan
68. fucoxanthin
69. fresh water
70. coenocyte

CHAPTER 24 Fungi

Learning Objectives

After mastering the material covered in chapter 24, you should be able to confidently do the following tasks:

- Explain what the fungi are and how they are different from other kinds of organisms.
- Describe the main modes of nutrition found among the fungi and outline their general reproductive cycle.
- Give an overview of the divisions of fungi and state the distinguishing characteristics of each division.
- Suggest some possible origins for the kingdom Fungi of today.
- Interpret multicellularity in the fungi and contrast this with prokaryotes, protists, plants, and animals.

Chapter Outline

I. What are the fungi?

A. Feeding in the fungi.
B. Reproduction in the fungi.
C. Fungal relationships.

II. Kingdom Fungi.

A. Zygomycota: Conjugating Molds.
1. *Rhizopus*, a bread mold.
2. Sexual reproduction.
B. Ascomycota: Sac Fungi.
1. Reproduction in the sac fungi.
2. Lichens.
3. Yeasts.
C. Basidiomycota: Club Fungi.
1. Sexual reproduction in the mushroom
2. Wheat rusts.
D. Deuteromycota: Fungi Imperfecti.

III. Fungal Ancestors

IV. Multicellularity

Key Words

mycelium (-ia)
hypha (-ae)
multinucleate
coenocytic
cytoplasmic streaming
haustoria
decomposers
cyclosporine
mycorrhiza (-ae)
spores
conjugation
delayed fertilization
dikaryon
dikaryotic
chitin
monophyletic
polyphyletic
division
Zygomycota
zygomycetes
conjugating molds
zygospores
saprobes
Pilobolus
bread mold
Rhizopus stolonifer
stolons
rhizoids
aerial hyphae
sporangium (-ia)
sporangiophores
gametangium (-ia)
mating types
Ascomycota
ascomycetes
sac fungi
septate
septum (-a)
morels
truffles
powdery mildews
Endotheia parasitica
chestnut blight
Claviceps purpurea
ergot
ergotism
Saint Anthony's fire
conidium (-ia)
conidiophores
ascospores
ascus (-i)
ascocarp
Ascogonium (-ia)
antheridium (-ia)
trichogyne
Peziza
lichens
crustose lichens
foliose lichens
fruticose lichens
reindeer moss
yeasts
budding
Basidiomycota
basidiomycete
club fungi
mushrooms
toadstools
shelf fungi
rust
smut
stalk
cap
clamp connection
basidiocarp
basidium (-ia)
gills
basidiospores
death cap
Puccinia graaminis
wheat rust
Berberis vulgaris
wild barberry
uredospore
teliospore
aeciospores
intermediate host
Deuteromycota
fungi imperfecti
Trichophyton mentagrophytes
athlete's foot
Candida albicans
thrush
Penicillium
Penicillium roquefortii
Penicillium camembertii
Aspergillus oryzae
Arthrobotrys
Pilobolus

Exercises

1. Provide the correct labels for the below diagram using words from the following list. To which division would this fungus be assigned?

hypha
rhizoid
sporangiophore
spore
sprorangium
stolon

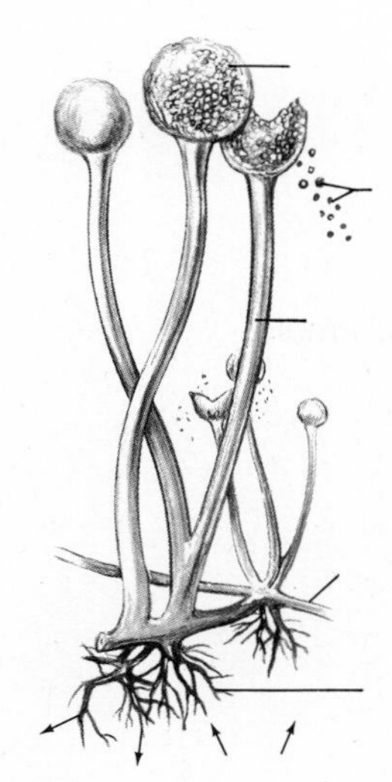

2. Provide the correct labels for the below diagram showing sexual reproduction in a fungus. You may use terms form the following list. To which division would this fungus be assigned?

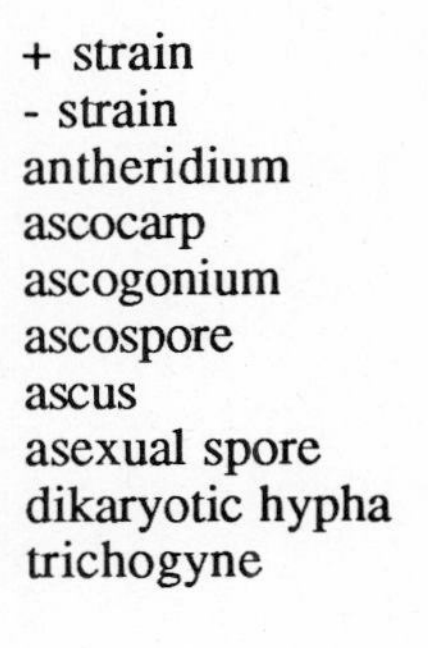

+ strain
- strain
antheridium
ascocarp
ascogonium
ascospore
ascus
asexual spore
dikaryotic hypha
trichogyne

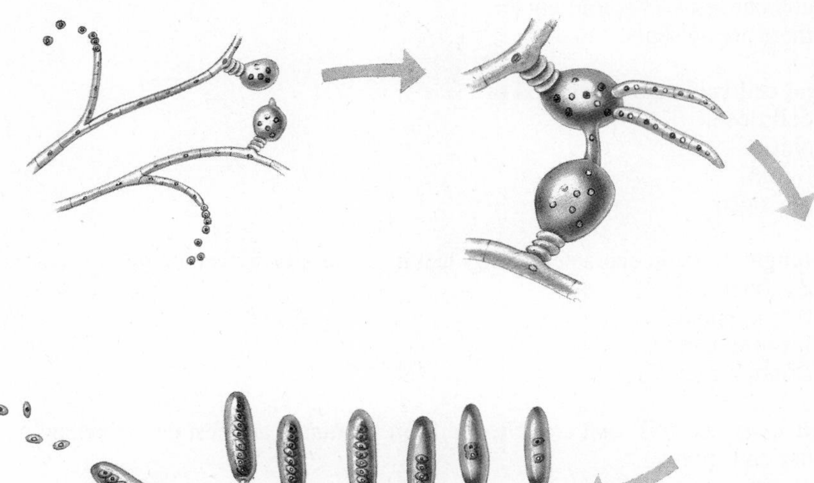

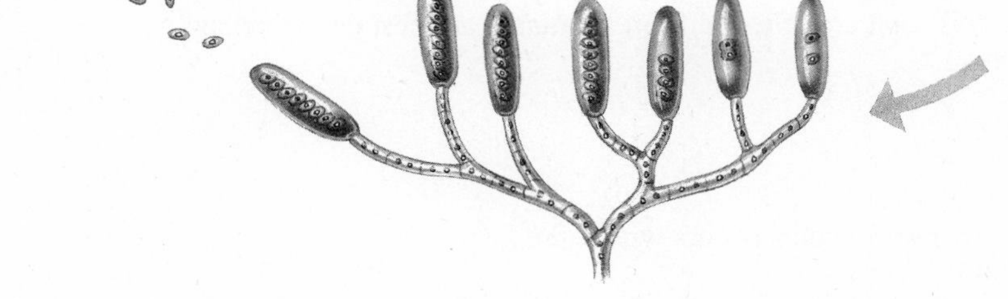

Self-Exam

You should be able to easily answer the following questions after learning the material in chapter 24. If you have difficulty with any question, study the appropriate section in the text and try again.

A. Multiple Choice Questions

Circle one alternative that best completes the statement or answers the question.

1. The nonreproductive part of a fungus is its
   a. cap.
   b. mycelium.
   c. basidium.
   d. conidium.

2. The cytoplasm of a coenocytic fungus is
   a. septate.
   b. imperfect.
   c. anucleate.
   d. multinucleate.

3. The hyphae through which parasitic fungi absorb their nutrients are termed
   a. mycorrhizae.
   b. trichogynes.
   c. haustoria.
   d. stolons.

4. Some higher plants and fungi can form mutualistic associations called
   a. mycorrhizae.
   b. haustoria.
   c. reindeer moss.
   d. sac fungi.

5. In fungi with dikaryotic hyphae,
   a. two cells share a nucleus.
   b. cells have diploid nuclei.
   c. cells have two haploid nuclei.
   d. there are no septa.

6. Fungal cell walls are composed of
   a. cellulose.
   b. chitin.
   c. lignin.
   d. glycogen.

7. The fungal division characterized by lacking regularly spaced complete crosswalls is
   a. Zygomycota.
   b. Ascomycota.
   c. Basidiomycota.
   d. Deuteromycota.

8. In *Rhizopus*, the cells that are released from sporangia are best described as
   a. haploid spores.
   b. zygospores.
   c. teliospores.
   d. gametes.

9. In a mold like *Rhizopus*, meiosis occurs within the
   a. sporangiophore.
   b. ascospore.
   c. zygospore.
   d. basidiocarp.

10. The so-called higher fungi, the ascomycetes and basidiomycetes, are distiguished from other fungi by
    a. being eukaryotic.
    b. being multinucleate.
    c. being septate.
    d. being asexual.

11. Ergotism can be attributed to which of the following organisms?
    a. *Claviceps purpurea.*
    b. *Endotheia parasitica.*
    c. *Puccinia graminis.*
    d. *Trichophyton mentagrophytes.*

12. In the sac fungi, meiosis occurs in which structure?
    a. the conidium.
    b. the ascus.
    c. the antheridium.
    d. the basidium.

13. In a fungus like *Peziza*, how many ascospores result from one zygote?
    a. one.
    b. four.
    c. eight.
    d. many.

14. A typical lichen might be an association between
    a. a fungus and an alga.
    b. a fungus and a plant.

c. two mycorrhizae.
d. haustoria and mycelia.

15. Reindeer moss is an example of a
    a. mycorrhiza.
    b. crustose lichen.
    c. foliose lichen.
    d. fruticose lichen.

16. Which of the following have been used to assess air quality?
    a. mushrooms.
    b. morels.
    c. molds.
    d. lichens.

17. Yeasts are single-celled fungi that are classified with the
    a. Zygomycota.
    b. Ascomycota.
    c. Basidiomycota.
    d. Deuteromycota.

18. Under anaerobic conditions, a typical brewer's yeast strain can carry out fermentation until the ethanol concentration reaches about
    a. 3 percent.
    b. 13 percent.
    c. 50 percent.
    d. 95 percent.

19. Shelf fungi and wheat rust are examples of fungi belonging to which division?
    a. Zygomycota.
    b. Ascomycota.
    c. Basidiomycota.
    d. Deuteromycota.

20. Basidiomycetes have a special way of maintaining the dikaryotic condition through mitotic divisions. This involves the
    a. ascus.
    b. sporangiophore.
    c. clamp connection.
    d. teliospore.

21. In mushrooms, meiosis occurs in which structure?
    a. basidium.
    b. ascocarp.
    c. conidium.
    d. antheridium.

22. The intermediate host of the wheat rust is infected by
    a. uredospores.
    b. teliospores.
    c. basidiospores.
    d. aeciospores.

23. The group of fungi for which sexual reproduction is not known is the division
    a. Zygomycota.
    b. Ascomycota.
    c. Basidiomycota.
    d. Deuteromycota.

24. Fungi of the genus *Penicillium* are important in producing
    a. athlete's foot.
    b. vaginal infections.
    c. cheese.
    d. soy sauce.

25. Which of these statements is best supported at present?
    a. Fungi are clearly monophyletic.
    b. Fungi are clearly polyphyletic.
    c. Fungi are clearly ancestral to plants.
    d. Fungi are clearly eukaryotic.

B. True or False Questions

Mark the following statements either T (True) or F (False).

_____ 26. The nonreproductive part of a fungus is usually called the mycelium

_____ 27. It is fungi but not bacteria that are the decomposers in ecosystems.

_____ 28. Although mycorrhizal associations have long been thought to be mutualistic associations, the fungus is probably actually a parasite.

_____ 29. In many fungi, the nuclear envelope remains intact during mitosis.

_____ 30. Fungi differ from higher plants in lacking centrioles and flagella.

_____ 31. Biologists believe the fungi to be either polyphyletic or monophyletic.

_____ 32. The cytoplasm of zygomycetes typically has a single nucleus per cell.

_____ 33. On sprorangia of *Rhizopus*, haploid spores arise through mitosis.

_____ 34. In general, sexual reproduction in the zygomycetes requires the presence of + and - mating types.

_____ 35. The powdery mildew are basidiomycetes.

_____ 36. The sac fungi reproduce asexually via conidia and sexually via ascospore formation.

_____ 37. Each ascospore will typically germinate and produce a diploid mycelium.

_____ 38. Spanish moss is a good example of a foliose lichen.

_____ 39. Common baker's yeast cells can be either diploid or haploid.

_____ 40. Toxic mushrooms are easily identified in the field.

_____ 41. Wheat rust produces two types of dikaryotic spores: red and black.

_____ 42. Complete eradication of the species *Berberis vulgaris* would eventually prevent wheat rust infections.

_____ 43. Sexual reproduction stage helps in identifying the various deuteromycetes.

_____ 44. The oldest clearly identified fossil fungi known are about 40 million years old.

_____ 45. The fungi differ from protista because all species are clearly multicellular.

C. Fill in the Blanks

Answer the question or complete the statement by filling in the blanks with the correct word or words.

46. In fungi, food and synthesized material are distributed throughout the mycelium via ________________.

47. The cell walls of fungi are composed of the carbohydrate ________________.

48. The fuzzy growth that is seen as bread mold is a mat of ________________.

49. The fungus *Claviceps purpurea* causes the disease ________________ on the cereal crop rye.

50. The ________________ is the spore-containing structure for which the sac fungi division is named.

51. Although the symbiotic association between fungi and algae to form lichens was usually considered mutualism, today some authorities consider it to be ________________.

52. Yeasts reproduce asexually by ________________.

53. Yeasts are placed in the division ________________.

54. The familiar fleshy mushroom is correctly termed a ________________.

55. In order to complete its sexual life cycle, *Puccinia graminis* must infect both ________________ and ________________.

56. The mouth infection called thrush is caused by the yeast-like fungus ________________.

57. Among other ideas, red algae have been proposed as ancestors of fungi, since both groups lack motile cells and ________________.

58. In fungi, the parts of the organism with the more specialized cells are usually the ________________.

59. Many fungal species that could be included among the sac fungi or club fungi are placed in the division ________________ because they have no known sexual phase.

60. The zygomycete *Pilobolus*, which is sometimes termed the "spitting mold," extracts its nutrition from ________________.

Questions for Discussion

1. For many years, plants and fungi were placed in the same kingdom, and even today mycology is often studied in botany courses. Why are plants and fungi separated today into separate kingdoms? Cite the characteristics of the two groups that support this classification.

2. What are the hypotheses for the origin of the Kingdom Fungi? What evidence supports each of these ideas? What are the weaknesses with each? State which of these hypotheses seems most reasonable to you and support your position.

3. Is the Division Deuteromycota a "natural" taxonomic group, or is it an assemblage of convenience? What other groups may be involved with the fungi imperfecti? State why a particular fungal species is placed in each of these divisions.

4. One area of active effort by both traditional plant breeders and the agricultural genetic engineers is the development of strains of crop plants that are genetically resistant to rusts and smuts. Are fungal pathogens

of plants a significant problem? In what ways? Why might this approach be preferable to other possible methods of control?

Answers to Self-Exam

Multiple Choice Questions

1. b
2. d
3. c
4. a
5. c
6. b
7. a
8. a
9. c
10. c
11. a
12. b
13. c
14. a
15. d
16. d
17. b
18. b
19. c
20. c
21. a
22. d
23. b
24. c
25. d

True or False Questions

26. T
27. F
28. F
29. T
30. F
31. T
32. F
33. T
34. T
35. F
36. T
37. F
38. F
39. T
40. F
41. T
42. T
43. F
44. F
45. F

Fill in the Blank Questions

46. cytoplasmic streaming
47. chitin
48. aerial hyphae
49. ergot
50. ascus
51. parasitsm
52. budding
53. ascomycota
54. basidiocarp
55. wheat, barberry
56. *Candida albicans*
57. centrioles
58. reproductive structures
59. Deuteromycota
60. manure

CHAPTER 25 Plant Evolution and Diversity

Learning Objectives

After mastering the material covered in chapter 25, you should be able to confidently do the following tasks:

- Outline the characteristics that unify the divisions of the plant kingdom.
- Distinguish, based on their characteristics and life cycles, the nonvascular plants from the vascular plants, the cryptogams from the phanerogams, and the gymnosperms from the angiosperms.
- Give the life cycle details for a typical moss, club moss, fern, gymnosperm, and angiosperm.
- List the distinguishing features of each plant division, and identify examples of species in each of them.
- Describe our knowledge of the evolutionary history of plants, including the relative sequence in which major plant groups have appeared in the fossil record.
- Discuss possible explanations that may account for the ascendancy of flowering plants, which have achieved such great diversity on earth today

Chapter Outline

I. What Are Plants?

A. Alternation of generations in plants.

II. Division Bryophyta: Nonvascular Plants.

A. Bryophyte characteristics.
B. Class Hepaticae: Liverworts.
C. Class Anthocerotae: Hornworts.
D. Class Musci: Mosses.
1. The gametophyte.
2. The sporophyte.
E. Bryophytes and the evolutionary history of the vascular plants.

III. The Vascular Plants.

A. Division Psilophyta: Whisk ferns.
B. Division Lycophyta: Club mosses.
1. *Selaginella*: A reproductively advanced lycophyte.
C. Division Spenophyta: Horsetails.
D. Division Pterophyta: Ferns.

IV. The Seed Plants.

A. Reproductive adaptations in the emerging seed plants.
   1. pollen.
   2. seeds.
B. The gymnosperms.
   1. The ginkgo.
   2. The cycads.
   3. The gnetophytes.
   4. The conifers.

V. The End of an Era.

A. Angiosperms: The rise of the flowering plants.
B. The Angiosperms today.
   1. Origin and phylogenetic relationships.

Key Words

chlorophyll *a*
chlorophyll *b*
carotenoids
amylose
cellulose
pectin
lignin
sporophyte
gametophyte
spore
gamete
nonvascular plants
bryophytes
vascular plants
homosporous
heterosporous
microspores
megaspores
microgametophytes
pollen grains
megagametophytes
ovules
ovary
seed
geological era
Precambrian
Paleozoic
Mesozoic
Cenozoic
geological period
Cambrian
Ordovician
Silurian
Devonian
Carboniferous
Permian
Triassic
Jurassic
Cretaceous
Tertiary
Quaternary
Hepaticae
liverwort
thalloid
Marchantia
rhizoid
fragmentation
gemma (-ae)
gemmae cups
gametangium (-ia)
archegonium (-ia)
antheridium (-ia)
archegoniophore
antheridiophore
monoecious
dioecious
homosporous
elaters
Anthocerotae
hornworts
Anthoceros
pyrenoid
meristem
Musci
moss
Sphagnum
peat
protonema (-ata)
epidermis
cortical cells
sterile jacket cells
seta (-ae)
vascular tissue
hydroid
leptoid
stoma (-ata)
guard cells
capsule
operculum
peristome
Coleochaete
Tracheophyta
tracheophyte
xylem
phloem
rhyniophytes
Rhynia
rhizome
tracheid
cryptogam
phanerogam
Psilophyta
whisk ferns
Psilotum nudum
Tmesipteris
epiphyte
Lycophyta
club moss
Lycopodium
ground pine
adventitious roots
microphylls
megaphylls
sporophylls
strobilus (-i)
Selaginella
resurrection plant
heterosporous
Lepidodendrales
Sphenophyta
horsetails
Equisetum
scouring rush
silica

Pteridophyta
fern
mycorrhiza
fiddleheads
compound leaf
tree fern
sorus (-i)
indusium
annulus
prothallus
antheridogen
gymnosperm
pollen grain
generative cell
tube cell
sperm cell
pollen tube
pollination
palynologist
seed
nucellus
integument
micropyle
ovule
cone
ovary
fruit
Ginkgo biloba
maidenhair tree
cycad
gnetophyte
Gnetum
Welwitschia
xylem vessels
conifer
taiga
Sequoia sempervirens
microspore mother cell
pollen cone
megaspore mother cell
ovulate cone
microsporogenesis
megasporogenesis
hypocotyl
apical meristem
cotyledon
angiosperm
Anthophyta
plate tectonics
continental drift
coevolution
Dicotyledonae
dicots
Monocotyledonae
monocots
Magnoliaceae
primitive feature
advanced feature
carpel
stamens
petals
regular flower
irregular flower
complete flower
perfect flower

Exercises

1. Complete the following chart which summarizes many of the characteristics of the different divisions of the plant kingdom. For each division, give the common name or names. List examples you know from your studies. Indicate whether conducting tissue is typically present (+) or absent (-), and whether the division has motile sperm cells (+) or nonmotile sperm (-). Give the dominant life cycle stage, either gametophyte or sporophyte. On a separate sheet of paper, describe the special characteristics of each division, such as presence or absence of true leaves, roots, seeds, and so forth. (To check you work, consult Table 25.2 in your textbook.)

Division	Common name	Examples	Conducting Tissue present?	Motile Sperm?	Dominant Life Cycle Stage
Bryophyta					
Psilophyta					
Lycophyta					
Sphenophyta					
Pterophyta					
Ginkgophyta					

Division	Common name	Examples	Conducting Tissue present?	Motile Sperm?	Dominant Life Cycle Stage
Cycadophyta					
Gnetophyta					
Coniferophyta					
Anthophyta					

2. Below are listed some geological eras and periods, as well as some groups of plants that became important at different times. You are provided with a "time line" extending more than 500 million years. Arrange these eras and periods in the correct order, and then indicate the relatives positions of the listed plant groups. (Check your work with Table 25.1 in your textbook.)

Eras
Cenozoic
Mesozoic
Paleozoic

Periods
Cambrian
Carboniferous
Cretaceous
Devonian
Jurassic
Ordovician
Permian
Quaternary
Silurian
Tertiary
Triassic

Plant groups
Conifers
Cycads
First bryophyte fossils
First conifer fossils.
First fern fossils
First plant fossils
Flowering plants
Ginkgos
Lycophytes
Rhyniophytes
Sphenophytes

Millions of years ago

570

500

400

300

200

100

Present

Self-Exam

You should be able to easily answer the following questions after learning the material in chapter 25. If you have difficulty with any question, study the appropriate section in the text and try again.

A. Multiple Choice Questions

Circle one alternative that best completes the statement or answers the question.

1. Which of the below best characterizes the alternation of generations typical of plants?
   a. alternation of a diploid gametophyte with a haploid sporophyte.
   b. alternation of a muticellular gametophyte with a single-celled sporophyte.
   c. alternation of a haploid gametophyte with a diploid sporophyte.
   d. alternation of a single-celled gametophyte with a multicellular sporophyte.

2. In flowering plants, megagametophytes arise from
   a. pollen grains.
   b. sporophytes.
   c. megaspores.
   d. gametes.

3. The conspicuous green organism that is typically considered to be a moss is
   a. a vascular sporophyte.
   b. a non-vascular gametophyte.
   c. a vascular gametophyte.
   d. a non-vascular sporophyte.

4. The majority of bryophytes are
   a. ferns.
   b. mosses.
   c. liverworts.
   d. hornworts.

5. A moss sporophyte can be recognized because it is
   a. diploid and nutritionally dependent on the gametophyte.
   b. diploid and able to live independently, since it is photosynthetic.
   c. haploid and nutritionally dependent on the gametophyte.
   d. haploid and able to live independently, since it is photosynthetic.

6. The trachaeophytes have special tissue for transport; the vascular elements are
   a. xylem for food and phloem for water.
   b. phloem for food and xylem for water.
   c. xylem and phloem, each transporting both food and water.
   d. xyem and phloem for water and tracheids for food.

7. Phanerogams produce seeds; the phanerogams include
   a. ground pine and cycads.
   b. ferns and conifers.
   c. horsetails and ginkgos.
   d. gnetophytes and pines.

8. The vast paleozoic forests were dominated by
   a. psilophytes, lycophytes, and sphenophytes.
   b. gnetophytes, pterophytes, and anthophytes.
   c. sphenophytes, conifers, and angiosperms.
   d. horsetails, ferns, and flowering plants.

9. With respect to the club moss *Lycopdium*,
   a. sporophytes are dominant, but they lack vascular roots, stems, and leaves.
   b. sporophytes are dominant, and they have vascular roots, stem, and leaves.
   c. gametophytes are dominant, but they lack vascular roots, stems and leaves.
   d. gametophytes are dominant, and they have vascular roots, stems, and leaves.

10. Which of the following statements about *Selaginella* is accurate?
   a. *Selaginella* is homosporous and has nonmotile sperm.
   b. *Selaginella* is homosporous and has motile sperm.
   c. *Selaginella* is heterosporous and has nonmotile sperm.
   d. *Selaginella* is heterosporous and has motile sperm.

11. Fern spores are typically produced in
   a. archegonia.
   b. microphylls.
   c. gametangia.
   d. sporangia.

12. A fern gametophyte is also referred to as a
   a. fiddlehead.
   b. protonema.
   c. prothallus.
   d. megaphyll.

13. A typical pollen grain of a seed plant might contain
   a. male and female gametophytes.
   b. a microspore mother cell.
   c. a generative cell and a tube cell.
   d. a megaspore and two sperm cells.

14. Which of the following seed plants have motile sperm?
   a. ginkgos and cycads.
   b. cycads and gnetophytes.
   c. gnetophytes and conifers.
   d. ginkgos and conifers.

15. The origin of the tissue that is found in a typical seed is
   a. the embryonic sporophyte alone.
   b. the parental sporophyte alone.
   c. both the microgametophyte and the embryonic sporophyte.
   d. both the parental sporophyte and the embryonic sprophyte.

16. The structure called a fruit in flowering plants typically originates as a(an)
   a. cone.
   b. ovary.
   c. ovule.
   d. strobilus.

17. The gymnosperm division with the fewest species is the
   a. Ginkgophyta.
   b. Cycadophyta.
   c. Gnoetophyta.
   d. Coniferophyta.

18. Pollen of conifers is typically transferred from plant to plant by
   a. wind.
   b. water.

c. insects.
d. pollinators.

19. The gymnosperm division that today occurs in nature mainly in the tropics is the
    a. Ginkgophyta.
    b. Cycadophyta.
    c. Gnoetophyta.
    d. Coniferophyta.

20. In an ovulate cone of a pine, meiosis occurs in the
    a. antheridium.
    b. nucellus.
    c. microspore mother cell.
    d. megaspore mother cell.

21. A typical pine embryo might have
    a. a hypocotyl and one cotyledon.
    b. a hypocotyl and two cotyledons.
    c. a hypocatyl and eight cotyledons.
    d. a hypocotyl, but no cotyledons.

22. In terms of the numbers of species that are placed in the plant divisions, the two most numerous are
    a. Anthophyta and Coniferophyta.
    b. Anthophyta and Pterophyta.
    c. Coniferophyta and Pterophyta.
    d. Pterophyta and Bryophyta.

23. Angiosperms probably arose during which period?
    a. Cambrian.
    b. Carboniferous.
    c. Cretaceous.
    d. Cenozoic.

24. To which group of flowering plants do cereal grains like wheat, rye, and maize belong?
    a. cryptogams.
    b. monocots.
    c. dicots.
    d. magnoliales.

25. Which of the following combinations is thought to illustrate the primitive condition in flowering plants?
    a. flower regular, complete, and perfect.
    b. carpels inferior and fused.
    c. petals separate, with bilateral symmetry.
    d. floral parts arranged in spirals, with few fused stamens.

B. True or False Questions

Mark the following statements either T (True) or F (False).

_____ 26. Mitosis in the haploid sporophyte yields single-celled haploid spores..

_____ 27. In mosses, the sporophyte is in many instances nutritionally dependent on the gametophyte.

_____ 28. Liverworts possess spore-forming archegonia and antheridia.

_____ 29. The sporophyte of a hornwort can be photosynthetic.

_____ 30. A monoecious moss can produce antheridia and archegonia on the same individual.

_____ 31. Although they are similar in many ways, bryophytes and green algae differ sharply in their types of photosynthetic pigments.

_____ 32. The woody portion in the stems of trees is xylem.

_____ 33. Some extinct ferns are known to have been seed plants.

_____ 34. The psilophyte *Psilotum nudum* has hairy rhizoids but lacks true roots.

_____ 35. A strobilus is a cluster of sporophylles.

_____ 36. *Selaginella* is thought to be reproductively advanced, since it is heterosporous and has nonmotile sperm.

_____ 37. The leaves of ferns are considered to be megaphylls.

_____ 38. A fern prothallus is nutritionally dependent on the sporophyte generation.

_____ 39. A seed plant's microgametophyte is the pollen tube and the cells within it.

_____ 40. Some seed plants are homosporous, and some are heterosporous.

_____ 41. The seeds of gymnosperms are formed in strobili.

_____ 42. The maidenhair tree *Ginkgo biloba* is one of the most primitive seed plants to have nonmotile sperm.

_____ 43. The greatest number of conifers is found in tropical regions of the earth.

_____ 44. Angiosperms are clearly linked to gymnosperm ancestors by the fossil record.

_____ 45. One hypothesis links the emergence of and proliferation of angiosperms on earth to global climatic changes.

C. Fill in the Blanks

Answer the question or complete the statement by filling in the blanks with the correct word or words.

46. The most common structural polysaccharide of plants is _________________.

47. On a liverwort like *Marchantia*, the antheridia occur within stalked structures called _________________.

48. An area of simple, undifferentiated, actively dividing cells in a plant is a _________________ region.

49. A young, threadlike moss gametophyte is referred to as a _________________.

50. The release of spores from a moss sporophyte capsule is moderated by the _________________, which is influenced by humidity.

51. In plants, as in the green alga *Coleochaete*, cell plates form between daughter cells via the coalescing of _________________.

52. The tube-like, non-living, water-filled parts of the xylem are termed _________________.

53. Roots that emerge from the stem, rather than from a primary root, are termed _________________ roots.

54. Relative to the types of spores they produce, *Lycopodium* is an example of a ________________ lycophyte, while *Selaginella* is an example of a ________________ species.

55. Horsetails are sometimes called "scouring rushes," since ________________ in the cell walls makes them quite abrasive.

56. On the underside of fertile fern leaves can be found clusters of sporangia called ________________, which are sometimes covered by a scale-like ________________.

57. A fern prothallus may produce the hormone ________________, which stimulates male development in neighboring gametophytes.

58. In seed plants, an ovule typically consists of a megasporangium surrounded by sporophytic tissue layers termed the nucellus and the ________________.

59. The gymnosperm divisions with motile sperm cells are the ________________ and the ________________.

60. The so-called seed leaves of conifers and flowering plants are termed ________________.

61. Angiosperms probably first arose during the ________________ period of the Mesozoic.

62. The coevolution between plants and pollen-carrying ________________ is believed to have been important in the rise of angiosperms.

63. The majority of angiosperms are placed in class ________________.

64. The plants of today that are thought to be most representative of the ancestral angiosperms are found in the plant family ________________ of the Ranales.

65. Individual traits of a species that are believed to have been present in the original founders of a group are termed ________________ characteristics.

Questions for Discussion

1. The bryophytes are nonvascular plants, while the other plant divisions are all considered to be vascular. How are the vascular and nonvascular plants believed to be related? What is known about their ancestors? Cite evidence that can be used in drawing conclusions about the evolutionary origins of vascular plants.

2. Discuss the various adaptations necessary as plants moved from aquatic to terrestrial habitats. What further adaptations developed to allow plants to inhabit drier and drier locales? Give examples of specific structural and reproductive characteristics of plant groups to illustrate the ties to the aquatic environment and ways of dealing with lack of water.

3. Dramatic change in the plant life of earth has been linked to global geologic and climatic changes between the late Paleozoic and the Mesozoic, and also between the Mesozoic and the Cenozoic. Outline the changes that occurred in the dominant plant life during these two episodes. What explanations have been offered for these global changes? Could such change occur again? How?

4. Plants typically exhibit a sporic life cycle and have an alternation of generations, to a greater or lesser extent. Animals, on the other hand, are typified by a gametic life cycle. What explanations can you offer for the alternation of generations being so well-established in plants? What extremes does it reach? Do any groups approach a functional life cycle that seems very similar to the animal gametic cycle? In what ways?

Answers to Self-Exam

Multiple Choice Questions

1. c
2. c
3. b
4. b
5. a
6. b
7. d
8. a
9. b
10. d
11. d
12. c
13. c
14. a
15. d
16. b
17. a
18. a
19. b
20. d
21. c
22. b
23. c
24. b
25. a

True or False Questions

26. F
27. T
28. F
29. T
30. T
31. F
32. T
33. F
34. T
35. T
36. F
37. T
38. F
39. T
40. F
41. T
42. F
43. F
44. F
45. T

Fill in the Blank Questions

46. cellulose
47. antheridiophores
48. meristematic
49. protonema
50. peristome
51. Golgi vesicles
52. tracheids
53. adventitious
54. homosporous, heterosporous
55. silica
56. sori, indusium
57. antheridogen
58. integuments
59. ginkgos, cycads
60. cotyledons
61. Cretaceous
62. insects
63. Dicotyledonae
64. Magnoliaceae
65. primitive

CHAPTER 26 Flowering Plant Reproduction

Learning Objectives

After mastering the material covered in chapter 26, you should be able to confidently do the following tasks:

- Describe the basic structure of the angiosperm flower and the variant forms that flowers sometimes take.

- Relate the various ways pollination can occur and explain some of the specializations flowers have acquired for various animal pollinators.

- Outline the events of megasporogenesis, microsporogenesis, and fertilization in a typical angiosperm.

- Compare embryogenesis and seed development in dicots and monocots.

- Distinguish apomixis and vegetative propagation as modes of asexual reproduction.

Chapter Outline

I. Sexual Reproduction

A. The flower.
B. Variations in flowers.
C. Flower structure and pollination.
   1. Animal pollinators and flower specialization.
   2. Wind pollination.
D. Sexual activity in flowers.
   1. Events in the ovary.
   2. Events in the anther.
   3. Pollination and double fertilization.
E. Development of the embryo and the seed.
   1. Dicot development.
   2. Monocot development.
F. Seed dormancy.
G. Seed dispersal.

II. Asexual Reproduction

A. Vegetative propagation.
B. Apomixis.

Key Words

whorls
sepals
calyx
petals
corolla
stamen
filament
anther
pollen sac
carpel
pistil
ovary
style
stigma
ovule
androecium
gynoecium
accessory parts
monocot
dicot
radially symmetrical
bilaterally symmetrical
irregular flower
incomplete flower
complete flower
imperfect flower
perfect flower
pistillate
staminate
inflorescence
composite flowers
ray flowers
disk flowers
nectar
nectary
pollinator
Yucca
Yucca moth
Symplocarpus
sporogenesis
placenta
funiculus
megasporangium
nucellus
integuments
micropyle
megaspore mother cell
megagametophyte
endosperm
central cell
endosperm mother cell
egg cell
embryo sac
double fertilization
microsporogenesis
microspore mother cell
microspore
microsporangium
microgametophyte
generative cell
tube cell
pollen grain
sperm cell
pollen tube
polar nuclei
primary endosperm nucleus
fruit
embryo
endosperm
coconut milk
seed
embryo axis
suspensor
basal cell
apical meristem
epicotyl
plumule
bud
hypocotyl
root cap
radicle
protoderm
procambium
ground meristem
coleoptile
coleorhiza
scutellum
aleurone
pericarp
wheat germ
flour
bran
corn silk
tassle
Lotus
Lupinus arcticus
annual plant
perennial plant
dissemination
winged fruit
bur
vegetative propagation
cutting
auxin
callus tissue
adventitious bud
sucker
leaf generation
Bryophyllum
runner
stolon
tuber
potato eye
rhizome
apomixis
clone
simple fruit
berry
pome
drupe
dehiscent
indehiscent
aggregate fruit
multiple fruit
achene

Exercises

1. List examples under each of the following kinds of fruit. Obtain as many of these as you can, and cut them up and eat them. Can you explain why each example is placed in its category?

Berry	Pome	Drupe	Aggregate Fruit	Multiple Fruit

2. Provide correct labels for this illustration that diagrams the events of pollination. Use terms from the following list. Check your work with figure 26.9 in the text.

Central cell nuclei
Egg nucleus
Micropyle
Nucellus
Ovule
Pistil
Pollen tube
Pollen
Sperm
Stigma
Style
Tube nucleus

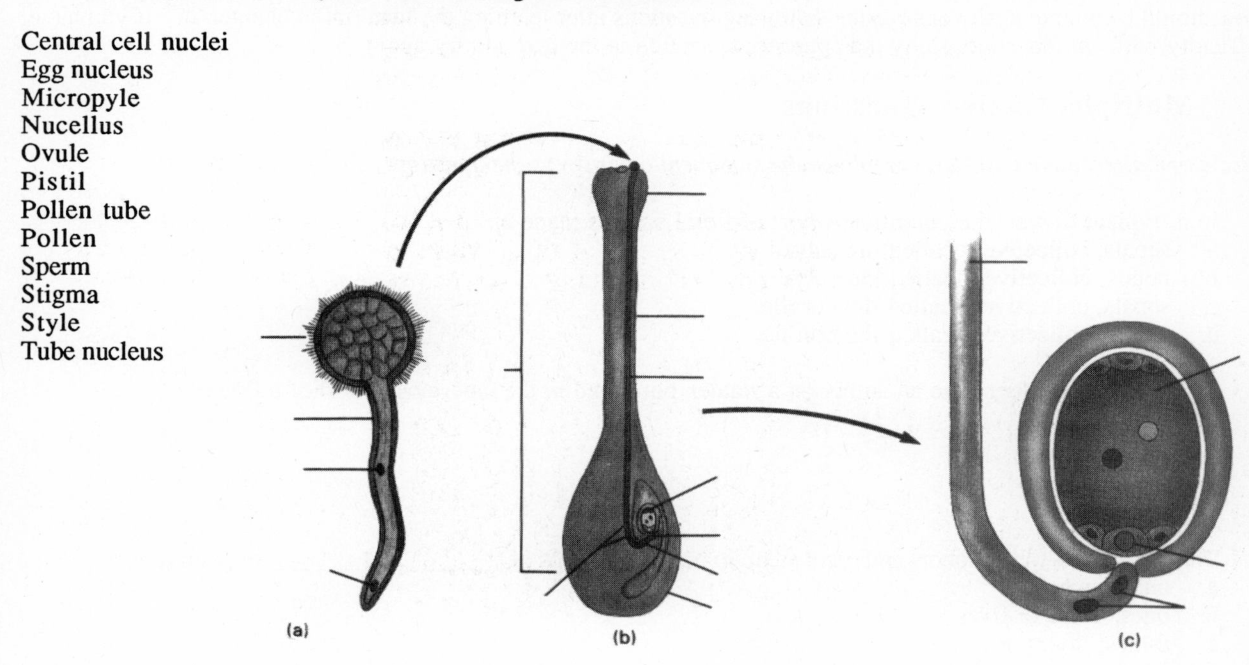

3. Two seeds are diagrammed below. Identify which is a monocot and which is a dicot. Label the diagrams appropriately using terms from the below list. Check your work with figure 26.11 in the text.

Aleurone
Coleoptile
Coleorhiza
Cotyledon
Endosperm
Epicotyl
Hypocotyl
Pericarp
Plumule
Radicle
Root apical meristem
Scutellum
Seed coat
Shoot apical meristem

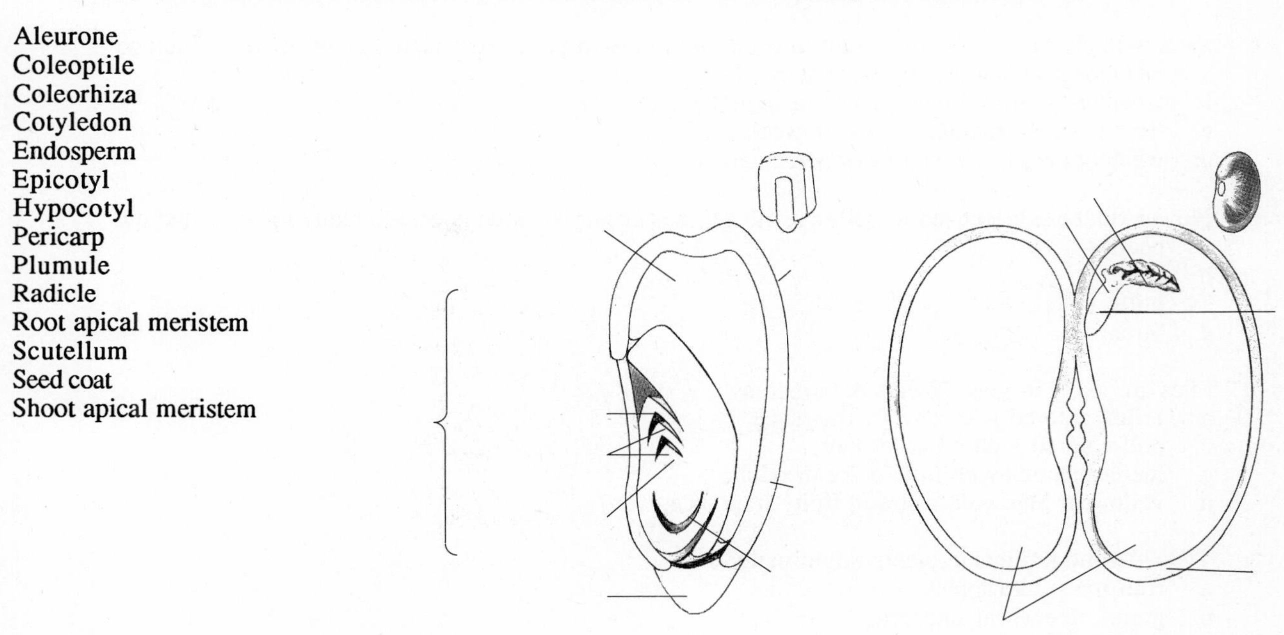

Self-Exam

You should be able to easily answer the following questions after learning the material in chapter 26. If you have difficulty with any question, study the appropriate section in the text and try again.

A. Multiple Choice Questions

Circle one alternative that best completes the statement or answers the question.

1. In a typical flower, the outermost whorl of floral parts is made up of
   a. sepals, collectively called the calyx.
   b. petals, collectively called the calyx.
   c. sepals, collectively called the corolla.
   d. petals, collectively called the corolla.

2. The structure analogous to an anther on a stamen but found in the innermost whorl of a flower is
   a. a filament.
   b. a style.
   c. a stigma.
   d. an ovary.

3. The floral parts of monocots and dicots tend to occur in groups of _______ and ______, respectively.
   a. ones; twos.
   b. ones; fours or fives.
   c. threes; fours or fives.
   d. fours or fives; threes.

4. Flowers that have pistils and stamens but lack a corolla and corona can be termed
   a. perfect complete flowers.
   b. perfect incomplete flowers.
   c. imperfect complete flowers.
   d. imperfect incomplete flowers.

5. Factors likely to be most important in determining which pollinators visit a plant's flowers include
   a. whether the plant is a monocot or dicot.
   b. whether the flowers are perfect or imperfect.
   c. the color and fragrance of the flowers.
   d. whether the plant is annual or perennial.

6. Flowers that are bright red or yellow, with copious nectar but little odor, are likely to be visited by
   a. beetles.
   b. bees.
   c. butterflies.
   d. birds.

7. Flies are likely to visit flowers described as
   a. bright-colored with a sweet fragrance.
   b. dull-colored with a putrid odor.
   c. inconspicuous with little color or odor.
   d. yellow or blue with a strong fruity fragrance.

8. Typical wind-pollinated plants might include
   a. fruit trees, like apples.
   b. grains, like wheat and corn.
   c. desert plants, like *Yucca*.
   d. crops, like peas and beans.

9. Although gymnosperms do not have flowers, their pollen is often carried by
   a. bees.
   b. flies.
   c. water.
   d. wind.

10. Within the ovary of a flower, the opening in the megasporangium that is ultimately penetrated by the pollen tube is called the
    a. placenta.
    b. funiculus.
    c. micropyle.
    d. nucellus.

11. Which of the following sequences best describes events in the pistillate plant?
    a. microspore mother cell ⇒ microspore ⇒ megagametophyte ⇒ egg cell.
    b. megagametophyte ⇒ microspore mother cell ⇒ microspore ⇒ egg cell.
    c. megagametophyte ⇒ egg cell ⇒ microspore mother cell ⇒ microspore.
    d. microspore mother cell ⇒ microspore ⇒ egg cell ⇒ megagametophyte.

12. In the angiosperm flower, microsporogenesis occurs via
    a. mitosis in the anthers.
    b. meiosis in the anthers.
    c. mitosis in the carpels.
    d. meiosis in the carpels.

13. The pollen grains produced by flowering plants are best described as
    a. microsporophytes.
    b. megasporophytes.
    c. microgametophytes.
    d. megagametophytes.

14. Germination of angiosperm pollen grains typically occurs
    a. in the anther.
    b. in the ovary.
    c. on the style.
    d. on the stigma.

15. The primary endosperm nucleus is the result of the fusion of
    a. two megaspore nuclei with a microspore nucleus.
    b. two polar nuclei with a sperm nucleus.
    c. two polar bodies with a sperm cell.
    d. the central cell with the egg cell.

16. In a mature bean seed, there is little or no endosperm; it is absorbed and most of the space is filled by
    a. the embryo axis.
    b. the radicle.
    c. the meristems.
    d. the cotyledons.

17. The portion of a dicot embryo below the level of cotyledon attachment is termed the
    a. epicotyl.
    b. hypocotyl.
    c. plumule.
    d. root cap.]

18. The tissue layers of an angiosperm embryo that give rise to epidermis and to vascular tissue are, respectively,
    a. protoderm and procambium.
    b. protoderm and ground meristem.

c. epicotyl and hypocotyl.
d. plumule and radicle.

19. The embryos of monocots, like corn, have protective sheaths that surround the plumule and the young root; these are termed the
   a. epicotyl and hypocotyl.
   b. scutellum and pericarp.
   c. coleoptile and coleorhiza.
   d. aleurone and bran.

20. Which of the following statements concerning food storage in seeds is most accurate?
   a. Most storage is in cotyledons in both beans and corn.
   b. Most storage is in endosperm in both beans and corn.
   c. Most storage is in endosperm in beans, but storage is in cotyledons in corn.
   d. Most storage is in cotyledons in beans, but storage is in endosperm in corn.

21. The component of grain called bran includes which of the following components?
   a. aleurone and pericarp.
   b. embryo and scutellum.
   c. coleoptile and coleorhiza.
   d. wheat germ.

22. The "tassels" on the tops of flowering corn plants are actually
   a. the styles of pistillate flowers.
   b. inflorescences of staminate flowers.
   c. devices to attract insect pollinators.
   d. the first stages of what will become the mature ear of corn.

23. The condition of dormancy in seeds is usually accompanied by
   a. increased water content and reduced metabolism.
   b. increased water content and increased metabolism.
   c. loss of water and reduced metabolism.
   d. loss of water and increased metabolism.

24. The edible portion of a potato is a tuber, which is a modified type of
   a. root.
   b. stem.
   c. rhizome.
   d. callus.

25. Apomictic seeds are produced by angiosperms in a way that avoids
   a. flowers.
   b. fruits.
   c. fertilization.
   d. seeds.

B. True or False Questions

Mark the following statements either T (True) or F (False).

_____ 26. The androecium is the innermost whorl of parts in the typical angiosperm flower.

_____ 27. Dicots typically have floral parts in fours or fives (or their multiples), while monocots more often have their parts arranged in threes or multiples of three.

_____ 28. A flower with only calyx, pistils, and stamens would be perfect but incomplete.

_____ 29. The most likely pollinators of bright yellow flowers with strong fragrances are birds.

_____ 30. Although ultraviolet light is not within human visual range, some insects can see it.

_____ 31. Although *Yucca* plants are usually pollinated by Yucca moths, the plants can be pollinated by several other less specialized moth and butterfly species.

_____ 32. Many grasses are similar to gymnosperms in that they are wind-pollinated.

_____ 33. The megagametophyte produces the egg cell via meiosis.

_____ 34. In plants, fertilization occurs when pollen is deposited on the receptive stigma of a flower.

_____ 35. In double fertilization, a single sperm fertilizes both the egg cell and the primary endosperm nucleus.

_____ 36. The mature structure called a fruit is actually derived from the ovary of the flower.

_____ 37. At early stages of development, endosperm often forms a multinucleate mass lacking cell walls.

_____ 38. In general, meristematic tissue can be recognized by its highly differentiated and specialized cells.

_____ 39. The plumule consists of the tiny embryonic leaves of the hypocotyl.

_____ 40. The protoderm layer of the plant embryo will differentiate into epidermis.

_____ 41. Food storage in corn kernels is in the form of starchy endosperm.

_____ 42. In wheat, the embryo is removed during milling and is included in the bran.

_____ 43. Viable *Lotus* seeds are known to be 20,000 years of age.

_____ 44. Tubers and rhizomes are both specialized types of stems.

_____ 45. The apomictically produced descendants of a plant are genetic clones of that plant.

C. Fill in the Blanks

Answer the question or complete the statement by filling in the blanks with the correct word or words.

46. The whorl of sepals and the whorl of petals are known as the ________________ and the ________________.

47. The accessory parts of a flower are the ________________ and ________________.

48. An irregular flower can be ________________ symmetrical.

49. The showy sunflower is actually an ________________ of individual ray and disk flowers.

50. The malodorous skunk cabbage is pollinated by ________________.

51. Plants with tiny, inconspicuous flowers that produce no nectar have pollen that is likely to be carried by ________________.

52. The significant pre-meiotic cell in an ovule is the ________________.

53. Another term for the completed megagametophyte is the ________________.

54. The process of microsporogenesis occurs within the _________________.

55. The pollen tube produces _________________, which digest maternal tissue as the pollen tube grows.

56. "Coconut milk" is best described as _________________.

57. A growing plant embryo is anchored in the embryo sac by a _________________.

58. _________________ is made up of simple, undifferentiated cells, and it is a site of cell proliferation that can give rise to new tissue throughout a plant's life.

59. The embryonic tissue that will form xylem and phloem is termed _________________.

60. The cotyledon in corn is called the _________________.

61. The indigestible substance in bran is the polysaccharide _________________.

62. Plants that live for many years rather than a single season are call _________________ plants.

63. The points where new plant growth can emerge from stems of damaged plants are _________________.

64. Strawberry plants tend to spread asexually by means of _________________.

65. A peach is a type of simple fruit termed a _________________.

Questions for Discussion

1. Some biologists believe that the relationship between angiosperm flowers and their animal pollinators is one of the most important classes of mutualisms. Explain why these relationships are so important. What are the ways in which a plant species benefits? What are the different ways that animals can benefit? How do you think the first pollinators came to visit plant reproductive structures?

2. A seed is a unit that can disperse or propagate a plant species. What are the tasks a seed must accomplish to be successful? What are some of the ways embryos are protected? Nourished? Disseminated?

3. Some plants reproduce exclusively by sexually produced seeds, while others are limited to vegetative propagation. Discuss the advantages and disadvantages to each of these approaches. Can you explain the advantage that might be gained by plants that can exploit both modes of reproduction?

Answers to Self-Exam

Multiple Choice Questions

1. a
2. d
3. c
4. b
5. c
6. d
7. b
8. b
9. d
10. c
11. a
12. b
13. c
14. d
15. b
16. d
17. b
18. a
19. c
20. d
21. a
22. b
23. c
24. b
25. c

True or False Questions

26. F
27. T
28. T
29. F
30. T
31. F
32. T
33. F
34. F
35. F
36. T
37. T
38. F
39. F
40. T
41. T
42. F
43. F
44. T
45. T

Fill in the Blank Questions

46. calyx, corolla
47. sepals, petals
48. bilaterally
49. inflorescence
50. flies
51. wind
52. megaspore mother cell
53. embryo sac
54. anthers
55. proteolytic enzymes
56. endosperm
57. suspensor
58. meristematic tissue
59. procambium
60. scutellum
61. cellulose
62 perennial
63. adventitious buds
64. runners
65. drupe

CHAPTER 27 Plant Growth and Structure

Learning Objectives

After mastering the material covered in chapter 27, you should be able to confidently do the following tasks:

- Explain what occurs during the process of germination.
- Distinguish the typical organization of newly emerged monocot seedlings and dicot seedlings.
- Outline the characteristic structures and functions of the various tissue types that occur in plants.
- Describe the patterns of primary growth in roots and in stems.
- Give the typical anatomy of a leaf and relate its structure to its function.
- Discuss secondary growth in plants, tell where it occurs, and explain its significance.
- Explain the meaning of totipotency, tell how it has been demonstrated for plant cells, and outline the potential importance of cloning plants from single cells.

Chapter Outline

I. Germination and Growth of the Seedling.

A. Germination.
B. The seedling.
C. Primary and secondary growth.

II. Tissue Organization in the Plant.

A. Protoderm and the epidermis.
B. Ground meristem.
C. Procambium and the vascular tissues.
1. xylem.
2. phloem.

III. Primary Growth in the Root.

A. The root tip.
1. elongation in the root tip.
2. differentiation in the young root.
3. the stele.
B. Root systems.

IV. Primary Growth in the Stem.

V. Growth of Leaves.

VI. Secondary Growth in the Stem.

A. Transition to secondary growth.
B. Secondary growth and the bark.
C. The older woody stem and annual rings.

Key Words

germination
Tambalacoque tree
Raphus cucullatus
dodo bird
gibberellins
alpha-amylase
radicle
shoot
hypocotyl
hypocotyl hook
epicotyl
plumule
primary root
lateral roots
Rhizophora mangle
red mangrove
primary growth
secondary growth
annual
perennial
open growth
indeterminate
bristlecone pine
cottonwood
creosote bush
dedifferentiation
primary meristems
protoderm
epidermis
cuticle
cutin
guard cells
stoma (-ata)
root hairs
periderm
suberin
ground meristem
parenchyma
collenchyma
sclerenchyma
chlorenchyma
lignin
fibers
sclereids
procambium
primary xylem
primary phloem
vascular cambium
tracheids
vessels
pits
pit pairs
vessel elements
xylem parenchyma
phloem parenchyma
sieve elements
sieve cells
sieve tube
sieve plates
companion cells
phloem fibers
root tip
root cap
region of cell division
region of cell elongation
region of maturation
cellulose fibers
auxin
cortex
endodermis
stele
pericycle
secondary xylem
secondary phloem
adventitious roots
aerial roots
prop roots
tap root system
diffuse root system
feeder roots
acacia tree
mesquite tree
pith
vascular bundle
leaf primordium
branch primordium
node
internode
leaf scar
lateral branch bud
axillary bud
bud scale scar
terminal bud
animal grazers
basal meristem
petiole
leaf blade
net venation
parallel venation
midrib
bundle sheath
upper epidermis
leaf mesophyll
palisade parenchyma
spongy parenchyma
lower epidermis
stomatal apparatus
leaf hairs
secondary cambium
cork cambium
cork
bark
lenticels
heartwood
sapwood
girdling
annual rings
vascular rays
totipotent
coconut milk

Exercises

1. Provide labels for the these diagrams of a bean seedling and a corn seedling using terms from the following list. Check your work with figure 27.1 in the text.

adventitious root
coleoptile
cotyledon
epicotyl
first leaf
hypocotyl
lateral root
primary root
radicle

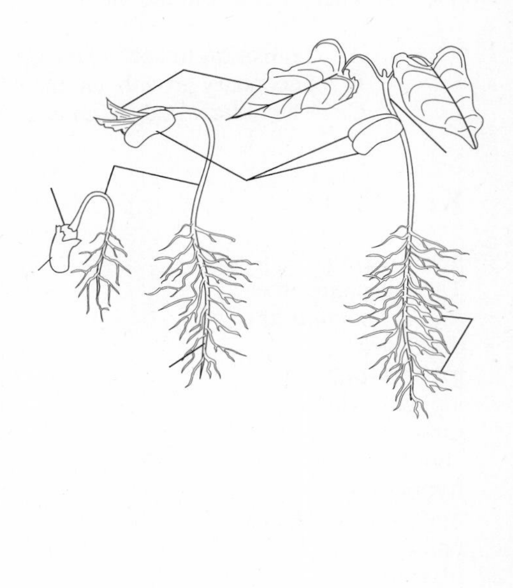

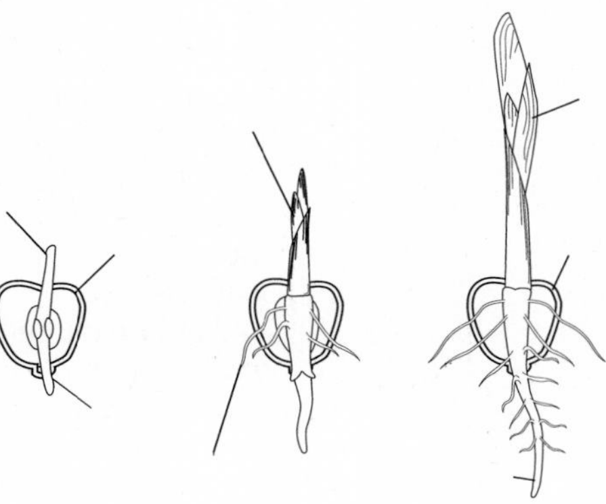

2. These diagrams are from stem cross-sections of two different plants. For each, indicate whether it is a monocot or a dicot and correctly label the indicated parts, using terms from the following list. Check your work with figure 27.12 in the text.

air space
cambium
collenchyma
companion cell
epidermis
parenchyma
phloem
sclerenchyma
sieve tube element
tracheid
vessel
xylem

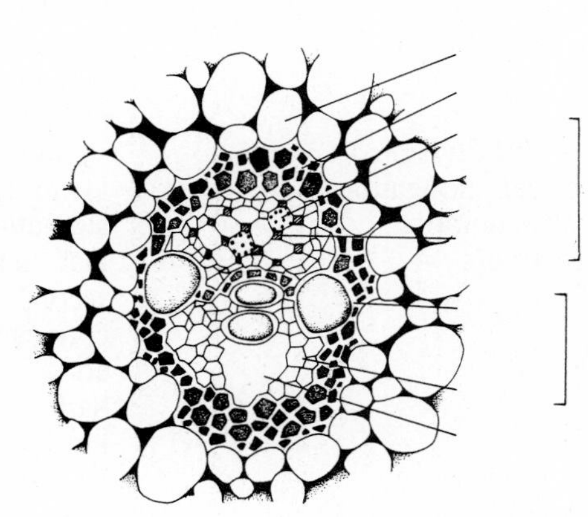

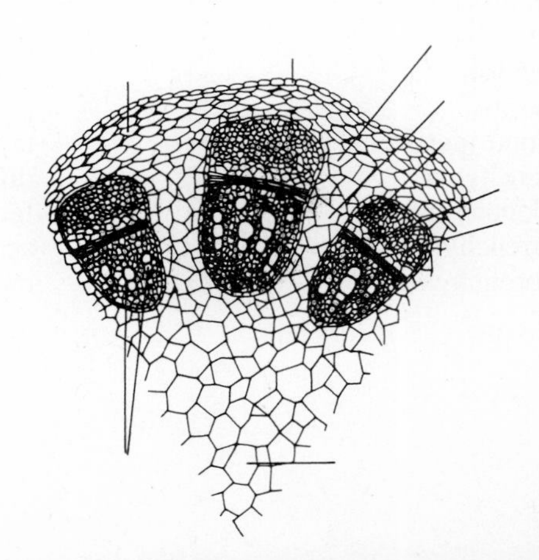

Self-Exam

You should be able to easily answer the following questions after learning the material in chapter 27. If you have difficulty with any question, study the appropriate section in the text and try again.

A. Multiple Choice Questions

Circle one alternative that best completes the statement or answers the question.

1. Which of the following sequences best describes events in germination of a corn seed?
   a. water uptake ⇒ root emergence ⇒ shoot emergence ⇒ starch mobilization.
   b. starch mobilization ⇒ water uptake ⇒ root emergence ⇒ shoot emergence.
   c. water uptake ⇒ starch mobilization ⇒ root emergence ⇒ shoot emergence.
   d. water uptake ⇒ starch mobilization ⇒ shoot emergence ⇒ root emergence.

2. In a bean seedling the foliage leaves arise from the
   a. epicotyl.
   b. hypocotyl.
   c. cotyledon.
   d. radicle.

3. Extensive secondary growth is generally lacking in which of the following groups?
   a. perennials.
   b. annuals.
   c. shrubs.
   d. trees.

4. Of the following, all could be correctly described as primary meristems except
   a. protoderm.
   b. collenchyma.
   c. ground meristem.
   d. procambium.

5. To help prevent water loss from internal tissue, epidermis is often covered by waxy
   a. cuticle.
   b. guard cells.
   c. sclerenchyma.
   d. stomata.

6. Collenchyma commonly forms a supporting tissue; it can be characterized as having cells that are
   a. no longer alive at maturity, with cell walls hardened by lignin.
   b. alive at maturity, with cell walls hardened by lignin.
   c. no longer alive at maturity, with cell walls that are quite thick but not hardened by lignin.
   d. alive at maturity, with cell walls that are quite thick but not hardened by lignin.

7. Sclerenchyma is a tissue type that includes both fibers and sclereids; its cells can be described as
   a. no longer alive at maturity, with cell walls hardened by lignin.
   b. alive at maturity, with cell walls hardened by lignin.
   c. no longer alive at maturity, with cell walls that are quite thick but not hardened by lignin.
   d. alive at maturity, with cell walls that are quite thick but not hardened by lignin.

8. Looking for the forms of sclerenchyma in natural products, we might expect to find
   a. fibers in rope and sclereids in string.
   b. fibers in rope and sclereids in nut shells.
   c. sclereids in rope and fibers in nut shells.
   d. both fibers and sclereids are characteristic of the soft tissue of fruit.

9. The types of cells found in xylem are called
   a. fibers and sclereids.
   b. vessels and tracheids.
   c. sieve tubes and companion cells.
   d. procambium and vascular cambium.

10. Water transport in plants typically passes
    a. through adjacent vessel elements via sieve plates.
    b. through the sieve tube elements of the phloem via sieve plates.
    c. through pit pairs between adjacent sieve tube elements.
    d. through pit pairs between adjacent vessel elements.

11. Among the cells of the phloem,
    a. sieve tube members are non-living, but the companion cells are living.
    b. sieve tube members and companion cells are living, but companion cells lack nuclei.
    c. sieve tube members and companion cells are living, but sieve tube members lack nuclei.
    d. neither sieve tube members nor companion cells are living at maturity.

12. One of the main tasks of the companion cells of the phloem is to
    a. transport sugars into and out of the sieve tubes.
    b. transport water up and down a plant's stem.
    c. filter the plant's sap.
    d. transport water up a plant's stem and transport sugars down the stem.

13. In an actively growing root tip, the most common site of frequent mitosis is
    a. the root cap.
    b. the apical meristem.
    c. the region of elongation.
    d. the region of maturation.

14. For cells to elongate and grow, the cell wall fibers are loosened
    a. under acidic conditions influenced by the hormone auxin.
    b. after being dissolved by auxin.
    c. under acidic conditions caused by the breakdown of cellulose.
    d. under acidic conditions, which leads to the breakdown of lignin.

15. The adventitious roots of corn
    a. emerge from the primary root of the plant as aerial roots.
    b. emerge from the primary root to form prop roots.
    c. emerge from the plant's stem and can act as prop roots.
    d. emerge from the plant's stem and can form tap roots.

16. When plants are characterized as having tap roots or diffuse roots,
    a. carrots and many grasses tend to have tap roots.
    b. carrots and many grasses tend to have diffuse roots.
    c. carrots have diffuse roots, while many grasses have tap roots.
    d. carrots have tap roots, while many grasses have diffuse roots.

17. Considering a vascular bundle in the stem of a dicot, we might typically find
    a. a bundle with xylem and phloem intermixed in the central portion of the stem.
    b. vascular cambium with xylem on its inner side and phloem on its outer side, all located peripherally.
    c. vascular cambium with phloem on its inner side and xylem on its outer side, all located peripherally.
    d. a centrally located ring of alternating xylem and phloem bundles.

18. The stem regions between patches of tissue that give rise to leaves are called
    a. nodes.
    b. internodes.

c. meristems.
d. leaf primordia.

19. The hormone auxin originates in
    a. the apical meristem and may inhibit lateral branch bud sprouting.
    b. the lateral branch bud and may promote lateral branch bud sprouting.
    c. the axillary bud and inhibits growth at the apical meristem.
    d. the basal meristem and may promote lateral branch development,

20. The cells that immediately surround the veins in typical dicot leaves are termed
    a. palisade parenchyma cells.
    b. guard cells.
    c. midrib cells.
    d. bundle sheath cells.

21. Carbon dioxide and water vapor primarily enter and exit typical leaves through
    a. stomata in the lower epidermis.
    b. stomata in the upper epidermis.
    c. the leaf hairs of the lower epidermis.
    d. the bundle sheaths of the upper epidermis.

22. Monocots can be characterized as having, in general,
    a. net venation in the leaves and a lack of secondary growth derived from vascular cambium.
    b. parallel venation in the leaves and a lack of secondary growth derived from vascular cambium.
    c. net venation in the leaves and extensive secondary growth derived from vascular cambium.
    d. parallel venation in the leaves and a lack of secondary growth derived from vascular cambium.

23. The older portion of the woody stem of a tree, the "heartwood," is composed of
    a. living xylem cells, which conduct water.
    b. living xylem and phloem cells.
    c. non-living xylem tissue that no longer conducts water.
    d. non-living xylem and phloem.

24. The so-called annual rings of trees in temperate seasonal climates are formed by
    a. different colors of xylem cells that are produced in the winter.
    b. alternating layers of xylem cells and phloem cells.
    c. a dark oxidized layer that occurs at the end of each season's growth.
    d. alternation of large xylem cell produced early with the smaller cells produced later in the season.

25. Which of the following statements concerning totipotency is most generally accurate?
    a. Almost any plant cell can easily be induced to produce a complete mature plant.
    b. Once plant cells are fully differentiated, it is virtually impossible to obtain any further growth from them.
    c. An excised plant embryo can be used to produce complete mature plants but not single cells.
    d. Under suitable conditions, single cells from mature plants can be induced to produce complete plants.

B. True or False Questions

Mark the following statements either T (True) or F (False).

_____ 26. In Mauritius, seeds of the Tambalacoque tree will not naturally germinate unless they have been eaten by a dodo bird.

_____ 27. One of the early signs of germination of a dicot is the swelling of the seed.

_____ 28. The corn embryo has food reserves stored as large, fleshy cotyledons, and it lacks endosperm.

_____ 29. In germination of the bean, unlike the pea, the cotyledons remain below the earth,s surface.

_____ 30. Plants that are grown from bulbs, like tulips, are considered annuals.

_____ 31. Procambium is one of the primary meristems found in plants.

_____ 32. Cutin and auxin are both waxy substances that help prevent water loss from plant tissues.

_____ 33. The leaf parenchyma that is highly photosynthetic is also called chlorenchyma.

_____ 34. Sclerenchyma cells are non-living, and the cell walls are usually hardened by lignin.

_____ 35. Primary xylem includes both sclereids and vessels.

_____ 36. Vessels are made up of elements lying end-to-end, and water easily moves through them.

_____ 37. There is evidence that companion cells transport sugars in and out of vessels.

_____ 38. On a root, the young cells of the epidermis will produce lengthy root hairs.

_____ 39. In dicots, the pericycle has the capacity to produce lateral roots.

_____ 40. In corn, the mature plant has adventitious roots, which arise from the primary root.

_____ 41. Carrots, beets, and sweet potatoes have in common a storage type of tap root.

_____ 42. Some plants produce roots that reach well over 30 meters below the surface.

_____ 43. In the typical monocot stem, the vascular bundles are arranged in a ring around the stem's perimeter.

_____ 44. Leaf growth in a typical grass occurs at an apical meristem in each leaf.

_____ 45. The leaves of monocots, such as grasses, have no petioles.

C. Fill in the Blanks

Answer the question or complete the statement by filling in the blanks with the correct word or words.

46. In a germinating seed, aleurone cells respond to the hormone _________________ by producing the starch-digesting enzyme amylase.

47. Both annual and perennials show _________________ growth, but _________________ growth is more characteristic of perennials.

48. Epidermis arises from the primary meristem termed _________________.

49. A widely distributed plant tissue that is often loosely arranged, irregularly shaped cells and that can serve as a food storage tissue is termed _________________.

50. The tough "strings" that are present in celery stems are made up of the tissue termed _________________.

51. The very thin cell walls between adjacent tracheids are _________________.

52. Cytoplasmic connections between adjacent sieve tube members are through _________________.

53. The hormone _________________ promotes cell elongation in a growing root.

54. In plants showing secondary growth, procambium produces a region of ________________ in the region between the xylem and phloem of the stem.

55. Adventitious roots arising from stems above ground level are ________________.

56. A leafy stem branch may be produced in the spring from an ________________ located above a leaf scar.

57. The large central vein of a typical dicot leaf is termed its ________________.

58. The tightly packed, vertically arranged cells of the leaf mesophyll make up the ________________.

59. Functioning to inhibit herbivorous insects, anti-feeding adaptations are known to occur among ________________ produced by leaf epidermis.

60. Monocot leaves have a blade that extends outward and a ________________ that surrounds the stem.

61. The cortex of a stem is composed of ________________ cells.

62. The outer cells of a stem's periderm are waterproof, being impregnated with suberin, and this layer of tissue is produced by the ________________.

63. In the woody stem of a tree, the non-conducting portion is the heartwood, and the recent layers of xylem that conduct water and minerals make up the ________________.

64. The radiating spokes or strands of parenchyma that are visible in cross-section of a woody stem are termed ________________.

65. In his early tissue culture studies, J. van Overbeek cultured cells from carrot embryos in a medium based on ________________.

Questions for Discussion

1. First identify the primary meristems that are derived from an apical meristem. Then classify and describe the various tissues that are derived from each of these primary meristems. Differentiation is involved in development of these tissues, but this process must be reversed in tissue culture. Which mature tissues would not be good choices for initiating tissue cultures? Why?

2. There are two major components of vascular transport system in angiosperms. Distinguish between these two components, describing the cells they comprise, the tissue from which they arise, and the specific functions they have in transport.

3. The two major groups of angiosperms are the dicots and the monocots, and they differ in a number of ways. On the basis of the discussion in chapter 27 of the text, outline the basic differences in the growth and development of these two groups. Be sure to consider embryos, germination, basic tissue organization, vascular system, and growth patterns.

4. Trace the path of the oxygen atom in a molecule of water from soil water, into the plant, and eventually to the atmosphere after it is liberated following photosynthesis. Specify each of the tissues that is encountered and name specific cell types through which the water molecule or oxygen atom must pass. In the same way, trace the path of the carbon atom of a carbon dioxide molecule from the atmosphere, into the plant, and to storage tissue in a root, after it is captured by photosynthesis.

5. What are the structural features of a terrestrial plant that function to prevent excessive water loss? What additional specializations are found in some desert plants to minimize water loss?

Answers to Self-Exam

Multiple Choice Questions

1. c
2. a
3. b
4. b
5. a
6. d
7. a
8. b
9. b
10. d
11. c
12. a
13. b
14. a
15. c
16. d
17. b
18. b
19. a
20. d
21. a
22. b
23. c
24. d
25. d

True or False Questions

26. T
27. T
28. F
29. F
30. F
31. T
32. F
33. T
34. T
35. F
36. T
37. F
38. T
39. T
40. F
41. T
42. T
43. F
44. F
45. T

Fill in the Blank Questions

46. gibberellin
47. primary, secondary
48. protoderm
49. parenchyma
50. collenchyma
51. pits
52. sieve plates
53. auxin
54. vascular cambium
55. aerial roots
56. axillary bud
57. midrib
58. palisade parenchyma
59. leaf hairs
60. leaf sheath
61. parenchyma
62 cork cambium
63. sapwood
64. vascular rays
65. coconut milk

CHAPTER 28 Plant Transport Mechanisms

Learning Objectives

After mastering the material covered in chapter 28, you should be able to confidently do the following tasks:

- Explain the mechanisms that are currently proposed to account for the observed movement of water in plants.
- Distinguish the roles of root pressure, transpiration, and water potential in explaining water transport.
- Describe the structural features of plants that are important for water movement in roots, stems, and leaves.
- Explain how the stomatal apparatus regulates the rate of transpiration, including the role of guard cells.
- Outline the ways plants transport food molecules from one tissue to another.
- List the mechanisms for gas exchange and movement in plants.
- Specify the mineral nutrient requirements common to plants and the specialized ways some plants acquire them.

Chapter Outline

I. The Movement of Water and Minerals.

A. Root pressure: Is the root a kind of pump?
B. Water potential and the vascular plant.
C. The leaf, transpiration, and the pulling of water.
D. Water movement in the xylem through the TACT mechanism.
E. Water transport in the root.
F. Guard cells and water transport.
1. The guard cell mechanisms.
2. CAM photosynthesis.

II. Food Transport in Plants.

A. Mechanisms of phloem transport..
1. A little help from aphids.
2. Flow from source to sink.
3. The pressure flow hypothesis.

III. Gas Transport in Plants.

IV. Mineral Nutrients and Their Transport.

Key Words

root hairs
root cortex
endodermis
stele
phloem
xylem
vessels
tracheids
spongy mesophyll
stoma (-ata)
transpiration
turgor pressure
root pressure
guttation
osmosis
water potential
solute
water potential gradient
upper epidermis
palisade parenchyma
spongy parenchyma
lower epidermis
plasmodesmata
free energy of evaporation
solar energy
TACT
adhesion
cohesion
tension
viscosity
tensile strength
apoplastic route
symplastic route
casparian strip
cuticle
guard cells
stomatal apparatus
potassium transport mechanism
water stress
abscisic acid
ABA
Crassulaceae
crassulacean acid metabolism
CAM
Calvin cycle
sap
hydrostatic pressure
honeydew
aphid
translocation
sink
source
pressure flow hypothesis
pneumatophores
airroot
black mangrove
Avicennia germinans
lenticels
mycorrhiza (-ae)
essential nutrients
macronutrients
micronutrients
trace elements
biogeochemical cycles
insectivorous
bladderwort
Utricularia
Venus's flytrap
Dionaea
pitcher plant
Sarracenia
sundew
Drosera

Exercises

1. Provide labels for this diagrammatic cross-section of a leaf, using terms from the following list. Show the route that water follows when this leaf is transpiring. Color the actively photosynthetic tissue lightly with a green pencil. Check your work with figure 28.4 in the text.

air space
guard cells
lower epidermis
palisade parenchyma
phloem
spongy parenchyma
stomatal apparatus
upper epidermis
xylem

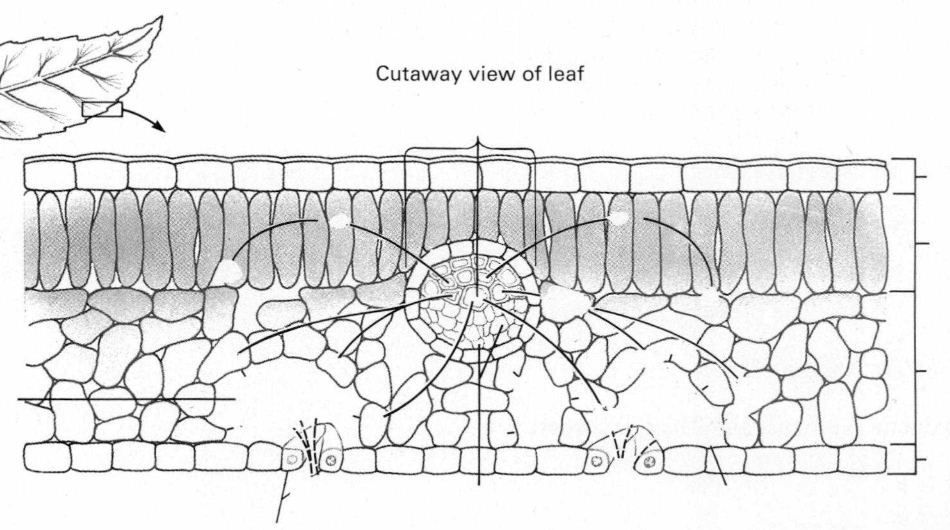

2. Two sections of a simple root are shown below. Indicate which is a longitudinal section and which is a transverse section. On each of the diagrams, label the appropriate parts, using terms from the following list. Using different color pencils or pens, show typical apoplastic routes and symplastic routes that water may follow in reaching the xylem. Check your work with figures 28.5 and 28.6 in the text.

epidermis
cortex
endodermis
pericycle
root hair
xylem
pit
casparian strip

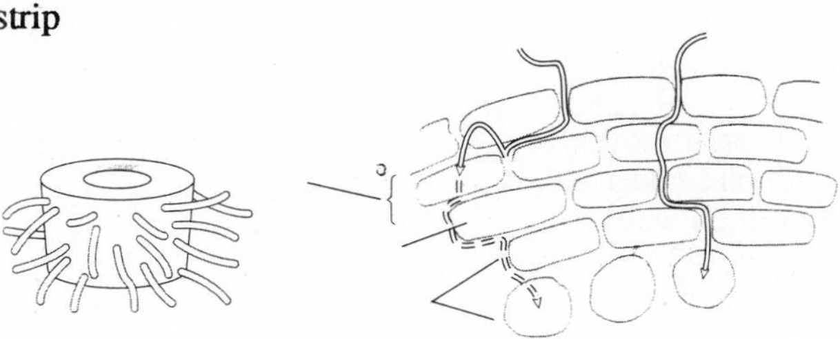

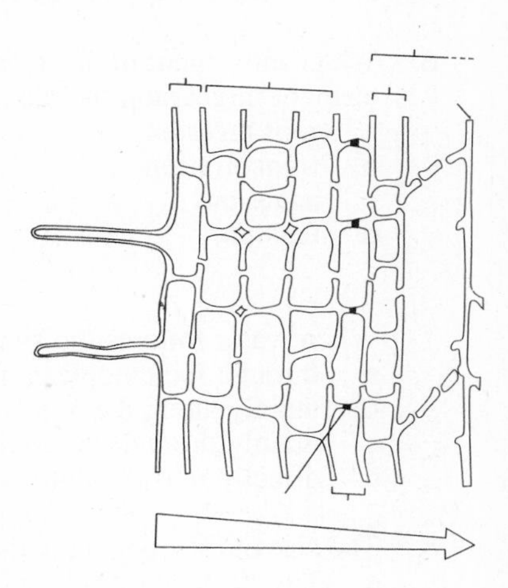

Self-Exam

You should be able to easily answer the following questions after learning the material in chapter 28. If you have difficulty with any question, study the appropriate section in the text and try again.

A. Multiple Choice Questions

Circle one alternative that best completes the statement or answers the question.

1. Which of the below sequences correctly describes the path water follows in entering the root?
   a. root hair ⇒ endodermis⇒ phloem ⇒ xylem.
   b. root hair ⇒ root cortex ⇒ endodermis ⇒ stele.
   c. root hair ⇒ endodermis ⇒ root cortex ⇒ stele.
   d. root hair ⇒ root cortex ⇒ stele ⇒ phloem.

2. The phenomenon of guttation in plants occurs when
   a. liquid water is lost from the plant, forced out by root pressure.
   b. water vapor is lost from leaves, forced out by root pressure.
   c. liquid water is lost from the plant, due to the water potential gradient established by transpiration.
   d. water vapor is lost from leaves, due to the water potential gradient established by transpiration.

3. In following a water potential gradient in a plant, water will tend to move from a given cell
   a. to a cell with a lower solute concentration and a higher water potential.
   b. to a cell with a lower solute concentration and a lower water potential.
   c. to a cell with a higher solute concentration and a higher water potential.
   d. to a cell with a higher solute concentration and a lower water potential.

4. The rate of water loss via transpiration should be greatest under conditions of
   a. high temperature, high winds, and high humidity.
   b. low temperature, low winds, and low humidity.
   c. low temperature, low winds, and high humidity.
   d. high temperature, high winds, and low humidity.

5. The energy required to lift many liters of water to great heights in trees is derived from
   a. the free energy of root pressure.
   b. the free energy of guttation, combined with siphoning.
   c. the free energy of evaporation of water.
   d. the free energy of passive water movement due to osmosis.

6. Water movement in the xylem involves several forces acting together. An air bubble introduced to the xylem element might stop the movement of water. Which of the following would be disrupted?
   a. root pressure.
   b. transpiration.
   c. adhesion.
   d. tension.

7. When water follows the symplastic route in a root, it moves
   a. through the cytoplasm from cell to cell via plasmodesmata.
   b. mainly along the highly porous parenchyma cell walls.
   c. mainly through the root hairs.
   d. directly into the phloem.

8. The size of the opening in the stomatal apparatus varies. Which is the general pattern we observe?
   a. Turgor in guard cells increases at night, and the stoma decreases in size.
   b. Turgor in guard cells increases at night, and the stoma increases in size.
   c. Turgor in guard cells increases during daylight, and the stoma decreases in size.
   d. Turgor in guard cells increases during daylight, and the stoma increases in size.

9. Which of the following sets of factors are known to decrease the opening of the stomatal apparatus?
   a. increased turgor in guard cells, increased solute concentration in guard cells, and presence of ABA.
   b. decreased turgor in guard cells, decreased solute concentration in guard cells, and absence of ABA.
   c. increased turgor in guard cells, increased solute concentration in guard cells, and absence of ABA.
   d. decreased turgor in guard cells, decreased solute concentration in guard cells, and presence of ABA.

10. In plants with CAM photosynthesis, carbon dioxide from the atmosphere is incorporated into
   a. the Calvin cycle at night, when the stomata are open.
   b. the Calvin cycle at night, when the stomata are closed.
   c. organic acids at night, when the stomata are open.
   d. organic acids at night, when the stomata are closed.

11. When sap moves through the phloem, it is
   a. pulled by the TACT forces.
   b. pushed by the TACT forces.
   c. pulled by the force of hydrostatic pressure.
   d. pushed by the force of hydrostatic pressure.

12. When sugars move from leaf cells, where sugar concentration may be relatively low, into the phloem, where the concentration of sugars may be much higher, the mechanism of movement is
   a. diffusion.
   b. active transport.
   c. TACT forces.
   d. osmosis.

13. The pressure flow hypothesis proposed to account for phloem sap flow suggests that
   a. active transport of solutes into and out of the phloem alters water potentials, and that water movement into the phloem creates sap-moving hydrostatic forces.
   b. the force of water flowing up in the xylem provides the pressure to move sap down in the phloem.
   c. the force of gravity is sufficient to generate the pressure leading to sap flow.
   d. water potential differences resulting from root pressure provide the force for sap movement.

14. The food molecule most likely to be carried in sap is
    a. starch.
    b. sucrose.
    c. cellulose.
    d. suberin.

15. In a typical dicot, gas transport in the plant is typically via
    a. active transport.
    b. vascular tissue.
    c. osmosis.
    d. diffusion.

16. A swamp-dwelling plant might be limited in its ability to exchange gas via its
    a. roots.
    b. leaves.
    c. pneumatophores.
    d. stems.

17. Mycorrhizae associated with plant roots
    a. are probably parasites, usually.
    b. are nutritionally independent of the plant associate.
    c. absorb and concentrate ions, such as phosphate.
    d. produce complex carbon compounds, such as sugars, for the plant.

18. Which of the following groups includes trace elements, as opposed to macronutrients?
    a. carbon, hydrogen, and oxygen.
    b. nitrogen, phosphorus, and potassium.
    c. calcium, magnesium, and phosphorus.
    d. chlorine, iron, and zinc.

19. Plants that are insectivorous use their insect prey as
    a. an energy source to supplement photosynthesis.
    b. a source of essential nutrients, including nitrogen.
    c. a means of insuring cross-pollination.
    d. a peculiar diversion of resources.

20. The trunk diameter of a tree is measured very carefully at midday; when this measurement is compared to the diameter of the same tree at midnight, we might discover that it is
    a. slightly decreased at midday, when transpiration is greatest.
    b. slightly increased at midday, when transpiration is greatest.
    c. slightly decreased at midday, when transpiration is least.
    d. slightly increased at midday, when transpiration is least.

B. True or False Questions

Mark the following statements either T (True) or F (False).

_____ 21. During transpiration, most evaporation of water occurs from the epidermal surface layer of leaves.

_____ 22. Because cacti typically occur in very arid locations, their transpiration rate is usually much greater than the transpiration rate of non-arid zone plants such as sunflowers.

_____ 23. Root pressure alone may be sufficient to move water from the roots to the leaves of some grasses.

_____ 24. Water will tend to move from an area of high water potential to a region of lower water potential.

_____ 25. The rate of water loss from leaves is greatly affected by relative humidity, but not by temperature.

_____ 26. The energy that moves water to the tops of tall trees is provided indirectly by sunlight.

_____ 27. The casparian strip found in roots is a waxy region in the cell walls of the xylem.

_____ 28. Since the photosynthetic stems of cacti are not leaves, they lack stomata.

_____ 29. Increased turgor pressure in guard cells will increase the stomatal opening.

_____ 30. In general, stomata are closed during daylight periods, when the risk of water loss is greatest.

_____ 31. Transport of potassium ions into guard cells is greater in a plant placed in direct blue light than in a plant placed in direct red light.

_____ 32. In the presence of the plant hormone abscisic acid, stomata will open.

_____ 33. The stomata of plants with crassulacean acid metabolism are characteristically closed in the daytime.

_____ 34. Sugar and other nutrients can move through the phloem in either direction, up or down.

_____ 35. Aphids feed on nutrient-laden sap by piercing vessels and sucking the fluid out.

_____ 36. We could characterize a plant's leaves as storage sinks for the food in phloem sap.

_____ 37. Accompanying phloem transport, water apparently circulates, moving from xylem to phloem at sources, and from phloem to xylem at sinks.

_____ 38. Lenticels provide pathways for gas exchange in the active tissues of plant bark.

_____ 39. Magnesium is a nutrient needed by plants for the production of chlorophyll molecules.

_____ 40. Although they are important for many plants, mycorrhizal relationships have been identified for only a small fraction of the plants studied.

C. Fill in the Blanks

Answer the question or complete the statement by filling in the blanks with the correct word or words.

41. Water loss from leaves through evaporation is called __________________.

42. If it is not dew, water on grass leaves in the early morning may be the result of __________________.

43. __________________ will tend to push water up a stem, while TACT forces tend to pull it up.

44. The tissue just below the upper epidermis in a leaf is the __________________.

45. The attractive force between molecules that allows water molecules to stick together and be pulled up a xylem column to great heights in trees is __________________.

46. In the root cortex, water tends to follow the __________________ route, but in the endodermis it follows the __________________ route.

47. The size of a stomatal opening is governed by the __________________ of the guard cells.

48. The presence of the plant hormone __________________ can lead to stomatal closing.

49. Plants placed in bright blue light will show an increase in the active transport of _________________ into guard cells.

50. The original source of carbon for photosynthesis in CAM plants is _________________.

51. In the early spring, for temperate trees such as maple we would expect the direction of transport in the phloem to be _________________.

52. The force that moves sap through the phloem is provided by _________________.

53. The movement of sap with its dissolved sugars, nutrients, hormones, and amino acids is referred to as _________________.

54. The principal sugar in sap is usually converted to _________________ for storage.

55. Where sugars are loaded into the phloem stream, the water potential in the phloem is decreased compared to the nearby _________________, and water moves from there into the phloem.

56. Among the angiosperms, the _________________ commonly have air channels in their vascular bundles, along with xylem and phloem.

57. For roots, gas exchange typically takes place across the surfaces of _________________.

58. The common association between a plant root and a fungal mycelium is called a _________________.

59. Copper, iron, and manganese are examples of _________________ needed for enzyme activity.

60. An instrument that can measure minute changes in the stem diameter of a tree is a _________________.

Questions for Discussion

1. Turgor in guard cells determines the size of the opening in a stomatal apparatus. This in turn regulates the rate of gas exchange via the stoma. What factors are known to influence guard cell turgor? How does each of these affect turgor? How might each of the factors be involved in regulating water loss in plants?

2. Plants that occur in very arid habitats have a number of adaptations to conserve water and minimize unnecessary water loss. Describe these adaptations, and explain how plants such as these can function perfectly well despite a chronic scarcity of water.

3. Explain how the TACT mechanism can move water from the ground to great elevations in a tree. What forces are involved in this movement? Identify the energy source that does this work.

4. For a brief period in the spring, the vascular tissue in a maple tree can be tapped, and it will exude a sugar-rich sap, which can be converted to maple syrup and maple sugar. Why does the sap flow out of the tree at this time? What are the forces that are moving the sap? What would have happened to this sap if it had been left in the tree?

Answers to Self-Exam

Multiple Choice Questions

1. b
2. a
3. d
4. d
5. c
6. d
7. a
8. d
9. d
10. c
11. d
12. b
13. a
14. b
15. d
16. a
17. c
18. d
19. b
20. a

True or False Questions

21. F
22. F
23. T
24. T
25. F
26. T
27. F
28. F
29. T
30. F
31. T
32. F
33. T
34. T
35. F
36. F
37. T
38. T
39. T
40. F

Fill in the Blank Questions

41. transpiration
42. guttation
43. root pressure
44. palisade parenchyma
45. cohesion
46. apoplastic, symplastic
47. turgor
48. abscisic acid or ABA
49. potassium ions
50. carbon dioxide
51. upward
52 hydrostatic pressure
53. translocation
54. starch
55. xylem elements
56. monocots
57. root hairs
58. mycorrhiza
59. micronutrients or trace elements
60. dendrometer

CHAPTER 29 Plant Regulation and Response

Learning Objectives

After mastering the material covered in chapter 29, you should be able to confidently do the following tasks:

- Explain the roles of plant hormones in regulating plant growth and response.
- Describe the chemical nature of the plant hormone auxin, where it is produced in plants, and how it brings about the elongation of the cells it acts upon.
- Identify the other major plant hormones and their effects.
- Explain the basis for plant movements and responses such as phototropism, gravitropism, thigmotropism, and solar tracking.
- Distinguish short-day, long-day, and day-neutral plants and explain how day length and flowering are related.

Chapter Outline

I. The Discovery of Auxin.

A. Later experiments on auxin.
B. The isolation of auxin.
C. Auxin: Its structure and roles.

II. Other Plant Growth Hormones.

A. Gibberellins.
B. Cytokinins.
C. Ethylene.
D. Abscisic acid.
E. Applications of plant hormones.

III. Plant Growth Responses and Movements.

A. Phototropism.
1. Auxin and apical dominance.
B. Roots and gravitropism.
C. Thigmotropism.
D. Nastic responses.
1. Solar tracking.
2. Thigmonastic response.

IV. Light and Flowering.

A. Photoperiodicity.
B. The dark clock and flowering.
1. Transmitting the stimulus.
2. Pineapples and auxin.

Key Words

Charles Darwin
Francis Darwin
coleoptile
plant hormone
Peter Boysen-Jensen
A. Paal
Fritz Went
auxin
indoleacetic acid
IAA
tryptophan
bioassay
abscission zone
gibberellins
Gibberella fujikuroi
foolish seedling disease
alpha-amylase
cytokinins
zeatin
kinetin
callus
totipotency
dedifferentiation
senescence
ethylene
abscisic acid
ABA
deciduous plants
2,4-dichlorophenoxyacetic acid
2,4-D
herbicide
biodegradability
2,4,5-trichlorophenoxyacetic acid
2,4,5-T
dioxin
agent orange
tropisms
phototropism
gravitropism
thigmotropism
nastic response
thigmonasty
photoreceptor
apical dominance
columella
amyloplasts
calmodulin
tendrils
sleep-movements
pulvinus (-i)
motor cells
Oxalis oregana
redwood sorrel
solar tracking
sunflower
heliotropism
Mimosa pudica
sensitive plant
Dionaea muscipula
Venus flytrap
photoperiodism
long-day plants
short-day plants
day-neutral plants
cocklebur
P_r
P_{fr}
phytochrome
dark clock
florigen

Exercises

1. Fill in the following table with some details for the hormones listed.

	Sites of Production	Sites of Action	Principal Effects
Auxin:			
Gibberellins:			
Cytokinins:			
Ethylene:			
Abscisic Acid:			

2. In the space below, sketch simple diagrams of the growing shoot and the growing root tip of a plant to show the effects of auxin. For the shoot, show what occurs to cause the shoot to bend toward light. For the root, indicate what occurs to cause the root to elongate downward toward the earth. Label the important tissues in each case, and show which cells auxin acts upon. Compare your diagrams with figures 29.5 and 29.13 in the text to make sure they are complete.

Self-Exam

You should be able to easily answer the following questions after learning the material in chapter 29. If you have difficulty with any question, study the appropriate section in the text and try again.

A. Multiple Choice Questions

Circle one alternative that best completes the statement or answers the question.

1. When Charles and Francis Darwin covered the tip of a canary grass seedling with foil,
   a. it bent toward the light source as it grew.
   b. it bent away from the light source as it grew.
   c. it bent neither toward nor away from its light source.
   d. it failed to grow in the absence of light.

2. In the Boysen-Jensen experiment, when a sliver of mica was placed in a slit only partway through the illuminated side of an oat coleoptile tip,
   a. the coleoptile bent toward the light source as it grew.
   b. the coleoptile bent away from the light source as it grew.
   c. the coleoptile bent neither toward nor away from the light source.
   d. the coleoptile failed to grow.

3. The growth-stimulating substance that Fritz Went collected from oat seedlings was
   a. agar.
   b. indoleacetic acid.
   c. an amino acid.
   d. a polysacharride.

4. Which of the following statements best describes the effects of auxin?
   a. Auxin stimulates plant growth in general.
   b. Auxin inhibits plant growth in general.
   c. In some cases auxin stimulates plant growth, while in other cases it inhibits plant growth.
   d. Auxin has no effect on plant growth.

5. The common triangular shape taken by many trees is often due to
   a. the inhibitory effects of auxin.
   b. the stimulatory effects of auxin.
   c. the inhibitory effects of gibberellin.
   d. the stimulatory effects of gibberellin.

6. The "foolish seedling disease" of rice, where stems are elongated and weakened, is due to
   a. *Gibberella fujikuroi.*
   b. *Dionaea muscipula.*
   c. ethylene.
   d. *Mimosa pudica.*

7. Dwarf corn can be induced to grow to the height of normal corn with application of the hormone
   a. auxin.
   b. gibberellin.
   c. zeatin.
   d. ethylene.

8. In tissue culture of tobacco pith cells, researchers found that under conditions of high auxin concentration but low cytokinin concentration
   a. there was very little growth and no differentiation.
   b. root growth was encouraged.
   c. shoot growth was encouraged.
   d. there was callus growth, but no differentiation.

9. In ripening warehouses, green bananas are ripened to yellow by using the hormone
   a. auxin.
   b. gibberellin.
   c. cytokinin.
   d. ethylene.

10. During periods of water stress, abscisic acid enters guard cells, K^+ is transported out of the guard cells, and,
    a. as water enters the guard cells, the stoma closes.
    b. as water enters the guard cells, the stoma opens.
    c. as water leaves the guard cells, the stoma closes.
    d. as water leaves the guard cells, the stoma opens.

11. The herbicide 2,4-D is an analog of which sort of plant hormone?
    a. auxin.
    b. gibberellins.
    c. cytokinins.
    d. ethylene.

12. The defoliant Agent Orange, widely used in the Vietnam War, relied on which hormone-like chemical?
    a. 2,4-D.
    b. 2,4,5-T.
    c. DDT.
    d. dioxin.

13. A plant growth in response to touching something, such as the tendril of a pea, is which type of response?
    a. phototropism.
    b. gravitropism.
    c. thigmotropism.
    d. thigmonasty.

14. Phototropic responses are mediated by the chemical
    a. abscisic acid.

b. indoleacetic acid.
c. 2,4-D.
d. calmodulin.

15. When a flowering shrub or fruit tree is pruned so that the growing tips are removed,
a. apical dominance leads to their familiar triangular shape.
b. further growth is inhibited.
c. flowering and fruit production are often reduced.
d. the number of lateral shoots may be increased.

16. In the root tips of plants, auxin
a. seems to stimulate elongation of the root, in general.
b. seems to inhibit elongation of some cells in a horizontal root tip.
c. accumulates in the uppermost cells of a horizontal root tip.
d. in unevenly distributed in a vertically oriented root tip.

17. Researchers have found that the gravitropic response in root tips is retarded by application of
a. calmodulin.
b. calmodulin inhibitors.
c. amyloplasts.
d. ethylene.

18. Pea tendrils coil around objects they touch by
a. unequal cell elongation, when they are exposed to light.
b. unequal cell elongation, but only in the dark.
c. unequal cell elongation, both in the the dark and when exposed to light.
d. by coiling unrelated to cell elongation.

19. The "sleep-movements" or folding and unfolding of leaves or leaflets are often caused by
a. unequal cell elongation, in response to touch.
b. coiling and uncoiling of tendrils.
c. changes in turgor in certain cells, caused by K^+ transport.
d. a phototropic response.

20. Solar tracking shown by a plant such as the sunflower is
a. a phototropic response due to differential growth.
b. a nastic response relying on changing turgor.
c. a response similar to tendril growth in peas.
d. a geotropic response.

21. In order to activate the "trap" of a Venus flytrap,
a. only one of three hairlike triggers must be touched.
b. two of triggers must be touched, or one must be touched twice.
c. all three of the hairlike triggers must be touched.
d. an edible insect must touch the hairlike triggers.

22. Plants referred to as short-day plants generally begin to flower
a. any time the days are short.
b. prior to the summer solstice, June 21, which is the longest day of the year.
c. when day length has shortened to some critical length, usually after the summer solstice.
d. shortly before long-day plants begin to flower.

23. Although there are exceptions, such as the pineapple,
a. auxin often has an inhibitory effect on flowering.
b. auxin stimulates flowering, for example, in roses.
c. the concentration of auxin usually goes up in plants about to flower.
d. auxin generally has no effect on flowering.

24. The cocklebur is a short-day plant; if, during the middle of 9 hours of darkness,
    a. only its leaves are exposed to a flash of white light, it will flower.
    b. its leaves and flowers are exposed to a flash of white light, it will flower.
    c. only its flowers are exposed to a flash of white light, it will not flower.
    d. its leaves and flowers are exposed to a flash of white light, it will not flower.

25. Under which of the following regimes will cocklebur plants flower?
    a. long day, short night, flash of far-red light.
    b. short day, long night, flash of red light.
    c. short day, long night, flash of far-red light.
    d. short day, long night, flash of far-red light followed by flash of red light.

B. True or False Questions

Mark the following statements either T (True) or F (False).

_____ 26. In general, a plant is able to respond to a stimulus.

_____ 27. When the Darwins covered the tip of a canary grass coleoptile with foil, it bent toward the light.

_____ 28. A bioassay for concentration of the hormone auxin is to measure bending induced in oat seedlings.

_____ 29. Indoleacetic acid is one of the naturally occurring amino acids found in protein.

_____ 30. Auxin leads to growth by causing cells to proliferate mitotically.

_____ 31. In all its natural forms, auxin always acts as a growth stimulant.

_____ 32. Gibberellins are produced not only by plants but also by certain fungi.

_____ 33. In a strain of dwarf corn in which dwarfism is genetically determined, application of gibberellins would be ineffective in stimulating growth to normal size.

_____ 34. In tissue culture of tobacco cells with high concentrations of auxin and cytokinin, there is callus growth but no differentiation of shoots or roots.

_____ 35. High concentrations of ethylene encourage germinating pea shoots to straighten prior to emergence from the ground.

_____ 36. In fruit storage warehouses, ethylene is used to prevent ripening of fruit like bananas.

_____ 37. Although it is a weed killer, 2,4-D promotes plant growth at low concentrations.

_____ 38. The indoleacetic acid molecule is probably the photoreceptor controlling phototropism.

_____ 39. With apical dominance, auxin can inhibit growth of lateral branches in some plants.

_____ 40. Roots of seedlings turn downward as a negative phototropic response mediated by auxin.

_____ 41. The gravitropic response in a root begins with events in the columella.

_____ 42. Calmodulin seems to be a necessary intermediate in the gravitropic response.

_____ 43. Coiling in the tendrils of a pea is due to unequal cell proliferation, with the greatest cell proliferation occurring on the side of the tendril opposite the touching surface.

_____ 44. Nastic responses, such as the folding of leaflets in *Mimosa pudica*, are usually irreversible.

_____ 45. Thigmonastic responses involve changes in turgor pressure in pulvinus motor cells.

C. Fill in the Blanks

Answer the question or complete the statement by filling in the blanks with the correct word or words.

46. From their experiments, Boysen-Jensen and Paal concluded that the source of the influence responsible for oat seedlings bending toward the light was the _________________ of the plant.

47. The chemical name of the hormone auxin is _________________.

48. Auxin may stimulate cell elongation by loosening the cellulose surrounding the cell and allowing _________________ to lengthen the cell.

49. Gibberellins produced by grain embryos stimulate the production of the enzyme _________________.

50. The storage life of some vegetable crops is extended by application of synthetic _________________.

51. The hormone that is important in fruit ripening is _________________.

52. The hormone believed to be responsible for inducing winter dormancy in plants is _________________.

53. The auxin-like compound that is the active ingredient in the defoliant agent orange is _________________.

54. Plant movements that are not related to growth are generally termed _________________ responses.

55. The initial photoreceptor for auxin-mediated phototropic responses is probably a yellow pigment, and it is most responsive to _________________ wavelengths of light.

56. The triangular shape of many conifers is due to the effects of _________________.

57. The coiling of tendrils around a support is an example of a _________________ response.

58. The day-by-day turning of leaves or flowers toward the sun is termed _________________.

59. In the thigmonastic response of the sensitive plant *Mimosa pudica*, _________________ ions are transported into cells that gain turgor pressure.

60. The energy required to operate the trap mechanism of *Dionaea muscipula* is provided by _________________.

61. Responses of plants and animals to particular day lengths or night lengths are termed _________________ responses.

62. The critical factor for determining the initiation of flowering in many plants species is the length of the _________________ portion of the 24-hour day.

63. Plants such as chrysanthemums, which flower after the nights lengthen following the summer solstice, can be termed _________________ plants.

64. Plants whose flowering seems indifferent to day and night lengths are termed _________________.

65. The pigment responsible for the photoperiod-based flowering behavior of cockleburs is _________________.

Questions for Discussion

1. Auxin has a role in a number of different plant responses. In some cases it promotes growth, while in others it seems to inhibit growth. Outline the roles that have been attributed to auxin and explain the proposed mechanism of action in each case.

2. The transport of K^+ ions is involved in different types of plant response. Identify the responses in which K^+ transport has been implicated and explain the mechanism that is believed to lead to these responses.

3. We know that photoperiod has many effects on plants, such as regulating time of flowering and dormancy in some cases. How would you evaluate the biological risks of installing very intense sodium vapor lighting in residential neighborhoods? What are the potential risks to flowering herbs, shrubs, and trees?

Answers to Self-Exam

Multiple Choice Questions

1. c
2. a
3. b
4. c
5. a
6. a
7. b
8. b
9. d
10. c
11. a
12. b
13. c
14. b
15. d
16. b
17. b
18. a
19. c
20. b
21. b
22. c
23. a
24. d
25. c

True or False Questions

26. T
27. F
28. T
29. F
30. F
31. F
32. T
33. F
34. T
35. F
36. F
37. T
38. F
39. T
40. F
41. T
42. T
43. F
44. F
45. T

Fill in the Blank Questions

46. tip
47. indoleacetic acid
48. turgor pressure
49. alpha amylase
50. cytokinins
51. ethylene
52. abscisic acid
53. 2,4,5-T
54. nastic
55. blue
56. apical dominance
57 thigmotropic
58. solar tracking
59. K^+
60. ATP reserves
61. photoperiodic
62. nighttime
63. short-day
64. day-neutral
65. phytochrome

CHAPTER 30 Invertebrates I

Learning Objectives

After mastering the material covered in chapter 30, you should be able to confidently do the following tasks:

- List the traits which we believe are generally characteristic of animals.
- Distinguish the Parazoa from the Metazoa.
- Distinguish the protostomes from the deuterostomes.
- Give characteristics which will separate the Phyla Porifera, Cnidaria, Ctenophora, Platyhelminthes, Nematoda, and Rotifera.
- Explain the relationship of the three germ layers and body cavities in the metazoans.
- Outline a representative life cycle for each of the Phyla.
- Among the Cnidarians, distinguish the hydrozoans, scyphozoans, and anthozoans.
- Among the Platyhelminthes, distinguish the free-living flatworms, the flukes, and the tapeworms.

Chapter Outline

I. What Is an Animal?

II. Animal Origin.

- A. The early fossil record.
- B. Phylogeny of the animal kingdom.

III. Levels of Organization and the Phylum Porifera.

- A. The biology of sponges.

IV. The Radial Plan and Higher Levels of Organization.

- A. Phylum Cnidaria.
  1. Class Hydrozoa: the hydrozoans.
  2. Class Scyphozoa: the jellyfish.
  3. Class Anthozoa: the corals and anemones.
- B. Phylum Ctenophora: the comb jellies.

V. The Bilateral Trend and the Versatile Mesoderm.

A. The three-layered embryo.
B. Phylum Platyhelminthes: the flatworms.
1. Class Turbellaria: the free-living flatworms.
2. Class Trematoda: the flukes.
3. Class Cestoda: the tapeworms.
C. Phylum Rhynchocoela: the ribbonworms.

VI. Body Cavities, A One-Way Gut, and a New Body Plan.

A. Phylum Nematoda.
B. Phylum Rotifera.

Key Words

multicellularity
chordates
Chordata
Arthropoda
Annelida
Mollusca
Nematoda
Cnidaria
Echinodermata
Porifera
Platyhelminthes
Parazoa
Metazoa
Ediacara fauna
Burgess Shale formation
deuterostome
protostome
tissue
organ
organ system
calcareous sponges
glassy sponge
sessile
spongocoel
osculum
porocytes
mesohyl
amebocytes
choanocytes
spicules
proteinaceous sponge
bath sponge
spongin
hermaphroditic
gemmules
bilateral symmetry
radial symmetry
cnidarian
ctenophores
pentaradial symmetry
planula larva
hydrozoans
jellyfish
coral
anemone
tentacle
cnidocyte
nematocysts
mesoglea
epidermis
gastrodermis
gastrovascular cavity
nerve net
neuron
polyp
medusa
Class Hydrozoa
budding
medusae
Obelia
hydra
Chlorhydra
Physalia
Class Scyphozoa
scyphomedusa
Cyanea
Aurelia
scyphistoma
Class Anthozoa
calcium carbonate
coral atoll
barrier reef
Phylum Ctenophora
comb jellies
bioluminescent
cephalization
germ layers
ectoderm
endoderm
mesoderm
diploblasts
triploblasts
Phylum Platyhelminthes
flatworms
acoelomate
Class Turbellaria
Dugesia
planarian
ganglia
eyespot
pharynx
osmoregulation
protonephridia
flame bulb
flame cell
Class Trematoda
fluke
oral sucker
ventral sucker
Clonorchis sinensis
human liver fluke
primary host
intermediate host
miracidia
sporocysts
redia
cercariae
Schistosoma mansoni
human blood fluke
schistosomiasis
Class Cestoda
tapeworm
scolex
proglottids
Phylum Rhynchocoela
ribbon worm
coelom
peritoneum
pseudocoelom
Phylum Nematoda
roundworm
Ascaris lumbricoides
Trichinella spiralis

trichinosis
Phylum Rotifera
gullet
gizzard
urinary bladder
parthenogenesis

Exercises

1. Below is a diagram of a hydrozoan and its life cycle. Provide labels for this diagram using terms from the list. Check your answers with figure 30.8 in the text.

asexually reproductive polyp
common gastrovascular cavity
egg
feeding polyp
medusa bud
planula larvae
sexually reproductive medusa
sperm
tentacle mouth
young polyp
zygote

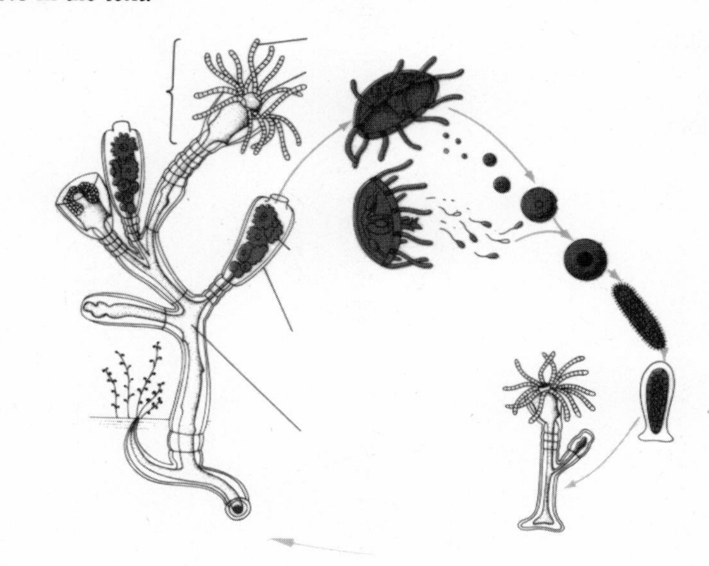

2. A diagram of a rotifer is reproduced below. Please provide labels for the various structures, using the terms in the accompanying list. Check you work with figure 30.20 in the text.

bladder
brain
cilia
egg
excretory tubule
eyespot
flame cell
foot
gizzard
intestine
mouth
ovary
oviduct
pharynx
pseudocoelom
spurs
stomach
ventral nerve cord

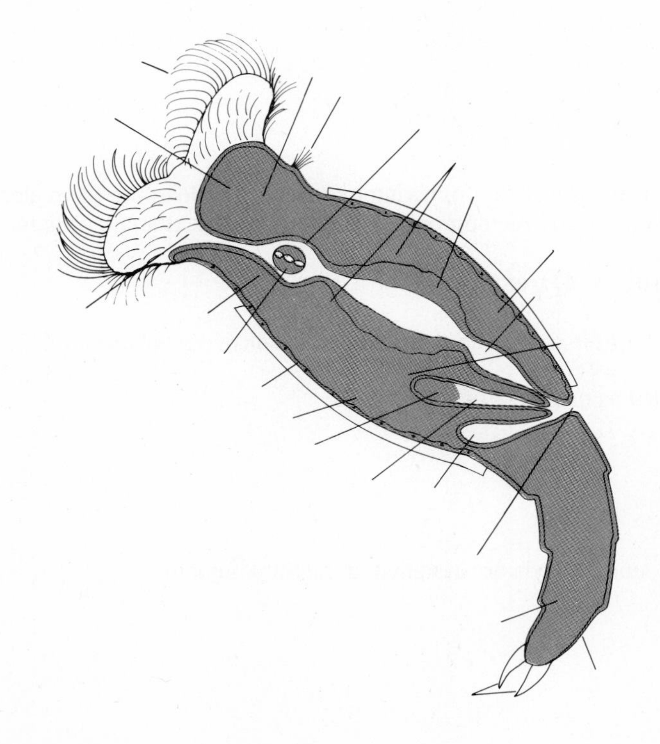

3. The diagrams below illustrate several animal body plans. Show which is acoelomate, which is pseudocoelomate, and which is coelomate. Use colored pencils to show which tissues are derived from ectoderm, mesoderm, and endoderm. Then label the diagrams to show each of the features in the following list. Check your work with figure 30.16 in the text.

coelom
ectodermal skin
endodermal lining
gastrovascular cavity
gut lumen
mesodermal layer
muscle
peritoneum
pesudocoelom

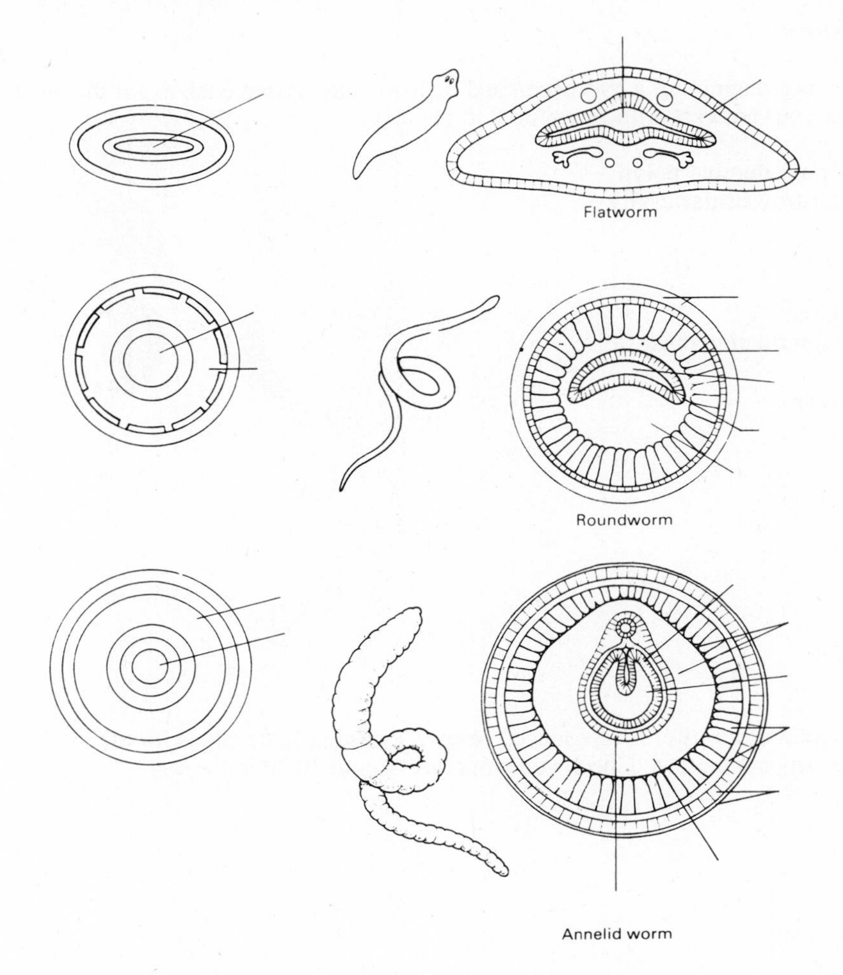

Self-Exam

You should be able to easily answer the following questions after learning the material in chapter 30. If you have difficulty with any question, study the appropriate section in the text and try again.

A. Multiple Choice Questions

Circle one alternative that best completes the statement or answers the question.

1. Humans are classified as members of the Phylum
   a. Chordata.
   b. Arthropoda.
   c. Anthozoa.
   d. Mammalia.

2. The majority of the animal phyla are assigned to the Subkingdom
   a. Parazoa.
   b. Porifera.
   c. Metazoa.
   d. Anthozoa.

3. Most invertebrate phyla first appeared
   a. in the Cenozoic, about 60 million years ago.
   b. in the Mesozoic, more than 65 million years ago.
   c. in the Paleozoic, more than 225 million years ago.
   d. in the Precambrian, more than 570 million years ago.

4. Humans can be correctly characterized as
   a. Parazoans.
   b. Protostomes.
   c. Deuterostomes.
   d. Hemichordates.

5. Which level of organization is the one that is the next most inclusive after the tissue level?
   a. Cell level of organization.
   b. Organ level of organization.
   c. Organ system level of organization.
   d. Organism level of organization.

6. The large central cavity in the body of a sponge is termed the
   a. spongocoel.
   b. osculum.
   c. porocyte.
   d. mesohyl.

7. In terms of nutrition, sponges are best described as
   a. photosynthetic.
   b. autotrophic.
   c. highly motile predators.
   d. filter-feeders.

8. Jellyfish and corals belong to the Phylum
   a. Porifera.
   b. Cnidaria.
   c. Ctenophora.
   d. Rotifera.

9. The sedentary form of Cnidarians is usually the
   a. medusa.
   b. polyp.
   c. cnidocyte.
   d. nematocyst.

10. A hydrozoan stage called a planula is
    a. a polyp.
    b. a medusa.
    c. a ciliated swimming larva.
    d. a scyphistoma.

11. The hard material secreted by corals is
    a. spongin.
    b. silica.
    c. calcium carbonate.
    d. bone.

12. The comb jellies are placed in the Phylum
    a. Porifera.
    b. Cnidaria.
    c. Ctenophora.

d. Rotifera.

13. In describing their germ layers, cnidarians and ctenophorans have
    a. an outer ectoderm and an inner endoderm.
    b. an outer endoderm and an inner ectoderm.
    c. from outside to inside, ectoderm, mesoderm, and endoderm.
    d, from outside to inside, endoderm, mesoderm, and ectoderm.

14. In development of animals which are triploblasts, muscle tissue
    a. originates from endoderm.
    b. originates from mesoderm.
    c. originates from ectoderm.
    d. originates from the skeletal support system.

15. Flatworms are
    a. parazoans.
    b. acoelomates.
    c. pseudocoelomates.
    d. coelomates.

16. Nutritionally, planarians can be described as
    a. autotrophs.
    b. filter-feeders.
    c. predators and scavengers.
    d. decomposers.

17. Unlike free-living flatworms, flukes are
    a. autotrophic.
    b. predatory.
    c. parasitic.
    d. independent.

18. The disease schistosomiasis is caused by
    a. a free-living flatworm.
    b. the human blood fluke.
    c. a human tapeworm.
    d. a pig bladderworm.

19. The tapeworms belong to which group?
    a. Class Cestoda.
    b. Class Trematoda.
    c. Class Turbellaria.
    d. Class Anthozoa.

20. A sexually mature tapeworm typically lives in the host's
    a. muscle.
    b. bladder.
    c. lungs.
    d. intestine.

21. The lining of a true coelom is
    a. the endoderm.
    b. the peritoneum.
    c. the gastrodermis.
    d. the ectoderm.

22. Roundworms belong to
    a. Class Trematoda.

b. Class Cestoda.
c. Phylum Nematoda.
d. Phylum Rotifera.

23. The human disease trichinosis is caused by
a. a flatworm.
b. a fluke.
c. a tapeworm.
d. a roundworm.

24. A true excretory system with protonephridia, ducts, and a urinary bladder is seen in
a. jellyfish.
b. hydrozoans.
c. flatworms.
d. rotifers.

B. True or False Questions

Mark the following statements either T (True) or F (False).

_____ 25. The Subkingdom Metazoa contains only one phylum, the Porifera.

_____ 26. In truly multicellular animals, the cells are interdependent and usually do not exist alone.

_____ 27. The fossils from the Burgess Shale formation are older than the fossils of the Ediacara fauna.

_____ 28. The flatworms and nematodes are deuterostomes.

_____ 29. A group of specialized cells with a common function can be considered a tissue.

_____ 30. Sponges can be separated into individual cells, and the cells can reorganize into new sponges.

_____ 31. Most adult sponges are highly motile and live in fresh water.

_____ 32. Among metazoans, bilateral symmetry is more common than is radial symmetry.

_____ 33. In the nerve cells of cnidarians, nerve impulses can travel in either direction.

_____ 34. Among the hydrozoans, the polyp state usually dominates.

_____ 35. A Portuguese man-of-war is not a jellyfish, but is rather a floating hydrozoan colony.

_____ 36. The dominant stage among the scyphozoans or jellyfish is the polyp stage.

_____ 37. Corals are colonial anthozoans whose polyps secrete walls of calcium carbonate.

_____ 38. The tentacles of comb jellies have glue cells to snare their prey.

_____ 39. Diploblasts have mesoderm and ectoderm.

_____ 40. In triploblasts, the nervous system arises from the ectodermal germ layer.

_____ 41. Flatworms have well-developed respiratory and circulatory systems.

_____ 42. Planarians have a true coelom.

_____ 43. Osmoregulation in flatworms is accomplished by protonephridia.

_____ 44. In general, the trematodes are parasites.

_____ 45. The human liver fluke causes schistosomiasis.

_____ 46. A tapeworm proglottid is a complete reproductive unit, with ovaries, testes, and genitalia.

_____ 47. Ribbonworms have a one-way gut, with a mouth at one end and an anus at the other.

_____ 48. Nematodes have a pseudocoelom rather than a true coelom.

_____ 49. The disease trichinosis is caused by a rotifer.

_____ 50. Diploid female rotifers may reproduce parthenogenetically to produce new diploid females.

C. Fill in the Blanks

Answer the question or complete the statement by filling in the blanks with the correct word or words.

51. The two subkingdoms of the animal kingdom are the __________________, with one Phylum, and the __________________, which includes the rest of the animal Phyla.

52. The earliest geological era is the Precambrian, and the most recent is the __________________.

53. The highest level of organization exhibited by the sponges is the __________________ level.

54. The large central cavity in the vaselike body of a simple sponge is the __________________.

55. The skeletal material of the proteinaceous sponges is the fibrous protein __________________.

56. Echinoderms such as starfish and sand dollars show a __________________ pattern of symmetry.

57. In Cnidarians, the stinging cells called __________________ release nematocysts.

58. Hydrozoans can reproduce asexually by __________________ and forming new polyps.

59. The swimming stage of a jellyfish (Class Scyphozoa) is called a __________________.

60. Coral barrier reefs are formed by members of the Class __________________.

61. Flatworms have solid bodies with no internal cavities, so they are called __________________.

62. Planarians take care of osmoregulation by means of __________________.

63. The human blood fluke, whose intermediate host is a snail, is the cause of the disease __________________.

64. The part of a tapeworm that is specialized for attaching to its host's intestinal wall is the __________________.

65. For parasitic worms with complex life cycles, sexual reproduction occurs in the __________________ host.

66. A true coelom is lined with the __________________, a tissue derived from embryonic mesoderm.

67. Humans risk contracting trichinosis by eating uncooked or undercooked __________________.

68. __________________ is an asexual reproduction process where diploid eggs skip meiosis and mitotically develop directly into new individuals.

Questions for Discussion

1. What are the basic traits we usually associate with animals? List these, and illustrate these traits with examples. Are there exceptions where animals lack one or more of these traits? Are any of these traits unique to animals, separating them from members of the other kingdoms.?

2. What are the levels of organization we see in the bodies of animals? Do all animals show all levels of organization? Outline the "progression" of levels of organization through the major animal phyla.

3. Distinguish the following groups of animals: the protostomes versus the deuterostomes; the parazoans versus the metazoans; the diploblasts versus the triploblasts. Can you characterize each of the major animal phyla with respect to these alternatives? What do these features tell us about the evolutionary relations among phyla?

4. What is a coelom? What sorts of coeloms are known, and how are they distinguished? Can you separate the animal phyla based on the presence or absence of a coelom, and the type of coelom, if present?

Answers to Self-Exam

Multiple Choice Questions

1. a
2. c
3. d
4. c
5. b
6. a
7. d
8. b
9. b
10. a
11. c
12. c
13. a
14. b
15. b
16. c
17. c
18. b
19. a
20. d
21. b
22. c
23. b
24. d

True or False Questions

25. F
26. T
27. F
28. F
29. T
30. T
31. F
32. T
33. T
34. T
35. T
36. F
37. T
38. T
39. F
40. T
41. F
42. F
43. T
44. T
45. F
46. T
47. T
48. T
49. F
50. T

Fill in the Blank Questions

51. Parazoa; Metazoa
52. Cenozoic
53. cellular
54. spongocoel
55. spongin
56. pentaradial
57. cnidocytes
58. budding
59. scyphomedusa
60. Anthozoa
61. acoelomates
62. protonephridia
63. schistosomiasis
64. scolex
65. primary
66. peritoneum
67. pork
68. parthenogenesis

CHAPTER 31 Invertebrates II

Learning Objectives

After mastering the material covered in chapter 31, you should be able to confidently do the following tasks:

- Describe the coelom and its use in defining the major phyla of animals.

- Contrast the protostome versus deuterostome characteristics of the major animal phyla, and explain the evolutionary implications of the contrast.

- Give the basic characteristics which distinguish the Phyla Mollusca, Annelida, Arthropoda, Onychophora, and Echinodermata.

- Distinguish the major classes of mollusks, annelids, and arthropods, and give examples of each.

- Describe the types of larvae that occur in the different animal phyla that have larval stages.

- Explain why the echinoderms, of all the invertebrate phyla, are believed to be the most closely related to the chordates.

Chapter Outline

I. The Coelom and the Coelomates.

A. Protostome and deuterostome lines.

II. The Protostome Line.

A. The lophophorate phyla.
B. Phylum Mollusca.
1. Class Polyplacophora: the chitons.
2. Class Gastropoda: snails.
3. Class Bivalvia (Pelycypoda): the bivalves.
4. Class Cephalopoda: the squids and octopuses.
C. The segmented body plan.
D. Phylum Annelida.
1. Class Oligochaeta: the earthworms.
2. Class Hirudinae: the leeches.
3. Class Polychaeta: the segmented marine worms.
E. Phylum Arthropoda.
1. Subphylum Trilobita: the trilobites.
2. Subphylum Chelicerata.
3. Subphylum Mandibulata.

a. Class Crustacea.
b. Classes Chilopoda and Diplopoda.
c. Class Insecta.
F. Phylum Onychophora.

III. The Deuterostome Line.

A. Phylum Echinodermata.
1. Class Asteroidea: the sea stars.

Key Words

coelom
coelomates
peritoneum
mesentery
protostome
deuterostome
lophophorate
blastopore
endoskeleton
adaptive radiation
Pangaea
continental drift
Proto-Laurasia
Proto-Gondwana
Ectoprocta
Brachiopoda
Phoronida
bryozoan
lophophore
plankton
Phylum Mollusca
mollusks
foot
shell
mantle
mantle cavity
gill
open circulatory system
closed circulatory system
radula
nephridia
trochophore larvae
Class Polyplacophora
chiton
Class Gastropoda
snail
torsion
operculum
Class Bivalvia (Pelycypoda)
valves
incurrent siphon
excurrent siphon
labial palps
Class Cephalopoda
squid
octopus
cuttlefish
nautilus
branchial heart
metamerism
metamere
Phylum Annelida
Class Oligochaeta
earthworm
segmentation
aortic arch
sperm receptacle
clitellum
typhlosole
ganglia
pharynx
seta (-ae)
Class Hirudinea
leech
ectoparasite
anticoagulant
hirudin
Class Polychaeta
parapodia
clamworm
Nereis virens
palolo worm
Eunice viridis
fan worm
feather worm
peacock worm
Phylum Arthropoda
exoskeleton
chiton
molt
Subphylum Trilobita
trilobites
Subphylum Chelicerata
horseshoe crab
sensory palps
chelicerae
Class Arachnida
arachnids
cephalothorax
abdomen
book lung
spinnerets
cocoon
web
fibroin
Subphylum Mandibulata
mandibles
Class Crustacea
ostrocod
copepod
compound eye
ommatidia
corneal lens
krill
Class Chilopoda
centipede
Class Diplopoda
millipede
Class Insecta
head
thorax
ovipositor
complete metamorphosis
larva (-ae)
pupa (-ae)
adult
incomplete metamorphosis
nymph
naiad
tracheal system
spiracle
trachea
tracheole
ocelli
labrum
labium
sensory palp
maxillae
Anoplura
Coleoptera
Dermaptera
Diptera
Ephemeroptera
Hemiptera
Hymenoptera

Isoptera
Lepidoptera
Neuroptera
Odonata
Orthoptera
Siphonaptera
Phylum Onychophora
Peripatus
Phylum Echinodermata
echinoderm
water vascular system
tube feet
Class Asteroidea
sea star
stone canal
madreporite
ampulla
bipinnaria larvae
eocrinoids
spiral cleavage
radial cleavage
blastopore
stomadeum
schizocoelous process
enterocoelous process
tornaria larva

Exercises

1. Using terms from the following list, provide labels for this diagram of a clam. Check your work with figure 31.7 in the text.

anus
digestive gland
excretory organ
excurrent siphon
foot
heart
incurrent siphon
gill
intestine
labial palps
mantle
mouth
muscle mass
stomach

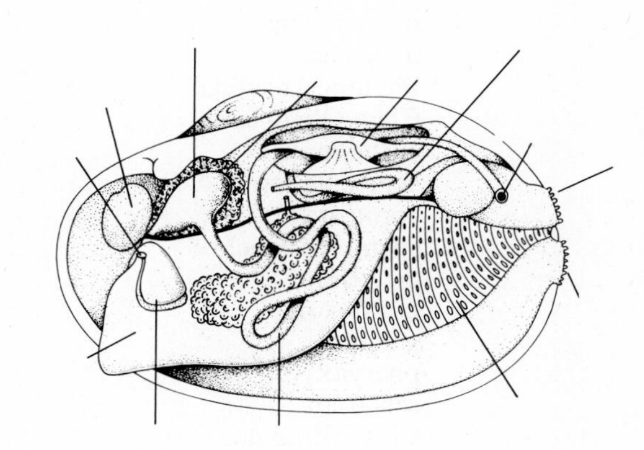

2. As in the previous exercise, use terms from the following list to provide labels for this diagram of the body plan of a grasshopper. Check your work with figure 31.17 in the text.

abdomen
antennae
compound eye
forewing
head
hindwing
legs
mouthparts
ovipositor
sensory palps
simple eye
spiracles
thorax

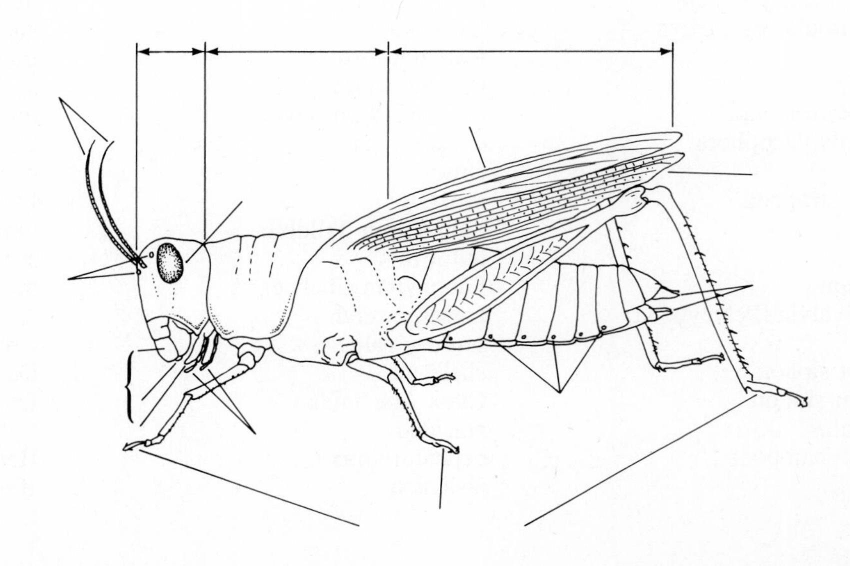

3. This is an illustration of an earthworm. Provide labels for this diagram, using terms from the following list. Check your work with figure 31.9 in the text.

anus
aortic arches
brain
circular muscle
clitellum
coelom
crop
dorsal blood vessel
epidermis
excretory pore
gizzard
intestine
longitudinal muscle
nephridia
setae
typhlosole
ventral blood vessel

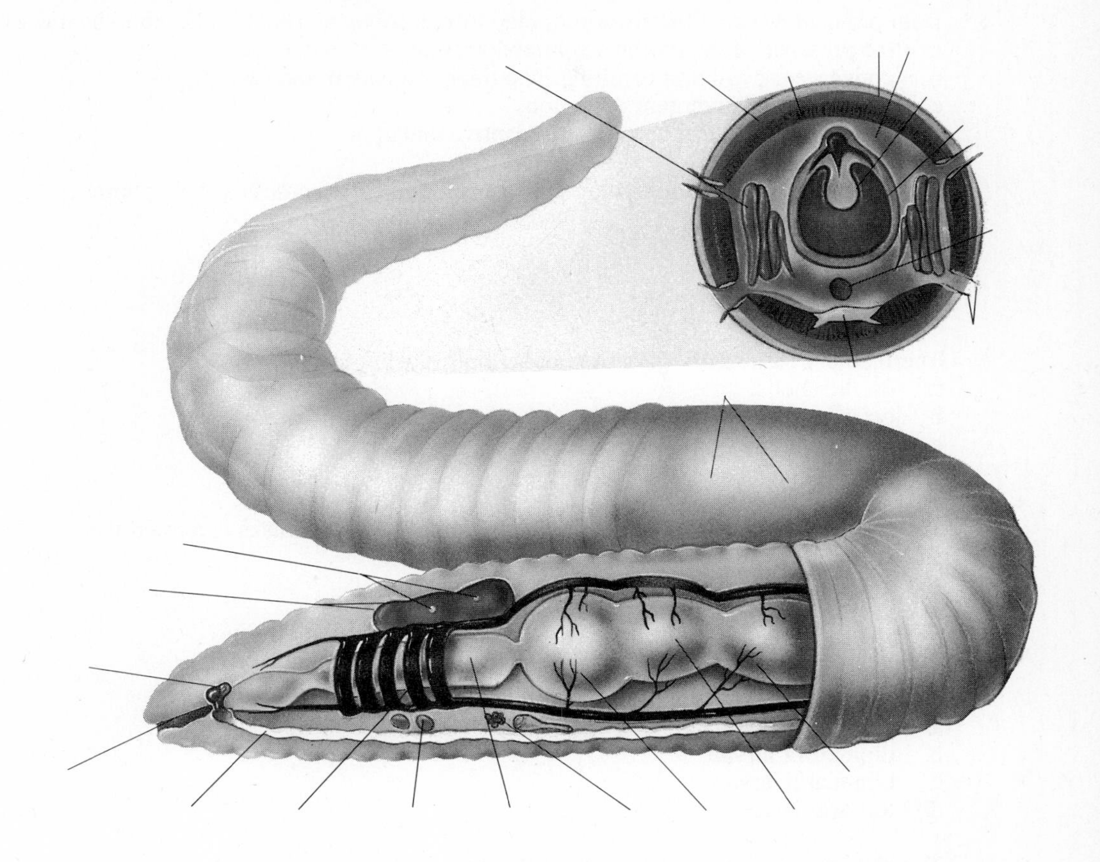

Self-Exam

You should be able to easily answer the following questions after learning the material in chapter 31. If you have difficulty with any question, study the appropriate section in the text and try again.

A. Multiple Choice Questions

Circle one alternative that best completes the statement or answers the question.

1. The peritoneum which lines the coelom is derived from
   a. endoderm.
   b. mesoderm.
   c. ectoderm.
   d. proctoderm.

2. Which of the following phyla *lacks* a coelom?
   a. Platyhelminthes.
   b. Mollusca.
   c. Annelida.
   d. Arthropoda.

3. Animals belonging to which of the following phyla are deuterostomes?
   a. Mollusca.
   b. Annelida.

c. Arthropoda.
d. Echinodermata.

4. Both protostomes and deuterostomes may form a coelom. This is believed to be due to
   a. the presence of the trait in a common ancestor.
   b. convergent evolution resulting in similar coeloms in the two groups.
   c. the actions of divergent evolution.
   d. the mechanisms that accompany adaptive radiation.

5. Species in which of the following phyla would typically possess a lophophore?
   a. Brachiopoda
   b. Polyplacophora.
   c. Gastropoda.
   d. Pelycypoda.

6. The feathery respiratory gills of aquatic mollusks are located
   a. on the shells.
   b. on the muscular foot.
   c. on part of the radula.
   d. in the mantle cavity.

7. The rasping, tongue-like structure of many non-bivalve mollusks is termed the
   a. foot.
   b. mantle.
   c. mollusk.
   d. radula.

8. Which type of larvae do some mollusks produce?
   a. trochophore larvae.
   b. dipleurula larvae.
   c. bipinnaria larvae.
   d. tornaria larvae.

9. The chitons are members of which of these mollusk classes?
   a. Class Cephalopoda.
   b. Class Polyplacophora.
   c. Class Bivalvia (Pelycypoda).
   d. Class Gastropoda.

10. Slugs are very much like snails; they belong to which class?
    a. Class Cephalopoda.
    b. Class Polyplacophora.
    c. Class Bivalvia (Pelycypoda).
    d. Class Gastropoda.

11. The mollusks which are filter feeders and circulate water through the mantle cavity via incurrent and excurrent siphons are members of
    a. Class Cephalopoda.
    b. Class Polyplacophora.
    c. Class Bivalvia (Pelycypoda).
    d. Class Gastropoda.

12. Which of the following mollusk classes typically has a closed circulatory system?
    a. Class Cephalopoda.
    b. Class Polyplacophora.
    c. Class Bivalvia (Pelycypoda).
    d. Class Gastropoda.

13. Highly developed segmentation of the body is characteristic of
    a. Phylum Ectoprocta.
    b. Phylum Mollusca.
    c. Phylum Annelida.
    d. Phylum Echinodermata.

14. Which of the following groups produces the anticoagulant hirudin?
    a. Predatory cephalopods.
    b. Scavenging earthworms.
    c. Parasitic leeches.
    d. Predatory polychaetes.

15. Members of the Phylum Arthropoda typically have
    a. endoskeletons of chitonous material.
    b. exoskeletons that are shells of calcium carbonate.
    c. exoskeletons of chitonous material.
    d. neither exoskeleton nor endoskeleton.

16. There are no living representative of which of the following arthropod groups?
    a. Trilobites.
    b. Arachnida.
    c. Crustacea.
    d. Chilopoda.

17. Horseshoe crabs are placed in the same subphylum with
    a. the nautilus.
    b. ticks.
    c. lobsters.
    d. insects.

18. The body of a member of which of the following groups would have a cephalothorax and an abdomen?
    a. Annelida.
    b. Arachnida.
    c. Chilopoda.
    d. Insecta.

19. Which of the following arthropod classes is *not* placed in the subphylum Mandibulata?
    a. Arachnida.
    b. Crustacea.
    c. Diplopoda.
    d. Insecta.

20. Which of the following sequences is correct for insects with complete metamorphosis?
    a. egg ⇒ larva ⇒ pupa ⇒ adult.
    b. egg ⇒ pupa ⇒ larva ⇒ adult.
    c. egg ⇒ larva ⇒ adult ⇒ pupa.
    d. egg ⇒ adult ⇒ larva ⇒ pupa.

21. In insects, the external openings of the tracheal system are the
    a. tracheae.
    b. tracheoles.
    c. spiracles.
    d. opercula.

22. The insect order Lepidoptera includes the
    a. beetles.
    b. moths.

c. grasshoppers.
d. lice.

23. Flies and mosquitoes have only a single pair of wings; they are placed in the order
a. Coleoptera.
b. Hymenoptera.
c. Diptera.
d. Isoptera.

24. Fleas are small ectoparasites of vertebrates. They are members of the same class as
a. ticks.
b. crabs.
c. butterflies.
d. leeches.

25. The onychophoran *Peripatus* is considered to be an evolutionary link between
a. protostomes and deuterostomes.
b. the invertebrates and the chordates.
c. the arthropods and the echinoderms.
d. the annelids and the arthropods.

26. Sea stars are frequently described as
a. ectoparasites of vertebrates.
b. filter feeders of krill.
c. scavengers of sea-bottom detritus.
d. predators of mollusks.

27. Sea stars have a life cycle with
a. trochophore larvae.
b. dipleurula larvae.
c. bipinnaria larvae.
d. tornaria larvae.

28. A radial cleavage pattern is seen in early embryos of
a. annelids.
b. mollusks.
c. arthropods.
d. echinoderms.

B. True or False Questions

Mark the following statements either T (True) or F (False).

_____ 29. The first embryonic opening, the blastopore, forms the mouth in deuterostomes.

_____ 30. Members of the Phylum Ectoprocta are also known as bryozoans.

_____ 31. There are relatively few species of mollusks, and it is one of the least common phyla.

_____ 32. All mollusks have a closed circulatory system.

_____ 33. Segmentation of body parts is not apparent in members of the Phylum Mollusca.

_____ 34. Although there are chiton fossils from hundreds of millions of years ago, some still exist today.

_____ 35. Bivalve mollusks are largely predators, capturing prey with a radula.

_____ 36. The segmented body plan of vertebrates is believed to have evolved independently of the segmentation found in several important invertebrate phyla.

_____ 37. Earthworms are hermaphroditic, with both male and female structures present in each individual.

_____ 38. Oligochaetes have a hollow dorsal nerve cord.

_____ 39. Leeches live primarily in fresh water, although marine and terrestrial forms are known.

_____ 40. Marine polychaetes produce swimming trochophore larvae.

_____ 41. In terms of the total number of known species, Arthropoda is the most diverse phylum.

_____ 42. There are living trilobites today very similar to those of 300 million years ago.

_____ 43. Spiders and mites are members of the arthropod subphylum Mandibulata.

_____ 44. Members of the subphylum Chelicerata have six pairs of appendages.

_____ 45. Centipedes have two pairs of appendages per segment.

_____ 46. Millipedes are generally harmless herbivores, while centipedes are predators.

_____ 47. Insect bodyies have two major regions: the cephalothorax and the abdomen.

_____ 48. Insects which proceed from egg to nymph to adult have an incomplete metamorphosis.

_____ 49. True bugs with piercing mouthparts are members of insect order Hemiptera.

_____ 50. The sea star, like all echinoderms, has a centralized nervous system.

C. Fill in the Blanks

Answer the question or complete the statement by filling in the blanks with the correct word or words.

51. Ectoprocts and brachiopods have a _________________, a ridge bearing ciliated tentacles.

52. Many mollusks have a rasping tongue-like structure termed a _________________.

53. Like marine annelids, some mollusks produce _________________ larvae.

54. Members of the mollusk class Polyplacophora are called _________________.

55. Attached to the foot of gastropods is a tough lid of horny material called the _________________.

56. Water is drawn into the mantle cavity of bivalve mollusks through the _________________.

57. Most segments of oligochaetes have paired excretory structures called _________________.

58. The _________________ is a prominent fold in the intestinal wall of an earthworm.

59. The anticoagulant secreted by parasitic leeches is _________________.

60. The fleshy setae-bearing extensions of each segment of a polychaete are _________________.

61. The "fangs" of members of subphylum Chelicerata are the appendages termed the _________________.

62. The silk of spiders is the protein ________________.

63. The compound eye of crustaceans has a large number of visual units called ________________.

64. In insects, the hollow appendage through which eggs are laid is the ________________.

65. Insects exchange air through a ________________.

66. Bees, wasps, and ants belong to the insect order ________________.

67. The insect order with the greatest number of species is the order ________________.

68. In sea stars, the water vascular system connects with the ________________, which often have suckers.

69. Sea stars produce ________________ larvae which are unlike other invertebrate larvae.

70. The embryonic invagination that becomes the mouth in protostomes is the ________________.

Questions for Discussion

1. Both protostomes and deuterostomes have the "tube-within-a-tube" body plan. What are the important differences between these two groups? If we conclude that the protostome body plan and the deuterostome body plan arose independently, how can we explain the continued coexistence of the two groups since the Precambrian?

2. Arthropods constitute one of the most successful and diverse groups of animals on earth today. What are some of the reasons that there are so many kinds of arthropods, and why are there so many individuals?

3. Among the phyla discussed in chapter 31 are seen several different approaches to respiration. What are the special structures for respiration found in the various groups? How do they work?

4. It is often concluded that of all the invertebrate phyla, the echinoderms are most closely related to the vertebrates, including humans. What is the evidence that supports this conclusion?

Answers to Self-Exam

Multiple Choice Questions

1. b
2. a
3. d
4. b
5. a
6. d
7. d
8. a
9. b
10. d
11. c
12. a
13. c
14. c
15. c
16. a
17. b
18. b
19. a
20. a
21. c
22. b
23. c
24. c
25. d
26. d
27. c
28. d

True or False Questions

29. F
30. T
31. F
32. F
33. T
34. T
35. F
36. T
37. T
38. F
39. T
40. T
41. T
42. F
43. F
44. T
45. F
46. T
47. F
48. T
49. T
50. F

Fill in the Blank Questions

51. lophophore
52. radula
53. trochophore
54. chitons
55. operculum
56. incurrent siphon
57. nephridia
58. typhlosole
59. hirudin
60. parapodia
61. chelicerae
62. fibroin
63. ommatidia
64. ovipositor
65. tracheal system
66. Hymenoptera
67. Coleoptera
68. tube feet
69. bipinnaria
70. blastopore

CHAPTER 32 The Chordates

Learning Objectives

After mastering the material covered in Chapter 32, you should be able to confidently do the following tasks:

- Describe the characteristics that distinguish the chordates from members of other phyla.

- Distinguish the three subphyla of chordates: urochordates, cephalochordates, and vertebrates.

- Specify the traits that characterize the vertebrates.

- Identify the seven living classes of vertebrates, and give the characteristics that are typical of each class.

- Give the unique characteristics of the mammal, and distinguish the three mammalian lines: prototherians, metatherians, and eutherians.

- Outline our current understanding of human evolutionary history, and state the position of *Homo sapiens* among the present day primates.

Chapter Outline

I. Phylum Hemichordata.

II. Phylum Chordata.

A. Subphylum Urochordata: tunicates and salps.
B. Subphylum Cephalochordata: lancelets.

III. Subphylum Vertebrata: Animal With Backbones.

A. Class Agnatha: the jawless fishes.
B. Class Placodermi: the extinct jawed fishes.
C. Class Chondrichthyes: the sharks and rays.
D. Class Osteichthyes: the bony fishes.
E. Class Amphibia: the amphibians.
F. Class Reptilia: the reptiles.
  1. reptile history.
G. Class Aves: the birds.
H. Class Mammalia: the mammals
  1. the mammalian brain.
  2. mammal origins.
  3. the primates.
  4. human specialization.

IV. Human Evolutionary History.

A. The Australopithecines.
B. The human line.
1. *Homo erectus.*
2. Neanderthals and us.

Key Words

Phylum Hemichordata
hemichordate
acorn worm
proboscis
collar
gill slit
phylum Chordata
dorsal hollow nerve cord
postanal tail
gill arches
ventral solid nerve cord
notochord
subphylum Urochordata
tunicate
ascidians
sea squirt
salp
gill sac
neoteny
subphylum Cephalochordata
lancelets
myotomes
atrium
subphylum Vertebrate
vertebral column
ventral heart
dorsal aorta
class Agnatha
agnathans
ostracoderms
lamprey
hagfish
subclass Cyclostomata
ammocoete
class Placodermi
class Chondrichthyes
cartilaginous fishes
shark
ray
placoid scale
spiral valve
cloaca
vent
lateral line organ
class Osteichthyes
bony fishes
swim bladder
gas gland
reabsorptive area
gill chamber
operculum
gill rakers
gill filaments
ventricle
Actinopterygii
ray-finned fish
Dipneustei
lungfish
Crossopterygii
lobe-finned fish
coelacanth
Latimeria chalumnae
class Amphibia
order Urodela
salamanders
order Anura
frogs
toads
order Apoda
caecilian
labyrinthodonts
spermatophores
class Reptilia
amniotic egg
amnion
extraembryonic membranes
uric acid
ectothermic
endothermic
Mesozoic era
age of the reptiles
cotylosaurs
pterosaurs
dinosaurs
Tyrannosaurus
Diplodocus
turtles
alligators
crocodiles
Sphenodon punctatum
tuatara
lizards
snakes
class Aves
Archaeopteryx
pulmonary circuit
systemic circuit
alveoli
urinary bladder
class Mammalia
mammary glands
mamma
placenta
ear ossicles
corpus callosum
Cenozoic era
age of mammals
prototherian
monotreme
metatherian
marsupial
eutherian
placental mammal
duckbilled platypus
echidna
pseudoplacenta
therapsids
opossum
Didelphis
order Primates
suborder Prosimii
tree shrew
tarsier
lemur
loris
suborder Anthropoidea
superfamily Ceboidea
New World monkeys
superfamily Cercopithecoidea
Old World monkeys
superfamily Hominoidea
apes
family Pongidae
chimpanzee
gorilla
family Hominidae
opposable thumb
bipedal
cultural evolution
molecular clock
Australopithecus robustus

Australopithecus africanus
Australopithecus boisei
Australopithecus afarensis
Homo habilis
cranial capacity
"Lucy"
Homo erectus
Java Ape Man
Peking Man
Lower Paleolithic
Homo sapiens
Neanderthal Man
Homo sapiens neandertalensis
Homo sapiens sapiens

Exercises

1. Provide labels for this illustration of larval and adult tunicates. Use the terms in the following list. Check your work with Figure 32.3 in the text.

anus
atrium opening
dorsal hollow nerve cord
heart
mouth
notochord
ovary
pharyngeal gill slits
testis
tunic

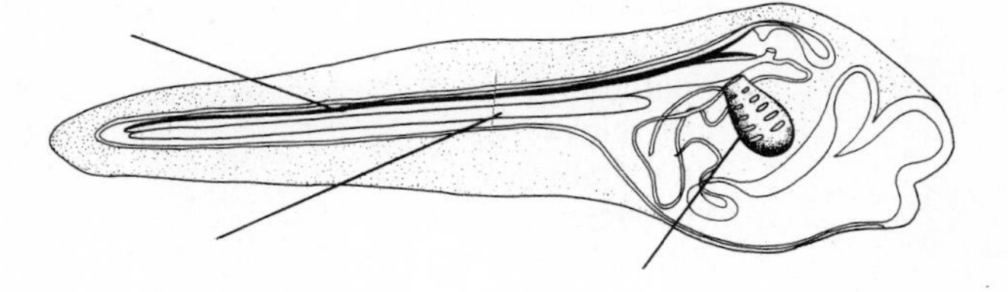

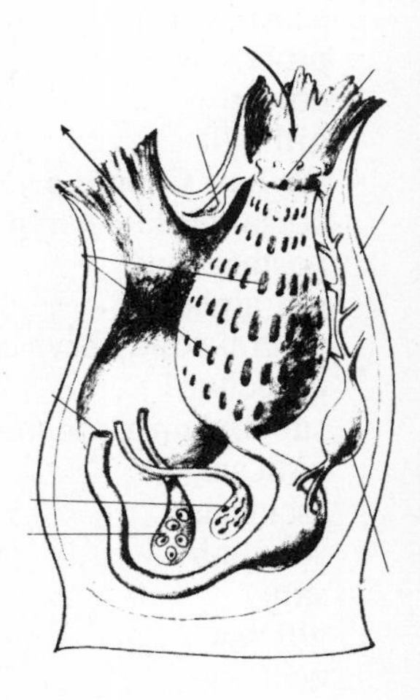

2.. The amniotic egg is a complex structure with both the embryo and extraembryonic membranes. Provide labels for the following diagram using the given terms. Check your work with Figure 32.18 in the text.

air space
albumen
allantois
amnion
amniotic chamber
blood vessels
chorioallantois
chorion
head of embryo
shell
shell membrane
yolk
yolk sac

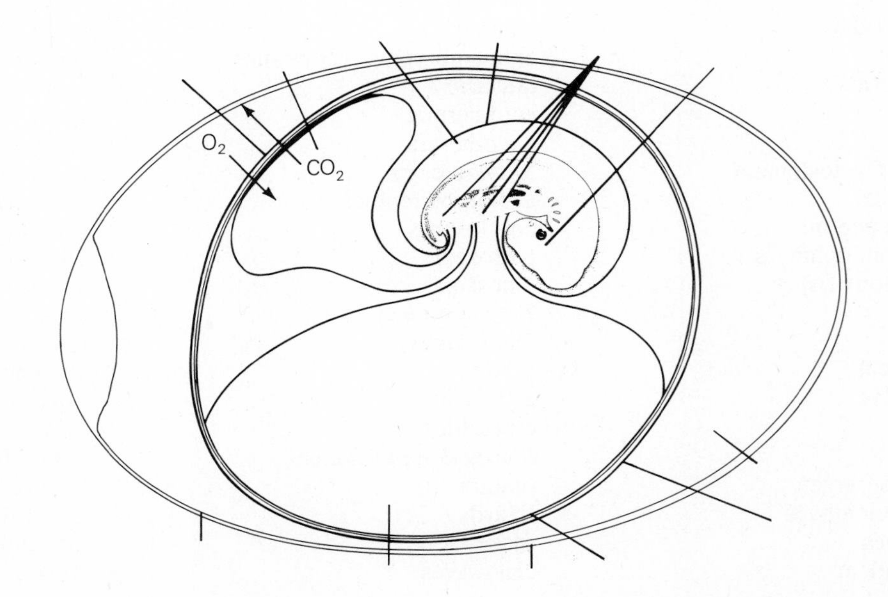

3. Below is an illustration of a cephalochordate. Use terms in the accompanying list to provide labels. Check your work with Figure 32.4 in the text.

anus
gill arches
gill basket
myotomes
nerve cord
notochord
opening of atrium
pharyngeal gill slits
tail
tentacles

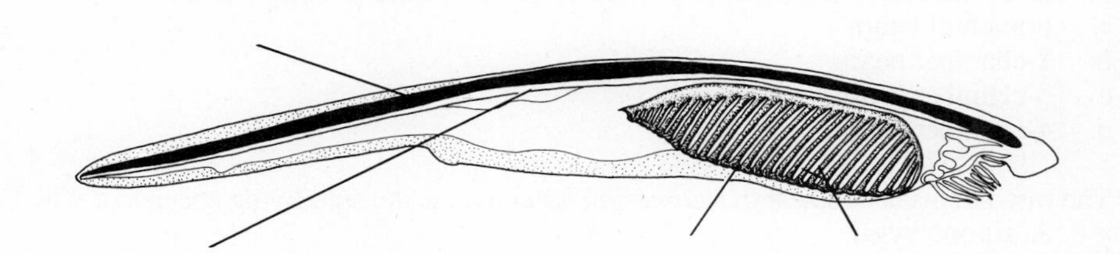

Self-Exam

You should be able to easily answer the following questions after learning the material in Chapter 32. If you have difficulty with any question, study the appropriate section in the text and try again.

A. Multiple Choice Questions

Circle one alternative that best completes the statement or answers the question.

1. A radial cleavage pattern is *not* seen in the embryos of
   a. arthropods.
   b. echinoderms.
   c. hemichordates.
   d. chordates.

2. Which of the following traits is *not* characteristic of members of the Phylum Chordata?
   a. ventral solid nerve cord.
   b. notochord.
   c. dorsal hollow nerve cord.
   d. postanal tail.

3. Traits such as a ventrally placed heart, a dorsal aorta, two pairs of limbs, and paired kidneys characterize
   a. hemichordates.
   b. tunicates.
   c. cephalochordates.
   d. vertebrates.

4. The filter-feeding larva of the parasitic lamprey is a(an)
   a. planophore.
   b. ammocoete.
   c. nymph or naiad.
   d. bipinnaria.

5. Which of the following groups is extinct, i.e., there are no known living representatives?
   a. agnathan fish.
   b. placoderm fish.
   c. dipneusteian (lungfish) fish.
   d. crossopterygian (lobe-fined) fish.

6. The so-called "bony-fishes" are members of the class

a. Osteichthyes.
b. Placodermi.
c. Chondrichthyes.
d. Agnatha.

7. Which of the following phrases best describes the heart of bony fishes?
   a. branchial heart.
   b. 2-chamber heart.
   c. 3-chamber heart.
   d. 4-chamber heart.

8. The rare deep-sea coelacanth *Latimeria chalumnae* is the sole living species of which group?
   a. Actinopterygii.
   b. Dipneustei.
   c. Crossopterygii.
   d. Placodermi.

9. Which of the following phrases best describes the heart of the typical amphibian?
   a. branchial heart.
   b. 2-chamber heart.
   c. 3-chamber heart.
   d. 4-chamber heart.

10. The period of time which we today refer to as the "age of reptiles" is the
    a. Paleozoic.
    b. Mesozoic
    c. Cenozoic.
    d. Pleistocene.

11. Among the reptiles were flying forms; the first flying reptile was probably a(an)
    a. cotylosaur.
    b. pterosaur.
    c. dinosaur.
    d. archaeopteryx.

12. Which of the orders of living reptiles is the least common?
    a. The order which includes turtles.
    b. The order which includes alligators.
    c. The order which includes tuataras.
    d. The order which includes lizards.

13. The egg-laying mammals, the monotremes, are identified with the term
    a. atherian.
    b. prototherian.
    c. metatherian.
    d. eutherian.

14. Taxonomically, humans are placed with
    a. prosimians.
    b. old world monkeys.
    c. new world monkeys
    d. apes.

15. Members of the species *Homo sapiens* are correctly classified in each of the following groups *except*
    a. Primates.
    b. Anthropoidea.
    c. Hominidae.
    d. Pongidae.

16. Which australopithecine species is considered to be ancestral to the human lineage?
    a. *Australopithecus robustus*
    b. *Australopithecus africanus*
    c. *Australopithecus boisei*
    d. *Australopithecus afarensis*

17. Which is today believed to be the most likely sequence of species leading to present-day humans?
    a. *Australopithecus afarensis* ⇒ *Homo habilis* ⇒ *Homo erectus* ⇒ *Homo sapiens.*
    b. *Homo erectus* ⇒ *Homo habilis* ⇒ *Homo sapiens neandertalensis* ⇒ *Homo sapiens sapiens.*
    c. *Australopithecus africanus* ⇒ *Homo habilis* ⇒ *Homo erectus* ⇒ *Homo sapiens.*
    d. *Australopithecus afarensis* ⇒ *Australopithecus afarensis* ⇒ *Homo erectus* ⇒ *Homo sapiens.*

18. The "Neanderthal Man" and other Neanderthals are today considered to have been
    a. a type of ape-man unrelated to the human lineage.
    b. an early form of human that became extinct.
    c. exactly the same and indistinguishable from present-day humans.
    d. one of the members of the genus *Australopithecus.*

19. The amniotic egg is most characteristic of which group?
    a. chordates.
    b. vertebrates.
    c. amphibians and birds.
    d. reptiles and birds.

20. With a few exceptions such as the opossum, marsupial mammals are largely restricted to which continent?
    a. North America.
    b. Eurasia.
    c. Africa.
    d. Australia.

21. The principal and unique specialization that distinguishes humans from other apes and primates is
    a. the ability to walk bipedally.
    b. the opposable thumb.
    c. the capacity for complex language.
    d. the absence of an adult tail.

B. True or False Questions

Mark the following statements either T (True) or F (False).

_____ 22. Hemichordates are members of the Phylum Chordata.

_____ 23. Vertebrates are characterized by a ventral solid nerve cord.

_____ 24. Although adult tunicates lack the postanal tail that is characteristic of vertebrates, it is present in their larval stage.

_____ 25. Tunicates are filter feeders, and their adult circulatory system is open.

_____ 26. Lancelets are assigned to the subphylum Vertebrata.

_____ 27. The cartilaginous fishes and the bony fishes arose from ancestral placoderm fishes.

_____ 28. Most evolutionists today believe that in fish bony skeletons have been derived from ancestral cartilaginous skeletons.

_____ 29. The very largest sharks are filter feeders that consume plankton.

_____ 30. Sharks and rays are considered primitive because they have external fertilization.

_____ 31. Some bony fishes have lungs and can breathe air.

_____ 32. The crossopterygian coelacanth *Latimeria chalumnae* is extinct.

_____ 33. Amphibians can carry out gas exchange across their moist skins.

_____ 34. Reptiles are generally considered to be endothermic.

_____ 35. Bird lungs have one-way air passages rather than the blind-ending alveoli found in most air-breathing vertebrate lungs.

_____ 36. Most birds lack a urinary bladder.

_____ 37. The early mammals probably branched from reptiles called therapsids.

_____ 38. The familiar marsupial opossum has inhabited North America continuously for at least 50 million years.

_____ 39. The bodies and limbs of very few primates are well-adapted for arborial life.

_____ 40. Humans are capable of cultural evolution in addition to genetic evolution.

_____ 41. Data from comparative study of DNA sequences indicate that humans are more closely related to chimpanzees and gorillas than to orangutans.

_____ 42. "Lucy," the oldest known hominid, is placed in the species *Homo sapiens*.

_____ 43. Fossil remains of Neanderthals have been found in Europe, Africa, and Asia.

C. Fill in the Blanks

Answer the question or complete the statement by filling in the blanks with the correct word or words.

44. Acorn worms are placed in the phylum _________________.

45. The flexible, turgid rod running along the backs of chordates is a _________________.

46. The repeating units of musculature that are evident in cephalochordates are termed _________________.

47. The lampreys and hagfish are members of the subclass _________________.

48. The miniature "teeth" present in the rough skin of sharks are actually _________________ scales.

49. The surface area of absorptive tissue in the shark's short digestive system is greatly increases by the structure called the _________________.

50. Some vertebrates have a _________________, a common chamber at the end of the digestive system that also receives ducts from the reproductive and excretory system.

51. A sensory organ that is characteristic of all fish is the _________________, which detects certain patterns of vibration and water movements.

52. In bony fish the hydrostatic organ that improves balance and allows an animal to maintain a particular depth is the ________________.

53. Both modern reptiles and modern amphibians are thought to have their origins among the ________________, an ancient terrestrial group.

54. The sperm of male salamanders is packaged in gelatinous packets called ________________.

55. Today's reptiles are considered to be ________________ or "cold-blooded."

56. Birds produce large, resistant ________________ eggs.

57. Eutherian mammals receive their nourishment during prenatal development via a ________________.

58. The duck-billed platypus and the echidna are the only living ________________.

59. In Australia, the only placental mammals found before Europeans visited were ________________.

60. Humans belong to the taxonomic family ________________.

61. The "old stone age" period when *Homo erectus* left crude stone tools is called the ________________.

Questions for Discussion

1. What is neoteny? What are the consequences of neoteny? What sorts of animals exhibit neoteny?

2. Discuss the various adaptations that accompanied the development of land-dwelling animals. What are typical differences in reproduction between aquatic and terrestrial groups? What are some of the structural differences? Are the adaptations related to gas exchange and water balance?

3. Dinosaurs were at one time a very successful group of large vertebrates, yet they have disappeared entirely. Why did they die off? Discuss the various explanations that have been offered. Which do you think are the most reasonable? Why?

4. Outline the structural adaptations for flight that are found in the bodies of birds.

5. Who were the Australopithecines? How long ago did they live? What are the various sorts of Australopithecines we know about, and how were they related to one another? What is thought to be the relation between Australopithecines and members of the genus *Homo*?

Answers to Self-Exam

Multiple Choice Questions

1. a
2. a
3. d
4. b
5. b
6. a
7. b
8. c
9. c
10. b
11. b
12. c
13. b
14. d
15. d
16. d
17. a
18. b
19. a
20. d
21. c

True or False Questions

22. F
23. F
24. T
25. T
26. F
27. T
28. F
29. T
30. F
31. T
32. F
33. T
34. F
35. T
36. T
37. T
38. F
39. F
40. T
41. T
42. F
43. T

Fill in the Blank Questions

44. Hemichordata
45. notochord
46. myotomes
47. Cyclostomata
48. placoid
49. spiral valve
50. cloaca
51. lateral line organ
52. swim bladder
53. labyrinthodonts
54. spermatophores
55. ectothermic
56. amniotic
57. placenta
58. monotremes or prototherians
59. bats
60. Hominidae
61. lower paleolithic

CHAPTER 33 Support and Movement

Learning Objectives

After mastering the material covered in Chapter 33, you should be able to confidently do the following tasks:

- Explain how a soft-bodied invertebrate can be supported by a hydrostatic skeleton.
- Describe the structure of the exoskeletons found among the invertebrates.
- Identify the principal types of tissues that vertebrates have.
- Describe the basic structure of vertebrate bone.
- Distinguish the major parts of the axial and appendicular skeletons of vertebrates.
- Describe and distinguish the types of vertebrate muscle: smooth, cardiac, and skeletal.
- Explain the gross structure and ultrastructure of skeletal muscle, and describe the ultrastructural and molecular events that accompany contraction.

Chapter Outline

I. Invertebrate Support and Movement.

A. Hydrostatic skeletons.
1. Nematodes.
2. Mollusks.
3. Annelids.

B. Exoskeletons.
1. Arthropods.
2. Mollusk shells.

C. Invertebrate endoskeletons.
1. Echinoderms.

II. Vertebrate Tissues.

A. Epithelial tissue.
B. Connective tissue.

III. Vertebrate Skeletons: Hardened Connective Tissue.

A. Structure of bone.

B. Organization of the vertebrate skeleton.
  1. the joints.
  2. the axial skeleton.
    a. the skull.
    b. the vertebral column.
  3. the appendicular skeleton.
    a. the pectoral and pelvic girdles.
    b. vertebrate limbs.

IV. Vertebrate Muscle: Its Organization and Movement.

A. Smooth muscle.
B. Cardiac muscle.
C. Skeletal muscle
  1. gross anatomy of skeletal muscle.
  2. ultrastructure of skeletal muscle.
  3. contraction at the ultrastructure level.
  4. molecular aspects of contraction.

Key Words

hydrostatic skeleton
antagonists
seta (-ae)
exoskeleton
ectoderm
endoskeleton
mesoderm
cuticle
chitin
calcium carbonate
muscle origin
muscle insertion
tendon
"cock-and-release"
direct flight muscles
indirect flight muscles
midge
mollusk shell
mantle
cuttlefish
cuttlebone
Nautilus
tissue
epithelium
connective tissue
muscle
nerve
epidermis
endothelium
glandular epithelium
connective tissue matrix
collagen
fibroblasts
scar tissue
elastin
elastic fibers
bony skeleton
periosteum
spongy bone
compact bone
red marrow
yellow marrow
calcium phosphate
Haversian system
osteon
central canal
lamella (-ae)
lacuna (-ae)
osteocytes
canaliculi
axial skeleton
appendicular skeleton
cranium
vertebral column
rib cage
sternum
pelvic girdle
pectoral girdle
joint
suture
ligament
synovial joint
gliding joint
pivotal joint
ball-and-socket joint
hinge
joint
maxilla
foramen (foramina)
foramen magnum
vertebra
coccyx
intervertebral disks
fibrocartilage
spinal cord
neural canal
carapace
keel
sacroiliac joint
sacrum
ilium
scapula (-ae)
clavicle
digit
humerus
radius
ulna
femur
tibia
fibula
patella
smooth muscle
cardiac muscle
skeletal muscle
muscle fibers
involuntary muscle
visceral muscle
autonomic nervous system
intercalated disks
striated muscle
voluntary muscle
somatic nervous system
fascicles
fascia
sphincters

aponeuroses
sarcolemma
motor neuron
neuromuscular junction
sarcoplasmic reticulum
transverse tubules
T-tubules
myofibril
fibril
myofilament
myosin
actin
Z line
sarcomere
I band
A band
H zone
myosin head
ATPase
tropomyosin
troponin
actin-myosin binding site
calcium ion channel

Exercises

1. Please provide labels for these diagrams of bone, using the terms listed below. Check your work with Figure 33.12 in the text.

blood vessels
cartilage
compact bone
cross canal
Haversian system
Haversian canal
lamellae
medullary cavity
periosteum
red marrow
spongy bone
yellow marrow

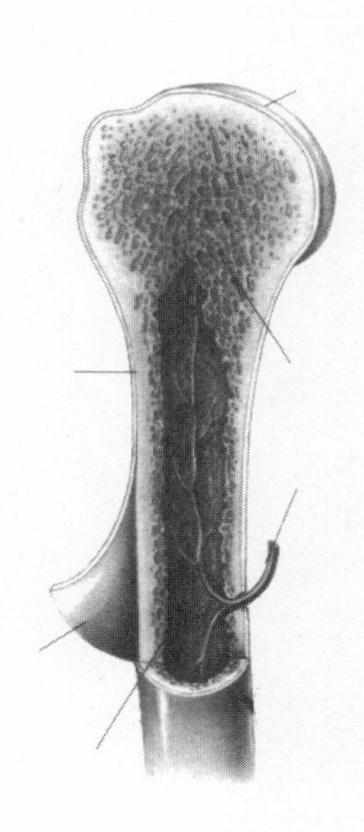

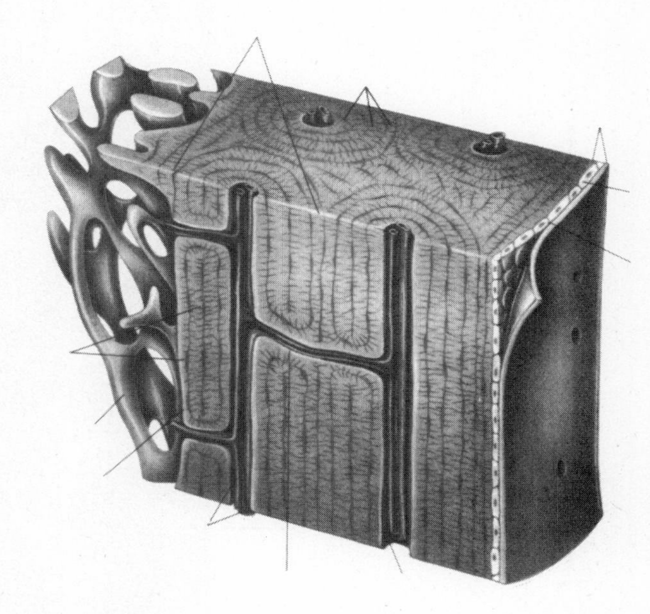

2. Below is a diagram of a muscle fiber at the myofibril level. Please label the various parts of this diagram using the following terms. Check your work with Figure 33.20 in the text.

A band
actin
H zone
I band
myosin
sarcomere
Z line

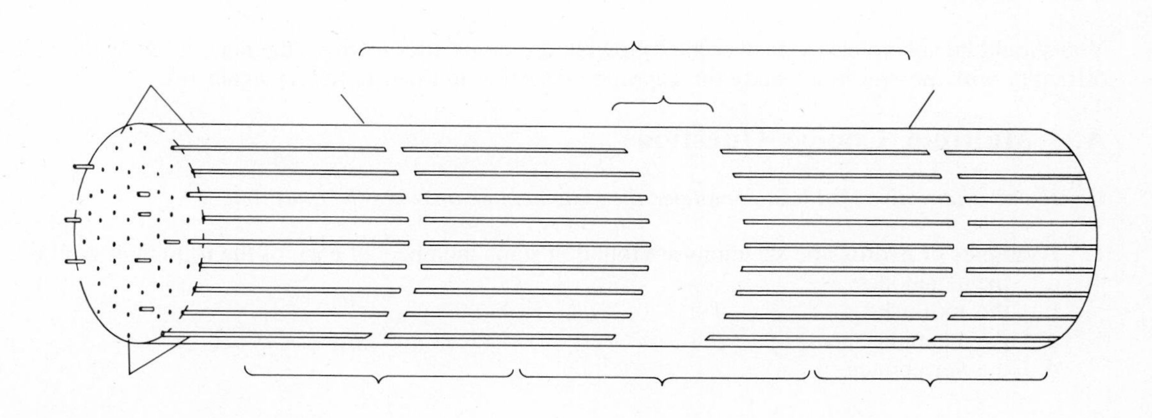

3. The major bones of the human skeleton are illustrated below. Provide the correct name for each of the bones indicated. Check your work with Figure 33.13 in the text.

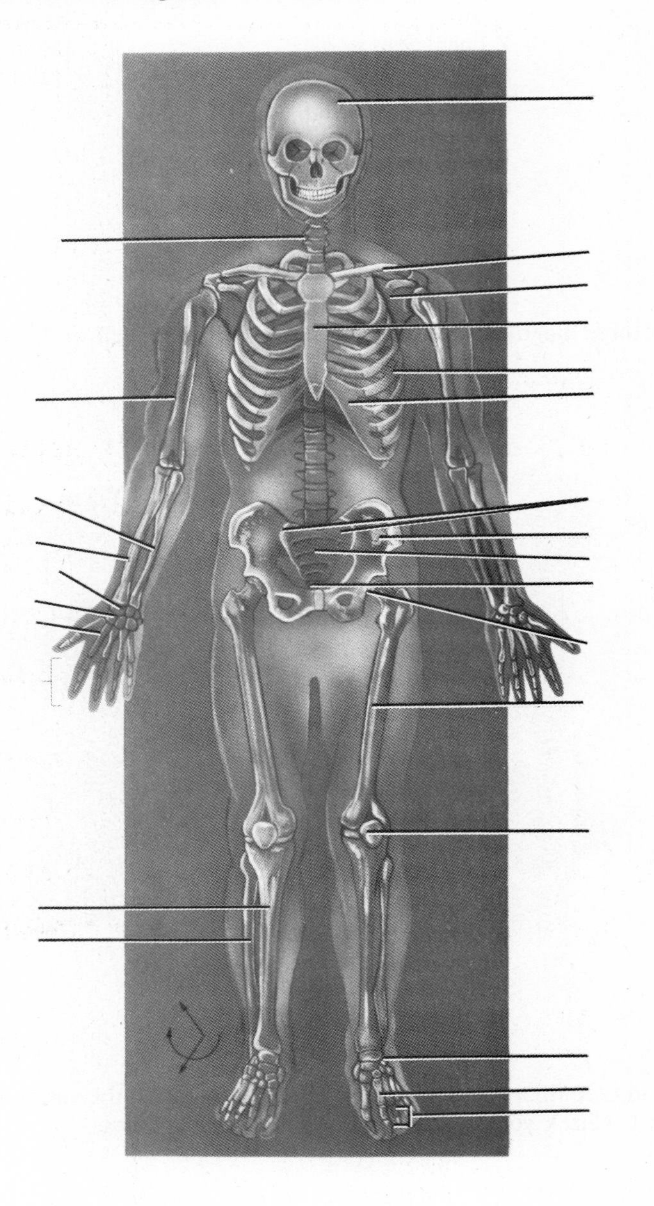

Self-Exam

You should be able to easily answer the following questions after learning the material in chapter 33. If you have difficulty with any question, study the appropriate section in the text and try again.

A. Multiple Choice Questions

Circle one alternative that best completes the statement or answers the question.

1. Examples of hydrostatic skeletons are found in some members of each of the following groups *except*
   a. the nematodes.
   b. the mollusks.
   c. the annelids.
   d. the vertebrates.

2. A nematode's body shape is maintained by pressure exerted by the cuticle against
   a. non-compressible fluids within.
   b. a chitonous exoskeleton.
   c. a bony endoskeleton.
   d. a cartilaginous skeleton.

3. The musculature of earthworm is best described as
   a. one layer of longitudinal muscles.
   b. one layer of circular muscles.
   c. an inner longitudinal layer and an outer circular layer.
   d. an inner circular layer and an outer longitudinal layer.

4. Exoskeletons are typically form by secretions from
   a. cells derived from ectoderm.
   b. cells derived from mesoderm.
   c. cells derived from endoderm.
   d. cells derived from gastroderm.

5. Endoskeletons in vertebrates are formed by
   a. cells derived from ectoderm.
   b. cells derived from mesoderm.
   c. cells derived from endoderm.
   d. cells derived from gastroderm.

6. The origin of a muscle refers to which part of a muscle?
   a. The end attached to a more stationary base.
   b. The end attached to a more moveable part of the skeleton.
   c. The central part of the muscle.
   d. The membrane surrounding the bundles of muscle fibers.

7. The shells of mollusks are secreted by
   a. epidermal tissue.
   b. endodermal tissue.
   c. mesodermal tissue.
   d. gastrodermal tissue.

8. Endoskeletons are found in which invertebrate group?
   a. annelids.
   b. arthropods.
   c. echinoderms.
   d. amphibians.

9. Which of the following tissue types is *not* an example of a connective tissue?
   a. bone.
   b. cartilage.
   c. blood.
   d. muscle.

10. Yellow marrow is often found
    a. in fatty tissue.
    b. in spongy bone tissue.
    c. in the central cavity of compact bone.
    d. surrounding cartilage.

11. The regions of concentric rings in calcified bone are termed
    a. lacunae.
    b. lamellae.

c. osteons.
d. canaliculi.

12. The axial skeleton is made up of
    a. the cranium, the vertebral column, the rib cage, and the sternum.
    b. the cranium, the vertebral column, the pelvic girdle, and the pectoral girdle.
    c. the limbs, the rib cage, the pelvic girdle, and the pectoral girdle.
    d. the limbs, the pelvic girdle, and the pectoral girdle.

13. Ligaments are composed of fibers of
    a. elastin.
    b. collagen
    c. myosin.
    d. myofibrils.

14. The human shoulder joint is an example of which type of joint?
    a. a gliding joint.
    b. a pivotal joint.
    c. a ball-and-socket joint.
    d. a hinge joint.

15. Which vertebrate structure has an important role in supporting the body?
    a. spinal cord.
    b. neural canal.
    c. vertebral column.
    d. notochord.

16. Which are the bones of the mammalian forelimb?
    a. humerus, radius, and ulna.
    b. humerus, tibia, and fibula.
    c. femur, radius, and ulna.
    d. femur, tibia, and fibula.

17. Which of the following is under control of the somatic nervous system?
    a. skeletal muscle.
    b. smooth muscle.
    c. involuntary muscle.
    d. visceral muscle.

18. In which of the following does smooth muscle characteristically occur?
    a. the muscular walls of the heart.
    b. the muscles of the tongue.
    c. the walls of blood vessels.
    d. the muscles which move bones.

19. Smooth muscle is typically under the control of
    a. the somatic nervous system.
    b. the autonomic nervous system.
    c. the automatic nervous system.
    d. the voluntary nervous system.

20. In skeletal muscle tissue, each cell is surrounded by
    a. the sarcoplasmic reticulum.
    b. the sarcomere.
    c. the sarcolemma.
    d. the neuromuscular junction.

21. If we examine myofibril ultrastructure, the light zone called the I band
    a. consists of actin only.
    b. is overlapping actin and myosin.
    c. consists of myosin only.
    d. is the boundary between sarcomeres.

22. Which of the following muscle proteins is characterized as a globular protein?
    a. myosin.
    b. actin.
    c. tropomyosin.
    d. troponin.

23. When a neural impulse reaches a neuromuscular junction, it initiates an electrical disturbance that
    a. causes calcium ion channels in the sarcoplasmic reticulum to close.
    b. causes calcium ion channels in the sarcoplasmic reticulum to open.
    c. causes calcium ion channels in the sarcolemma to close.
    d. causes calcium ion channels in the sarcolemma to open.

B. True or False Questions

Mark the following statements either T (True) or F (False).

_____ 24. The shell of a clam is an example of a hydrostatic skeleton.

_____ 25. In an earthworm, each segment has its own bundle of circular muscles.

_____ 26. It is likely that the first type of animal skeleton was the hydrostatic skeleton.

_____ 27. Hard skeletons function only in support; they are not involved in movement.

_____ 28. In animals, some muscles pull while other push.

_____ 29. Hardened, jointed exoskeletons are characteristic of the phylum Mollusca.

_____ 30. Arthropod skeletons are essentially just hollow, thin-walled, jointed tubes.

_____ 31. The small midge insects have been observed to beat their wings at more than 1,000 beats per second.

_____ 32. In general the shell of a marine snail is much thinner and lighter than that of a land snail.

_____ 33. Blood is classified as a connective tissue; the matrix of blood is plasma.

_____ 34. In general, compact bone is not considered to be a living tissue.

_____ 35. Unlike humans, in fish both the top and bottoms jaws move.

_____ 36. Snakes, like all other vertebrates, have exactly 33 vertebrae in their vertebral columns.

_____ 37. Among humans, males have 11 pairs of ribs while females have 12 pairs of ribs.

_____ 38. Smooth muscle tends to occur in sheets rather than in the dense bundles seen in cardiac and skeletal muscle.

_____ 39. Cardiac muscle cells, like smooth muscle cells are typically multinucleate.

_____ 40. In vertebrates, all skeletal muscles move bones.

_____ 41. When skeletal muscle is fully contracted, the H zone tends to disappear.

_____ 42. The thin actin filaments consist of three different protein components.

_____ 43. A muscle cell will contract only when calcium ions are absent from the contractile unit.

C. Fill in the Blanks

Answer the question or complete the statement by filling in the blanks with the correct word or words.

44. Soft-bodied animals are often supported by a _________________ skeleton.

45. Opposing muscles are termed _________________.

46. In insects, the _________________ helps retard water loss in terrestrial species.

47. The leaping ability of insects like grasshoppers is based on a _________________ mechanism.

48. Some cephalopod mollusks have a specialized internalized shell, the _________________, which is used to modify the animal's buoyancy.

49. Collagen is produced and secreted by cells known as _________________.

50. A major component of elastic fibers in connective tissue is the protein _________________.

51. The thin membrane of connective tissue that surrounds long bones of the limbs is the _________________.

52. The _________________ are repeating units of bone structure consisting of a central canal and lamellae.

53. Where the ligaments form an enclosing capsule they produce a _________________ joint.

54. The largest opening in the vertebrate skull is the _________________.

55. The pectoral girdle consist of two _________________ and two _________________.

56. The cells of _________________ muscle have single nuclei and lack striations.

57. Skeletal muscles are organized into bundles or _________________ surrounded by a sheath of connective tissue.

58. In skeletal muscle, the region between two Z lines is known as a _________________.

59. In resting muscle, calcium ions are stored in the _________________.

60. Tropomyosin and troponin control muscle contraction because they block the _________________.

Questions for Discussion

1. In general we see exoskeletons more commonly with small animals, while large animals typically have endoskeletons. Why would an exoskeleton not work well for animals as large as some large mammals?

2. A skeleton is important for supporting the body of many vertebrates. Discuss the additional ways that a bony skeleton is important for vertebrates.

3. Outline the three types of muscle found in vertebrates. Compare their structures at gross and ultrastructural levels, and specify where each type occurs.

4. Discuss the mechanism of neural stimulation of skeletal muscle, and describe the role of calcium ions in contraction.

Answers to Self-Exam

Multiple Choice Questions

1. d
2. a
3. c
4. a
5. b
6. a
7. a
8. c
9. d
10. c
11. b
12. a
13. b
14. c
15. c
16. a
17. a
18. c
19. b
20. c
21. a
22. d
23. b

True or False Questions

24. F
25. T
26. T
27. F
28. F
29. F
30. T
31. T
32. F
33. T
34. F
35. T
36. F
37. F
38. T
39. F
40. F
41. T
42. T
43. F

Fill in the Blank Questions

44. hydrostatic
45. antagonists
46. cuticle
47. "cock-and-release"
48. cuttlebone
49. fibroblasts
50. elastin
51. periosteum
52. Haversian systems
53. synovial
54. foramen magnum
55. scapulae, clavicles
56. smooth
57. fascicles
58. sarcomere
59. sarcoplasmic reticulum
60. actin-myosin binding sites

CHAPTER 34 Neural Control I: The Neuron

Learning Objectives

After mastering the material covered in Chapter 34, you should be able to confidently do the following tasks:

- Describe the basic structure of a neuron, and identify the different types of cells in the nervous system.
- Explain how an action potential is used to send a signal along a neuron.
- Show the role of ion channels and gates in the neuronal membrane in propagation of an action potential.
- Explain how myelinated neurons are able to conduct impulses faster than nonmyelinated neurons.
- Distinguish electrical and chemical synapses between neurons.
- Describe the events that occur as an impulse is transmitted from one neuron to another via the neurotransmitter acetylcholine.
- Give an illustration of a reflex arc in action.

Chapter Outline

I. The Neuron.

   A. Cellular structure.
   B. Cells of the nervous system.
      1. Nerves.

II. Neural Signals.

   A. The action potential.
      1. the resting state.
      2. generating action potentials.
      3. reestablishing the resting potential.
   B. Ion channels and gates.
      1. the sodium gates.
      2. the potassium gates.
   C. Myelin and impulse velocity.

III. Communication Among Neurons.

   A. Electrical synapses.
   B. Chemical synapses.

1. action at the synapse.
2. inhibitory synapses.
3. recovery at the synapse.

C. The reflex arc.

Key Words

neuron
detection
integration
response
stimulus
cell body
cell process
neurotransmitter
effector
dendrite
axon
action potential
hillock
axon tree
neuromuscular junction
myelin sheath
nodes of Ranvier
internode
Schwann cell
oligodendrocyte
neuroglia
glial cell
astrocyte
blood-brain barrier
sensory neuron
interneuron
motor neuron
afferent neuron
sensory receptor
peripheral axon
central axon
efferent neuron
motor impulse
nerve
graded signal
resting state
sodium/potassium ion exchange pump
resting potential
millivolt (mV)
depolarizing wave
all-or-none principle
threshold voltage
refractory period
absolute refractory period
relative refractory period
ion channel
ion gate
sodium ion gate
potassium ion gate
activation gate
inactivation gate
potassium leak channels
saltatory propagation
synapse
electrical synapse
chemical synapse
synaptic cleft
presynaptic neuron
postsynaptic neuron
axonal knobs
acetylcholine
acetylcholine receptor site
excitory synapse
inhibitory synapse
hyperpolarization
chloride ion channel
choline group
acetyl group
acetylcholinesterase
organophosphate pesticide
reflex arc
knee jerk reflex

Exercises

1. Please provide labels for the following diagram of a motor neuron, using terms provided below. Check your work with Figure 34.2 in the text.

axon
axonal tree
cell body
dendrites
hillock
internodes
myelin sheath
node
nucleus

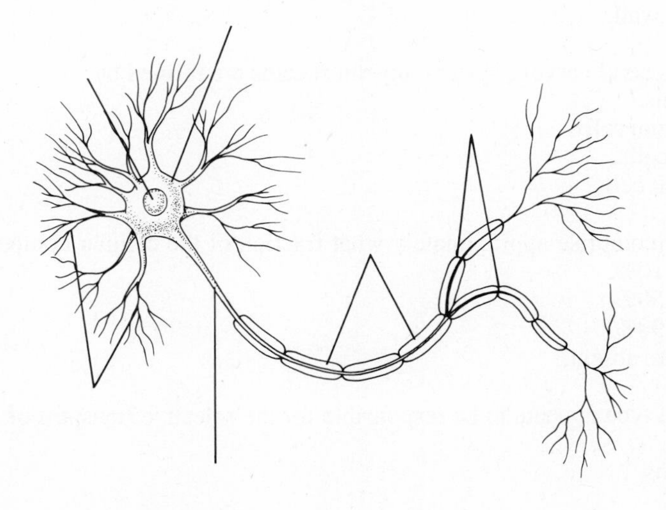

2. When a neuron "fires", the changes in polarization of an axonal membrane can be visualized by the tracing on an oscilloscope. On the axes below, plot the polarization and changes you might observe in an axon as an action potential is generated.. Begin at the resting potential and plot the entire event. Label the X-axis as time in milliseconds. Indicate the periods of Na^+ inflow and outflow. Show the absolute and relative refractory periods. Indicate the correct values for maximum and minimum membrane potentials.

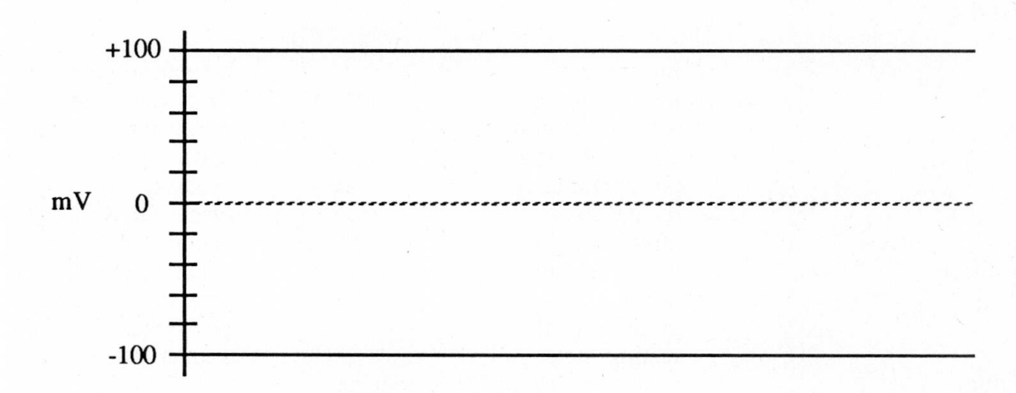

Self-Exam

You should be able to easily answer the following questions after learning the material in chapter 34. If you have difficulty with any question, study the appropriate section in the text and try again.

A. Multiple Choice Questions

Circle one alternative that best completes the statement or answers the question.

1. Which of the following is *not* one of the things accomplished by a functioning neural network?
   a. detection.
   b. integration.
   c. response.
   d. stimulation.

2. Neurons are cells whose structure typically includes all of the following *except*
   a. a cell body.
   b. dendrites.
   b. axons.
   d. a cell wall.

3. In the peripheral nervous system, myelin sheaths are formed by
   a. neurons.
   b. Schwann cells.
   c. glial cells.
   d. muscle cells.

4. Glial cells comprise approximately what fraction of the cellular component of the vertebrate brain?
   a. about 10%.
   b. about 50%.
   c. about 90%.
   d. they are absent.

5. Which cell type appears to be responsible for the selective transport of material across the blood-brain barrier?
   a. neurons.

b. astrocytes.
c. Schwann cells.l
d. smooth muscle cells of the capillaries.

6. Which type of cell is responsible for *integration* in the nervous system?
   a. sensory neurons.
   b. afferent neurons.
   c. interneurons.
   d. motor neurons.

7. Motor impulses are transmitted to effectors via
   a. afferent neurons.
   b. efferent neurons.
   c. sensory neurons.
   d. interneurons.

8. Which processes are involved in movement of ions across axonal membranes during generation of action potentials?
   a. active transport and diffusion.
   b. active transport, but not diffusion.
   c. diffusion, but not active transport.
   d. active transport and osmosis.

9. In a neuron during the resting state,
   a. there are more sodium ions outside the axonal membrane than inside.
   b. there are more sodium ions inside the axonal membrane than outside.
   c. there are more potassium ions outside the axonal membrane than inside.
   d. the potassium ions and sodium ions have the same distribution gradient.

10. Ion distribution in a resting axon is maintained by sodium/potassium ion exchange pumps which move
    a. sodium ions into the axon and potassium ions out of the axon.
    b. sodium ions out of the axon and potassium ions into the axon.
    c. both sodium and potassium ions into the axon.
    d. both sodium and potassium ions out of the axon.

11. The electrical charge difference across the axonal membrane during the resting state is called the resting potential, and the value of the resting potential is
    a. about -70 mV.
    b. about +70 mV.
    c. about -30 mV.
    d. about +30 mV.

12. The time period of an action potential's passing, from depolarization to repolarization, is about
    a. 2 seconds.
    b. 2/100 second.
    c. 2 milliseconds.
    d. 2 millionths of a second.

13. After a neuron "fires," the time period of repolarization is termed
    a. the polarization period.
    b. the repolarization period.
    c. the resting period.
    d. the refractory period.

14. When an axon is at resting potential,
    a. both the sodium activation gates and the sodium inactivation gates are open.
    b. the sodium activation gates are open and the sodium inactivation gates are closed.
    c. the sodium activation gates are closed and the sodium inactivation gates are open.

d. both the sodium activation gates and the sodium inactivation gates are closed.

15. Neurotransmitters move across synaptic clefts by
    a. active transport.
    b. diffusion.
    c. the sodium/potassium ion exchange pump.
    d. osmosis.

16. When an action potential reaches an axonal knob,
    a. the calcium ion channels open and allow calcium ions to diffuse inward.
    b. the potassium ion channels open and allow potassium ions to diffuse inward.
    c. the calcium ion channels close and prevent calcium ions from diffusing inward.
    d. the potassium ion channels close and prevent potassium ions from diffusing inward.

17. Which elements of the nervous system are involved in a knee-jerk reflex, at a minimum?
    a. sensory neurons, interneurons, and motor neurons.
    b. sensory neurons and motor neurons.
    c. sensory neurons and interneurons.
    d. interneurons and motor neurons.

B. True or False Questions

Mark the following statements either T (True) or F (False).

_____ 18. Neuron cell bodies must receive all their stimuli via dendrites; they cannot be stimulated directly.

_____ 19. Action potentials usually originate in the hillock region of a neuron.

_____ 20. A neuron often has many dendrites, but it usually has only one axon.

_____ 21. The messages from receptors responding to stimuli like touch, light, and heat are carried by efferent neurons.

_____ 22. Like action potentials, passive graded signals are quite suitable for long distance communication.

_____ 23. An action potential does not gain or lose in strength as it travels the length of a neuron.

_____ 24. Resting axons are more negative inside than outside.

_____ 25. At rest, an axon's plasma membrane is quite permeable to sodium ions.

_____ 26. The difference in membrane potential from the resting state to depolarization is about 100 mV.

_____ 27. In general, the stronger a stimulus is, the greater the magnitude of the action potential.

_____ 28. After an action potential passes, during repolarization potassium ion movement restores the membrane potential to -70 mV.

_____ 29. The maximum rate of firing for a neuron is limited by the refractory period for that neuron.

_____ 30. Sodium ion gates are voltage sensitive, while potassium ion gates are not voltage sensitive.

_____ 31. Myelin sheathes are produced by Schwann cells and by oligodendrocytes.

_____ 32. The giant axons of such invertebrates as squids, cockroaches, and crayfish are not myelinated.

_____ 33. A neuron typically has a single neuron, and therefore can synapse with only one subsequent neuron.

_____ 34. At a chemical synapse, an action potential can pass directly from one neuron to the next.

_____ 35. Neurotransmitters are often stored in vesicles located in the axonal knobs.

_____ 36. Neural impulses can be transmitted in either direction across a typical synapse.

_____ 37. The patellar response is a reflex arc involving just two neurons.

C. Fill in the Blanks

Answer the question or complete the statement by filling in the blanks with the correct word or words.

38. The chemical messengers that are ordinarily responsible for relaying impulses from one neuron to another are termed _________________.

39. The lengthy neuron processes specialized for transmitting signals over long distances are termed the _________________.

40. Some neurons stimulate responding organs termed _________________.

41. On myelinated axons, the myelin sheath is interrupted at intervals by non-myelinated _________________.

42. Many cells of the neural component of the brain and spinal cord are _________________, which communicate only with other neurons.

43. The ions most important to neural activity are _________________ ions and _________________ ions.

44. When the membrane potential of a neurons is about -70 mV, it is termed the _________________.

45. An _________________ begins with a sudden increase in permeability of an axon's plasma membrane to sodium ions.

46. Axons are governed by an "all-or-none" characteristic; a _________________ voltage must be attained to initiate an action potential.

47. New action potentials cannot be initiated along an axon during the absolute _________________.

48. A resting axon's membrane is impermeable to sodium ions because the _________________ gates are closed.

49. A neural impulse moves rapidly ("jumps")along a myelinated axon via _________________.

50. The points of communication between two neurons are called _________________.

51. _________________ synapses occur at gap junctions between neurons.

52. The space between neurons at a chemical synapse is termed the _________________.

53. The common neurotransmitter _________________ causes chemical gates to open and admit sodium ions into the postsynaptic neuron.

54. This common neurotransmitter is dismantled into choline and acetyl groups by the enzyme _________________.

55. The patellar response involves only two neurons; it is an example of a _________________.

Questions for Discussion

1. Describe the generation of an action potential along an axon. Be sure to include discussion of establishing the polarized, resting state, discussion of generating and propagating an action potential, and discussion of restoring to polarized resting state. Specify the value of the membrane potential during the complete "firing" cycle.

2. Discuss the role of ion channels and gates in resting potentials and action potentials. Relate how they work and how they are controlled during the resting state, during the propagation of an action potential, and during repolarization.

3. How does the neurotransmitter acetylcholine relay a neural impulse from one neuron to another? Describe the events in both the presynaptic neuron and the postsynaptic neuron. What is the role of acetylcholinesterase?

4. Draw a diagram of a simple reflex arc. Show the involvement of receptors, neurons, and effectors. Is the central nervous system involved at all? When a reflex arc reaction occurs, can the individual be aware of the reaction?

Answers to Self-Exam

Multiple Choice Questions

1. d
2. d
3. b
4. c
5. b
6. c
7. b
8. a
9. a
10. b
11. a
12. c
13. d
14. c
15. b
16. a
17. b

True or False Questions

18. F
19. T
20. T
21. F
22. F
23. T
24. T
25. F
26. T
27. F
28. T
29. T
30. F
31. T
32. T
33. F
34. F
35. T
36. F
37. T

Fill in the Blank Questions

38. neurotransmitters.
39. axons
40. effectors
41. nodes of Ranvier
42. interneurons
43. sodium, potassium
44. resting potential.
45. action potential
46. threshold
47. refractory period
48. sodium activation
49. saltatory propagation
50. synapses
51. Electrical
52. synaptic cleft
53. acetylcholine
54. acetylcholinesterase
55. reflex arc

CHAPTER 35 Neural Control II: Nervous Systems

Learning Objectives

After mastering the material covered in Chapter 35, you should be able to confidently do the following tasks:

- Contrast the nervous systems found in invertebrates with the highly organized vertebrate nervous system.

- Identify the major parts of the vertebrate central and peripheral nervous systems.

- Show the locations of the major potions of the human brain, and tell the functions associated with each.

- List the principal kinds of neurotransmitters in the human brain, and describe their functions.

- Explain how the sympathetic and parasympathetic divisions of the autonomic nervous system work to promote homeostasis.

- Contrast the sensory receptors in invertebrates with those of vertebrates.

- Explain how the receptors of the vertebrate retina respond to light and send visual information to the brain.

Chapter Outline

I. Invertebrate Nervous Systems.

    A. Simple nerve nets and ladders.
    B. Echinoderms and nerve rings.
    C. Massed ganglia to organized brain.
        1. cephalopods and the appearance of a brain.
        2. arthropods and a segmented nervous system.

II. Vertebrate Nervous Systems.

III. The Human Nervous System.

    A. The human brain.
        1. the cerebrum.
        2. the thalamus.
        3. the hypothalamus.
        4. the limbic system.
        5. the midbrain and hindbrain.
    B. The spinal cord.
    C. Chemicals in the brain.
    D. Electrical activity in the brain.

1. sleep

E. The peripheral nervous system.

1. the autonomic nervous system

IV. The Senses.

A. Neural codes.

B. Tactile reception.

1. touch in invertebrates.
2. touch in vertebrates.

C. Thermoreception.

1. heat receptors in invertebrates.
2. heat and cold in vertebrates.

D. Chemoreception.

1. chemical receptors in invertebrates.
2. chemical receptors in vertebrates.

E. Proprioception.

1. proprioceptors in invertebrates.
2. proprioceptors in vertebrates.

F. Auditory reception.

1. hearing in invertebrates.
2. hearing in vertebrates.

G. Gravity and movement.

H. Visual reception

1. vision in invertebrates.
2. vision in vertebrates: the eye and retina.

Key Words

central nervous system (CNS)
cephalization
nerve net
ganglionic masses
nerve ring
giant axons
cerebrum
cortex
brain
gray matter
white matter
meninges
cerebrospinal fluid
circle of Willis
blood brain barrier
astrocytes
forebrain
cerebrum
cerebral hemispheres
cerebral cortex
occipital lobe
temporal lobe
lateral fissure
frontal lobe
prefrontal area
parietal lobe
central fissure
gyrus (gyri)
handedness
aphasia
corpus callosum
language centers
Broca's area
Wernicke's area
angular gyrus
thalamus
third ventricle
nuclei
cerebellum
reticular system
hypothalamus
pituitary
limbic system
amygdala
hippocampus
midbrain
hindbrain
medulla oblongata
cerebellum
pons
cranial nerves
spinal cord
foramen magnum
vertebral canal
sensory root
motor root
acetylcholine
norepinephrine
monamines
dopamine
histamine
serotonin
neuropeptides
enkephalins
endorphins
opioid neurotransmitter
amphetamine
cocaine
LSD
mescaline
psilocybin
electroencephalograph
electroencephalogram
paradoxical sleep
REM sleep
peripheral nervous system (PNS)
somatic nervous system
motor division
sensory division
autonomic nervous system
homeostasis
sympathetic division
parasympathetic division

arrector pili muscles
cardioaccelerator
vagus nerve
"fight or flight" response
sympathetic ganglia
celiac plexus
solar plexus
sensory organs
transducers
receptor potential
tactile receptors
mechanoreceptors
distance receptors
lateral line organ
contact receptors
Pacinian corpuscles
Meissner's corpuscles
baroreceptor
piezoelectric crystal
thermoreception
pit viper
Ruffini's corpuscles
Krause's corpuscles
Krause's end bulbs
chemoreceptors
bombykol
sensillum
gustatory receptors
olfactory receptors
Jacobson's organ
olfactory epithelium
olfactory bulb
taste buds
proprioception
proprioceptors
audition
tympanal organs
inner ear
Weberian ossicles
auditory canal
middle ear
tympanic membrane
external ear
pinna
tympanum
middle ear
malleus
incus
stapes
cochlea
vestibular apparatus
oval window
round window
organ of Corti
basilar membrane
hair cells
tectorial membrane
statocysts
semicircular canal
saccule
utricle
vision
simple eye
compound eye
ommatidium (-ia)
eyeball
sclera
cornea
choroid
iris
pupil
lens
retina
anterior chamber
aqueous humor
posterior chamber
vitreous humor
ciliary muscles
rods
cones
bipolar cells
ganglion cells
optic nerve
night vision
fovea centralis
rhodopsin
opsin
vitamin A
retinal

Exercises

1. Provide labels for this diagrams of the human ear, using the terms provided. Check your work with Figure 35.26 in the text.

auditory canal
cochlea
eustachian tube
incus
lateral semicircular canal
malleus
oval window
pinna
posterior semicircular canal
round window
stapes
superior semicircular canal
tympanic membrane
vestibule

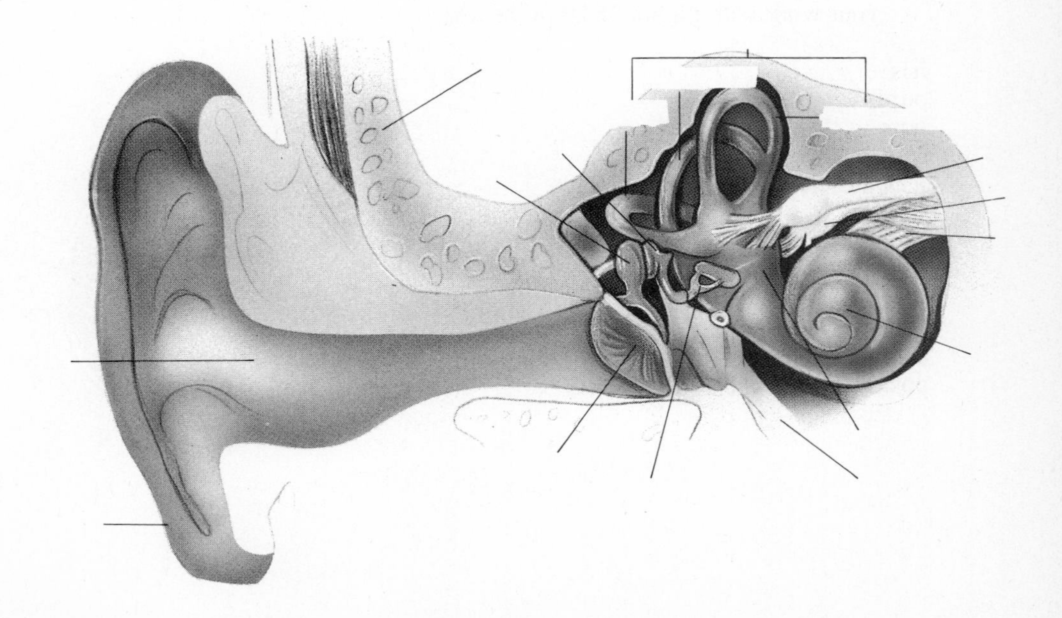

2. Many organs are affected by the autonomic nervous system. Often, the effects on an organ by the sympathetic and parasympathetic divisions are opposite. Complete the below chart to show the main effects of these two parts of the autonomic nervous system on a variety of major organs.

Organ	Sympathetic Effect	Parasympathetic Effect
Heart	accelerates heart rate	
Blood vessels of heart		
Lungs		
Blood vessels of respiratory system		
Stomach glands		
Blood vessels of stomach		
Intestine		
Salivary glands		
Pupil of eye		
Rate of metabolism		

3. Please provide labels for the following diagram of a human eye. Use the terms in the list provided. Check your work with Figure 35.29 in the text.

iris
pupil
lens
cornea
lens muscle
sclera
choroid coat
retina
fovea
optic nerve
anterior chamber
posterior chamber

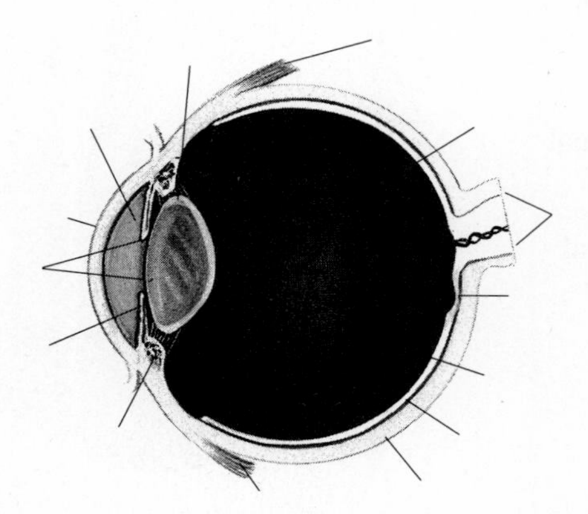

4. This diagrammatic section of the human brain shows the major components of the limbic system, as well as the cerebrum. Provide labels for this diagram using the terms listed below. Also, indicate the relative locations of the frontal, parietal, occipital, and temporal lobes of the cerebrum. Check your work with Figure 35.15 in the text.

amygdala
cerebellum
corpus callosum
hippocampus
hypothalamus
medulla oblongata
pituitary
pons
spinal cord
thalamus

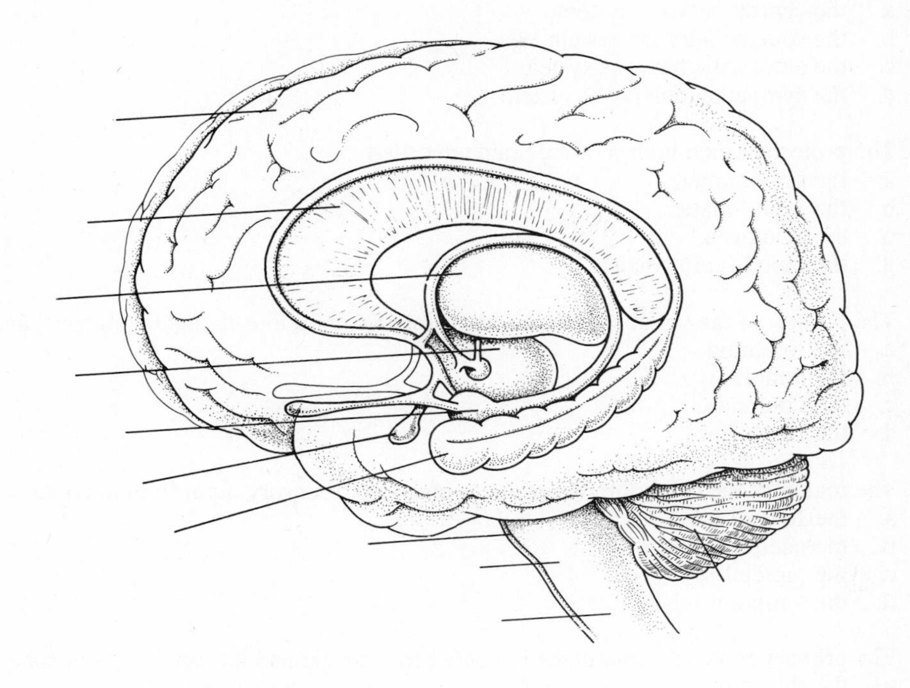

Self-Exam

You should be able to easily answer the following questions after learning the material in Chapter 35. If you have difficulty with any question, study the appropriate section in the text and try again.

A. Multiple Choice Questions

Circle one alternative that best completes the statement or answers the question.

1. The organized neurons of animals belonging to phylum Cnidaria form a
   a. nerve net.
   b. nerve ladder.
   c. nerve ring.
   d. central nervous system.

2. Seastars are echinoderms. The principal part of the seastar nervous system includes a
   a. nerve net.
   b. nerve ladder.
   c. nerve ring.
   d. central nervous system.

3. Which statement best describes the brain of an octopus?
   a. It is a nerve net
   b. It is a ganglionic mass of tissue around its esophagus, showing differentiation.
   c. It has the typical vertebrate cerebrum and cerebellum.
   d. It is largely a single giant axon.

4. The forebrain, midbrain, and hindbrain are characteristic parts of
   a. the Cnidarian brain.
   b. the insect brain.

c. the cephalopod brain.
d. the vertebrate brain.

5. The spinal cord of vertebrates is considered to be a major part of
   a. the central nervous system.
   b. the somatic nervous system.
   c. the autonomic nervous system.
   d. the sympathetic nervous system.

6. The protective membranes of the brain are called
   a. the gray matter.
   b. the white matter.
   c. the meninges.
   d. the blood-brain barrier.

7. The portion of the vertebrate brain responsible for conscious thought, language, and sensory reception is
   a. the forebrain.
   b. the midbrain.
   c. the hindbrain.
   d. the cerebellum.

8. The region of the cerebrum that receives raw, visual sensory input from the optic nerve is contained in
   a. the frontal lobe.
   b. the occipital lobe.
   c. the parietal lobe,
   d. the temporal lobe.

9. The primary route of communication between the right and left cerebral hemisphere is via
   a. the thalamus.
   b. the speech center.
   c. the gyrus.
   d. the corpus callosum.

10. Major components of the limbic system include
    a. the amygdala, hippocampus, thalamus, and hypothalamus.
    b. the cerebrum and the cerebellum.
    c. the frontal lobe, parietal lobe, occipital lobe, and temporal lobe.
    d. the pons, the medulla oblongata, and the cerebellum.

11. The pituitary is an endocrine gland; much of its hormone secretion is prompted by
    a. the thalamus.
    b. the hypothalamus.
    c. the corpus callosum.
    d. the hippocampus.

12. In humans, the hindbrain consists of
    a. the amygdala, hippocampus, thalamus, and hypothalamus.
    b. the cerebrum and the cerebellum.
    c. the frontal lobe, parietal lobe, occipital lobe, and temporal lobe.
    d. the pons, the medulla oblongata, and the cerebellum.

13. All voluntary movement in the limbs and body of humans is coordinated in
    a. the cerebrum.
    b. the midbrain.
    c. the corpus callosum.
    d. the cerebellum.

14. The spinal cord emerges from the skull through

a. the vertebral canal.
b. the foramen magnum.
c. the motor root.
d. the sensory root.

15. Familiar examples of neuropeptides are
a. acetylcholine and norepinephrine.
b. dopamine and serotonin.
c. enkephalins and endorphins.
d. amphetamines and cocaine.

16. The so-called "fight or flight" response in an emergency is instigated by
a. the sympathetic nervous system.
b. the parasympathetic nervous system.
c. the somatic nervous system.
d. the automatic nervous system.

17. In humans and other mammals, the pressure and touch receptors are encapsulated nerve endings called
a. Ruffini's corpuscles and Krause's corpuscles.
b. Pacinian corpuscles and Meissner's corpuscles.
c. Jacobson's organs and the olfactory epithelium.
d. the cochlea and the organ of Corti.

18. Among the sensory structures responding to odor in vertebrates are
a. Ruffini's corpuscles and Krause's corpuscles.
b. Pacinian corpuscles and Meissner's corpuscles.
c. Jacobson's organs and the olfactory epithelium.
d. the cochlea and the organ of Corti.

19. Which of the four basic tastes is often characteristic of poisonous alkaloids?
a. sweet.
b. sour.
c. salty.
d. bitter.

20. In the mammalian ear, the malleus, incus, and stapes bones are located
a. in the outer ear.
b. in the middle ear.
c. in the inner ear.
d. in the vestibular apparatus.

21. Many invertebrates detect body position through sensory organs called
a. semicircular canals.
b. statocysts.
c. organs of Corti.
d. ommatidia.

22. Which animals have image-forming eyes much like those of vertebrates?
a. flatworms.
b. arthropods, such as insects and crustaceans.
c. echinoderms, such as sea stars.
d. cephalopods, such as the octopus.

23. As light enters the human eye, it passes through, in order
a. the cornea, aqueous humor, lens, and vitreous humor.
b. the lens, cornea, vitreous humor, and aqueous humor.
c. the lens, cornea, aqueous humor, and vitreous humor.
d. the cornea, vitreous humor, lens, and aqueous humor.

24. In the human eye, the photoreceptors are
    a. the bipolar cells.
    b. the ganglion cells.
    c. the rods and cones.
    d. the optic nerves.

25. In retinal rods, the photoactive pigment that breaks down in the presence of light is
    a. rhodopsin.
    b. retinal.
    c. opsin.
    d. vitamin A.

B. True or False Questions

Mark the following statements either T (True) or F (False).

_____ 26. In a highly visual insect species like the housefly, visual centers may occupy as much as 80% of the brain.

_____ 27. In vertebrates, the cerebrum is part of the midbrain.

_____ 28. The autonomic nervous system has motor and sensory divisions.

_____ 29. The white matter of the human brain represents myelinated axons.

_____ 30. The brain utilizes the majority of the body's intake of the sugar glucose.

_____ 31. A monkey has a more highly developed cerebrum than does a rat.

_____ 32. In humans, the occipital lobe is located at the front of the cerebrum.

_____ 33. Species such as rats and parrots show left- and right-handedness.

_____ 34. The human brain's "great relay station' through which most sensory input must pass is the thalamus.

_____ 35. The amygdala can produce docility if it is stimulated and rage if it is removed.

_____ 36. Cranial nerves are large nerves that emerge directly from the brain, rather than from the spinal cord.

_____ 37. The cell bodies of motor neurons are clumped in large ganglia just outside the spinal cord.

_____ 38. Acetylcholine belongs to the neuropeptide class of neurotransmitters.

_____ 39. During REM sleep, an electroencephalogram recording is similar to that during wakefulness.

_____ 40. As a generalization, we can state that the somatic nervous system promotes homeostasis.

_____ 41. Very often, sympathetic neurons secrete norepinephrine and parasympathetic neurons secrete acetylcholine.

_____ 42. Receptor potentials, like action potentials, show an "all-or-none" behavior when stimulated.

_____ 43. A baroreceptor can respond to blood pressure within vessels.

_____ 44. Pit vipers can detect prey location by sensing the prey's metabolic heat.

_____45. In land vertebrates, the keenest olfactory senses are found among the birds.

_____46. Insects are unable to detect sound.

_____47. Humans detect differences in the pitch (highness or lowness) of sounds by the number of auditory neurons that fire.

_____48. Some animals can detect ultraviolet light visually.

_____49. In the human retina, the rods and cones are beneath the bipolar cell layer and the ganglion cell layer.

_____50. In humans, night blindness can result from a deficiency of dietary vitamin A.

C. Fill in the Blanks

Answer the question or complete the statement by filling in the blanks with the correct word or words.

51. The evolutionary trend toward development of the anterior end of organisms as sites of sensory receptors and brain development is termed ________________.

52. The principal part of the nervous system of the seastar includes a ________________ around the mouth.

53. The wrinkled outer region of the mammalian cerebrum is the ________________.

54. The brain and spinal cord together make up the ________________ nervous system.

55. The human brain contains many more of the supporting ________________ cells than it has neurons.

56. Many toxic substances are unable to reach brain cells from the bloodstream due to the ________________.

57. Broca's area, Wernicke's area, and the angular gyrus are all important to human ________________ ability.

58. The ________________ is located at the base of the human forebrain, and is sometimes called the brain's "great relay station."

59. The human brain's ________________ has a major function of coordination the nervous system with the endocrine system.

60. Large nerves that emerge directly from the brain are termed the ________________.

61. The cell bodies of ________________ neurons are found in large ganglia just outside the spinal cord.

62. Enkephalins and endorphins belong to the ________________ class of neurotransmitters.

63. The record of electrical activity in the brain recorded via electrodes is called a ________________.

64. The rate of the human heartbeat will speed up when the heart is stimulated by a ________________ nerve.

65. A special kind of mechanoreceptor activated by blood pressure within vessels is a ________________.

66. A silk moth "smells" with its ________________.

67. The olfactory receptors in the roofs of the mouths of some vertebrates are the ________________.

68. The inner ear of mammals consists of the ________________ and the ________________.

69. In insects, the compound eye is made up of a number of visual units, the ________________.

70. In the human eye, vision is most acute in the ________________, which contains a concentration of cone cells.

Questions for Discussion

1. In outline form, list the major portions and subdivisions of the human brain. Identify the known functions for each part.

2. Contrast the types of nervous system organization found in invertebrates with the basic vertebrate nervous system. The vertebrate nervous system is often considered to be more complex or more organized than most invertebrate nervous systems. Why might this be?

3. What is the major role of the autonomic nervous system in humans? What are its two divisions? What are the typical neurotransmitters of each division. Compare the main effects of these divisions on the major organ systems.

4. Consider the reaction when a person touches something very hot, draws his hand away, and comes to realize what has happened. Trace the neural events from stimulus to reaction to awareness and pain. Identify each portion of the nervous system involved.

5. How does the human eye detect light and convert the sensory stimulus into a neural message? Outline the photochemical events of vision, and trace the neural pathway from receptor cell to the brain.

Multiple Choice Questions

1. a
2. c
3. b
4. d
5. a
6. c
7. a
8. b
9. d
10. a
11. b
12. d
13. d
14. b
15. c
16. a
17. b
18. c
19. d
20. b
21. b
22. d
23. a
24. c
25. a

True or False Questions

26. T
27. F
28. F
29. T
30. T
31. T
32. F
33. T
34. T
35. F
36. T
37. F
38. F
39. T
40. F
41. T
42. F
43. T
44. T
45. F
46. F
47. F
48. T
49. T
50. T

Fill in the Blank Questions

51. cephalization
52. nerve ring
53. cerebral cortex
54. central
55. glial
56. blood-brain barrier
57. language
58. thalamus
59. hypothalamus
60. cranial nerves
61. motor
62. neuropeptide
63. electroencephalogram
64. sympathetic
65. baroreceptor
66. antennae
67. Jacobson's organs
68. cochlea, vestibular apparatus
69. ommatidia
70. fovea centralis

CHAPTER 36 Thermoregulation, Osmoregulation, and Excretion

Learning Objectives

After mastering the material covered in Chapter 36, you should be able to confidently do the following tasks:

- Explain the meaning of homeostasis, tell how positive and negative feedback loops work, and give appropriate examples.

- Contrast endothermic and ectothermic pathways of thermoregulation.

- Provide examples of mechanisms of thermoregulation in invertebrates and in vertebrate organisms.

- Explain the problems of osmoregulation encountered in marine, freshwater, and terrestrial habitats.

- Diagram the anatomy of the human excretory system.

- Outline the functional aspect of the human kidney, relating structure to function.

- Describe the mechanisms of control for the human kidney.

Chapter Outline

I. Homeostasis.

A. Negative and positive feedback loops.

II. Thermoregulation.

A. Why thermoregulate?
B. Categories of thermoregulation.
C. Endothermic regulation.
1. endothermy in cold-adapted moths.
2. endothermy in the bluefin tuna.
3. thermoregulation in birds and mammals.
4. behavioral adaptations in endotherms.
5. internal thermoregulation.
D. Ectothermic regulation.
1. some cold adaptations.
2. behavioral adaptations in ectotherms.

III. Osmoregulation and Excretion.

A. Producing nitrogen wastes.

B. The osmotic environment.
   1. the marine environment.
   2. the freshwater environment.
   3. the terrestrial environment.

IV. The Human Excretory System.

A. Anatomy of the human excretory system.
B. Microanatomy of the nephron.
C. The work of the nephron.
   1. Bowman's capsule.
   2. the proximal convoluted tubule.
   3. the loop of Henle.
   4. the distal convoluted tubule and collecting ducts.
   5. tubular secretion.
D. Control of nephron function.

Key Words

homeostasis
feedback
negative feedback loop
positive feedback loop
homeotherms
poikilotherms
"warm-blooded"
"cold-blooded"
heterotherms
endothermy
thermiogenesis
ectothermy
core temperature
hibernate
owlet moth
countercurrent heat exchanger
bluefin tuna
rete mirable
metabolic rate
shrew
oxidative respiration
shivering
brown fat
frostbite
blubber
underhair
guard hair
radiation
conduction
convection
pant
gular flutter
adrenal gland
epinephrine
pituitary gland
thyroid-stimulating hormone
thyroid gland
thyroxin
cryogenic surgery
Bracon cephi
glycerol
osmoregulation
excretion
deamination
amine group
NH_2
ammonia
NH_3
ammonium ion
NH_4^+
uric acid
urea
hyperosmotic
hypoosmotic
isoosmotic
osmoconformer
osmoregulator
Artemia salina
flame cell system
protonephridium
flame bulb
nephridium (-ia)
nephridiopore
Malpighian tubule
kidney
ureter
urinary bladder
urethra
renal circuit
renal arteries
renal veins
renal cortex
renal medulla
renal pelvis
nephron
collecting ducts
pyramids
calyx (calyces)
urine
Bowman's capsule
glomerulus
renal artery
afferent arteriole
efferent arteriole
peritubular capillaries
renal vein
proximal convoluted tubule
loop of Henle
distal convoluted tubule
juxtaglomerular complex
force filtration
tubular secretion
antidiuretic hormone (ADH)
diuretic
antidiuretic
aldosterone
renin

Exercises

1. Provide labels for the below diagram of a nephron. Use the terms provided in the list. Show which portions of the nephron are in the renal cortex, and which are in the medulla. Check your work with Figure 36.22 in the text.

afferent arteriole
Bowman's capsule
capillary bed
collecting duct
distal convoluted tubule
efferent arteriole
glomerulus
juxtaglomerular complex
loop of Henle
peritubular capillaries
proximal convoluted tubule

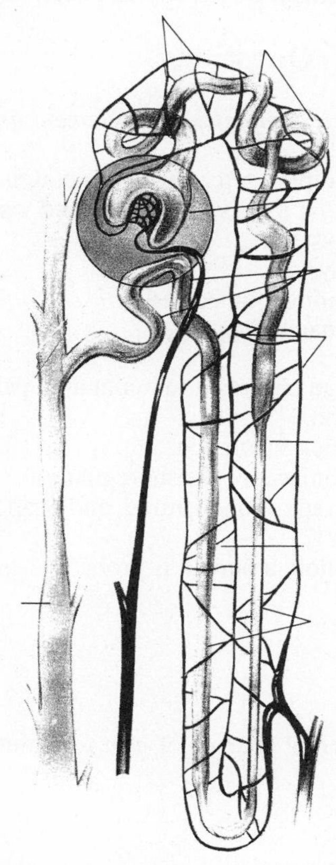

2. Complete the table below to summarize the functional performance of the human kidney's nephrons. For each portion of the nephron, give the process that takes place there and list the substance that move in or out of the bloodstream or the filtrate. Check your work with the appropriate sections of Chapter 36.

Nephron Region	Process Occurring	Substances Moving In or Out
Bowman's capsule	force filtration	
Proximal convoluted tubule		
Descending loop of Henle		
Ascending loop of Henle		
Distal convoluted tubule		
Collecting duct		

Self-Exam

You should be able to easily answer the following questions after learning the material in Chapter 36. If you have difficulty with any question, study the appropriate section in the text and try again.

A. Multiple Choice Questions

Circle one alternative that best completes the statement or answers the question.

1. Consider this homey example: after flushing a toilet, the tank inlet valve opens and water flows into the tank until it is refilled. Then the inlet valve closes and water flow stops. The valve control system illustrates
   a. a positive feedback system.
   b. a negative feedback system.
   c. a mechanism for avoiding homeostasis.
   d. an analogy of the human kidney.

2. In general, the ability of an organism to maintain a particular body temperature is
   a. due to osmoregulation.
   b. always a result of thick insulation.
   c. due to various mechanisms of thermoregulation.
   d. always a result of perspiration, panting, and evaporation.

3. In terms of thermoregulation, a human is probably best described as
   a. a heterotherm.
   b. a poikilotherm.
   c. cold-blooded.
   d. warm-blooded.

4. Animals who utilize external sources of heat to maintain body temperature are
   a. endothermic.
   b. ectothermic.
   c. thermiogenic.
   d. always cold-blooded.

5. For a homeotherm, the temperature is usually most constant
   a. deep within the body.
   b. in the extremities, like the limbs.
   c. in the surface layers of the trunk.
   d. in all body parts uniformly; they are "warm-blooded."

6. Which of the following statements is most generally true concerning endothermic animals?
   a. Because of heat loss constraints, endothermy tends to limit how small an animal can be.
   b. Because of heat loss constraints, endothermy tends to limit how large an animal can be.
   c. Across many species, endothermic animals have a very constant surface area to volume ratio.
   d. The rate of heat loss from very large animals is proportionately much greater than from small animals.

7. Which statement best describes heat maintenance in owlet moths of the family Noctuidae?
   a. They are exclusively ectothermic.
   b. They are warm-blooded.
   c. They are heterothermic, being endothermic only at certain times.
   d. They are exclusively endothermic.

8. The bluefin tuna is a cold-adapted fish. Which statement is most accurate about bluefins?
   a. Bluefins are cold-adapted by being cold-blooded.
   b. Bluefins use a countercurrent heat exchanger to rid their bodies of excess heat.
   c. Bluefins use a countercurrent heat exchanger to retain body heat and keep the body core warm.
   d. Bluefins conserve energy by swimming very slowly.

9. Which of the following mammals would have the highest metabolic rate?
   a. an elephant.
   b. a human living in the arctic.
   c. a human living in the tropical rain forest.
   d. a shrew.

10. Which of the following is *not* considered to be a mechanism of thermiogenesis?
    a. basking in the sun.
    b. oxidative respiration of carbohydrates.
    c. shivering.
    d. utilization of brown fat.

11. Why do anthropologists believe that humans have so little body hair which might aid in heat retention?
    a. Most humans have a good layer of blubber.
    b. Humans rely on brown fat for heat production, rather than hair to retain heat.
    c. Human skin has evolved to retain as much heat as a furry coat might retain.
    d. Humans probably evolved in hot climates where a furry coat might have retained too much heat.

12. In birds, the gular flutter is involved in accelerating
    a. heat loss from their wings as they flutter up and down.
    b. evaporative cooling in their mouths.
    c. the take-off speed, as they embark on cooling flights.
    d. the cooling of aquatic birds, as they flutter their webbed feet in cooler water.

13. In terms of internal thermoregulation, the heat-monitoring portion of the human brain is
    a. the cerebrum.
    b. the cerebellum.
    c. the thalamus.
    d. the hypothalamus.

14. The effect of thyroxin in humans is
    a. to increase cellular respiration and thereby produce body heat.
    b. to decrease cellular respiration and thus reduce body heat production.
    c. to stimulate the pituitary gland to secrete thyroid stimulating hormone.
    d. to stimulate the thyroid gland to release TSH.

15. Desert lizards regulate their body temperatures by basking in the sun or minimizing sun exposure. This is
    a. endothermic thermoregulation.
    b. a behavioral means of ectothermic thermoregulation.
    c. an adaptation for thermiogenesis.
    d. not really involved in thermoregulation, but rather is an aspect of behavior.

16. The primary nitrogen waste of earthworms and all mammals is
    a. amino groups.
    b. ammonia.
    c. urea.
    d. uric acid.

17. Oceanic fish live in a marine environment. The ocean is typically
    a. hyperosmotic to their bodies.
    b. isoosmotic to their bodies.
    c. hypoosmotic to their bodies.
    d. endoosmotic to their bodies.

18. *Artemia salina* can exist in environments up to 30% salinity. Its osmoregulatory organ is
    a. its kidney, which actively transports salt out of the blood.
    b. its skin, which has salt glands that actively transport salt out of the blood.

c. its gill, which has salt glands that actively transport salt out of the blood.
d. its nasal lining, which has salt glands that actively transport salt out of the blood.

19. The chief osmoregulatory device of freshwater planarians is
    a. the gill, which has salt glands that actively transport salt out of the blood.
    b. the kidney, which actively transports salt out of the blood.
    c. the flame cell system.
    d. the Malpighian tubules.

20. Which of the following sequences correctly gives the urine-forming pathway in humans?
    a. kidney ⇒ ureters ⇒ urinary bladder ⇒ urethra.
    b. ureters ⇒ kidney ⇒ urinary bladder ⇒ urethra.
    c. kidney ⇒ ureters ⇒ urethra ⇒ urinary bladder.
    d. kidney ⇒ urinary bladder⇒ ureters ⇒ urethra.

21. Filtration of blood takes place in which portion of the human kidney?
    a. the loop of Henle.
    b. Bowman's capsule.
    c. the proximal convoluted tubule.
    d. the collecting duct.

22. After blood is force filtered, which of the following might remain in the blood rather than in the filtrate?
    a. glucose.
    b. amino acids.
    c. large proteins.
    d. urea.

23. How do sodium ions move out of the proximal convoluted tubule into surrounding cells?
    a. osmosis.
    b. diffusion.
    c. active transport.
    d. leakage.

24. Water is reclaimed from the filtrate in the nephron and moves back into the capillaries via
    a. osmosis.
    b. countercurrent exchange.
    c. active transport.
    d. leakage.

25. The glomeruli and convoluted tubules of the nephrons lie in which portion of the kidney?
    a. the renal cortex.
    b. the renal medulla.
    c. the renal pelvis.
    d. the renal plexus.

26. What is the activity of antidiuretic hormone (ADH)?
    a. Its targets are the epithelial cells of the collecting ducts, and it increases urine flow.
    b. Its targets are the epithelial cells of the collecting ducts, and it decreases urine flow.
    c. Its targets are the epithelial cells of the loop of Henle, and it increases urine flow.
    d. Its targets are the epithelial cells of the loop of Henle, and it decreases urine flow.

27. What is the activity of aldosterone?
    a. Its targets are the distal tubules, and it increases sodium transport back to the blood.
    b. Its targets are the distal tubules, and it decreases sodium transport back to the blood.
    c. Its targets are the glomeruli, and it increases sodium transport back to the blood.
    d. Its targets are the glomeruli, and it decreases sodium transport back to the blood.

B. True or False Questions

Mark the following statements either T (True) or F (False).

_____ 28. A negative feedback loop is inherently stabilizing.

_____ 29. Thermoregulation is a specialized adaptation found only in certain groups of animals.

_____ 30. Animals that actively thermoregulate are termed poikilotherms.

_____ 31. A flea moving to a warmer portion of a dog is an example of ectothermic thermoregulation.

_____ 32. Bats are examples of animals that can hibernate, where their core temperatures are decreased.

_____ 33. Under warm conditions, a cold-adapted moth may overheat and fall into a metabolic stupor.

_____ 34. A cold-adapted bluefin tuna should be considered a cold-blooded animal.

_____ 35. In general, the large the body size of an animal, the greater its metabolic rate.

_____ 36. During hibernation, some mammals use brown fat thermiogenesis to maintain body temperature.

_____ 37. When the human body becomes chilled, venous blood is shunted to veins near the skin surface.

_____ 38. Blubber on animals like seals and whales is an efficient insulator.

_____ 39. On mammals, the long coarse guard hairs provide more insulation than do the fine underhairs.

_____ 40. Heat can escape from animal surfaces by radiation, conduction, and convection.

_____ 41. For mammals, the real thermostat of the body is the brain.

_____ 42. Through the presence of antifreeze substances like glycerol, some insect larvae can resist freezing at temperatures well below 0° C.

_____ 43. Urea is a semisolid insoluble nitrogenous waste product of animals.

_____ 44. The body fluids of marine organisms are typically hyperosmotic to the surrounding salt water.

_____ 45. Because it can tolerate salinities up to 30%, *Artemia salina* is considered an osmoconformer.

_____ 46. Amphibians can actively transport salts back into their bodies, via the skin.

_____. 47 The excretory system of insects consists of the Malpighian tubules.

_____ 48. The primary nitrogen waste of mammals is soluble urea.

_____ 49. Arterial blood is subjected to force filtration in the loop of Henle.

_____ 50. A diuretic is an agent that increases urine flow.

C. Fill in the Blanks

Answer the question or complete the statement by filling in the blanks with the correct word or words.

51. In a _________________ loop, a stimulus evokes a response that further increases the stimulus.

52. Homeotherms are sometimes called ________________ and poikilotherms are ________________.

53. Metabolic production of heat is termed ________________.

54. A circulatory arrangement where heat is transferred from one vessel to another is a ________________.

55. The complex heat exchanger in the bluefin tuna has been termed the ________________.

56. Birds rely on ________________ for body insulation.

57. For humans, sweat glands can reduce body heat via ________________.

58. The hormone ________________ can lead to higher body temperature by increasing cellular respiration.

59. The nitrogenous waste of mammals is water-soluble ________________.

60. Some limpets match the salinity of their intertidal habitat. They are called osmo-________________.

61. The blood vessels leading to the human kidney from the aorta are the ________________.

62. The functional units of the human kidney are the ________________.

63. In Bowman's capsule, the glomerular capillaries arise from the ________________ arterioles.

64. In humans, urine volume and the osmotic consistency of blood are controlled by ________________.

Questions for Discussion

1. Why is it necessary for animals to thermoregulate? What are the adaptive solutions that have been used by various animals to achieve thermal homeostasis? Outline the different mechanisms that are used by animals to produce, maintain, acquire, retain, and dissipate heat.

2. How does body size affect the heat balance of an organism? What is the volume to surface area relationship that one expects in terms of maintaining body temperature? Illustrate your discussion with appropriate mammalian examples.

3. Make a general contrast between thermoregulation in insects and thermoregulation in birds. Which mechanisms are more common in each. What are some specific adaptations? Use as many examples as you can.

4. Living in the marine environment presents an osmotic challenge to many marine organisms. What are the difficulties such organisms encounter? List the mechanisms by which animals achieve osmotic homeostasis in the saltwater environment.

5. How does the human kidney rid the body of wastes and yet maintain proper water balance? Trace the urine-forming pathway from the blood of the aorta to the release of urine. Identify the structures involved and the process which occur at each stage. You may find drawing a diagram helpful in explaining the complete process.

Answers to Self-Exam

Multiple Choice Questions

1. b
2. c
3. d
4. b
5. a
6. a
7. c
8. c
9. d
10. a
11. d
12. b
13. b
14. a
15. b
16. c
17. a
18. c
19. c
20. a
21. b
22. c
23. c
24. a
25. a
26. b
27. b

True or False Questions

28. T
29. F
30. F
31. T
32. T
33. T
34. F
35. F
36. T
37. F
38. T
39. F
40. T
41. T
42. T
43. F
44. F
45. F
46. T
47. T
48. T
49. F
50. T

Fill in the Blank Questions

51. positive feedback
52. warm-blooded, cold-blooded
53. thermiogenesis
54. countercurrent heat exchanger
55. rete mirable
56. feathers
57. evaporative cooling
58. thyroxin
59. urea
60. osmoconformers
61. renal arteries
62. nephrons
63. afferent
64. antidiuretic hormone (ADH)

CHAPTER 37 Hormonal Control

Learning Objectives

After mastering the material covered in Chapter 37, you should be able to confidently do the following tasks:

- Explain how hormones work to coordinate control of function in animals.
- Distinguish hormonal control from neural mechanisms of control.
- Identify the two large classes of messenger molecules. the peptide hormones and the steroid hormones.
- Explain the second messenger concept and give an example showing the role of second messengers in hormonal control.
- Illustrate the role of steroid hormones in controlling expression of genes.
- Give examples showing the hormonal control of development in arthropods.
- Identify the major components of the vertebrate endocrine system.
- For each of the endocrine glands, identify the hormones produced, their target cells, and the actions they elicit.

Chapter Outline

I. The Chemical Messengers.

A. Neural and hormonal control compared.
B. Molecular structure of hormonal messengers compared.
C. Characteristics of hormonal control.
D. Identifying hormones.

II. Chemical Messengers and the Target Cell.

A. Peptide hormones and second messengers.
B. Steroid hormones and gene control.

III. Invertebrate Hormones.

A. Hormonal activity in arthropods.
1. hormones and development in insects.

IV. The Vertebrate Endocrine System.

A. The pituitary.
   1. control by the hypothalamus.
   2. hormones of the anterior pituitary.
   3. hormones of the posterior pituitary.
   4. evolutionary relationships of the neurohypophyseal hormones.
B. The thyroid.
C. The parathyroid glands.
D. The pancreas: islets of Langerhans.
E. The adrenal gland.
   1. hormones of the adrenal medulla.
   2. adrenal hormones and physiological stress.
F. The gonads: ovaries and testes.
G. The thymus.
H. The pineal body.
I. Prostaglandins

Key Words

hormones
endocrines hormones
paracrine hormones
autocrine hormones
ductless glands
ducted glands
neurohormones
neurosecretions
peptide hormones
lipid hormones
steroids
target cells
endocrinology
radioimmunoassay
second messenger
cyclic AMP
cyclic GMP
inositol triphosphate
adenylate cyclase
protein kinase
phosphorylase kinase
glycogen phosphorylase *a*
glucose-1-phosphate
glucose-6-phosphate
glycogen synthetase
cascade
cytoplasmic binding protein
chromatin
ecdysis
molt-inhibiting hormone (MIH)
molting hormone (MH)
X-organ
Y-organ
Hyalophora cecropia
complete metamorphosis
diapause
ecdysone
prothoracic gland
brain hormone (BH)
juvenile hormone (JH)
corpus allatum
hypothalamus
portal circuit
releasing hormone
inhibiting hormone
growth hormone releasing hormone (GHRH)
growth hormone release inhibiting hormone (GHRIH)
corticotropin releasing hormone (CRH)
thyrotropin releasing hormone (TRH
prolactin releasing hormone (PRH)
prolactin release inhibiting hormone (PRIH)
melanocyte-stimulating hormone releasing hormone (MSHRH)
melanocyte-stimulating hormone release inhibiting hormone (MSHRIH)
hypophysis
pituitary gland
adenohypophysis
anterior pituitary
adrenocorticotropic hormone (ACTH)
growth hormone (GH)
somatotropin
giantism
pituitary giant
pituitary dwarf
acromegaly
thyroid-stimulating hormone (TSH)
thyrotropin
prolactin
crop milk
follicle-stimulating hormone (FSH)
luteinizing hormone (LH)
melanocyte-stimulating hormone (MSH)
neurohypophysis
posterior pituitary
oxytocin
pitocin
vasopressin
antidiuretic hormone (ADH)
arginine vasotocin (AVT)
thyroid
thyroxin
T_4
triiodothyronine
T_3
hyperthyroidism
hypothyroidism
Grave's disease
goiter
myxedema
cretinism
calcitonin
parathyroid gland
parathyroid hormone (PTH)
osteoclasts
osteoblasts
vitamin D
pancreas
islets of Langerhans
islet cells
alpha islet cells
glucagon

beta islet cells
insulin
diabetes mellitus
adrenal gland
adrenal cortex
corticosteroid
mineralocorticoid
aldosterone
glucocorticoid
cortisol
corticosterone
sex hormones
adrenal androgen
adrenal medulla
epinephrine
norepinephrine
adrenalin
noradrenalin
"fight or flight" response
physiological stress
gonad
ovary
estrogen
estradiol
estriol
estrone
progesterone
testis
testosterone
thymus
lymphocyte
thymosin
pineal body
melatonin
melanocyte
circadian rhythm
"third eye"
tuatara
Sphenodon punctatus
prostaglandins
thromboxane
prostacyclin
hyperglycemia
insulinase
juvenile-onset diabetes
hypoglycemia

Exercises

1. The pituitary has sometimes been called the "master gland." Many different number of hormones are secreted by the pituitary. Organize the functions of the pituitary by completing the following table by listing the pituitary hormone that affects the listed targets, and by stating the action caused by each hormone at the target.

Anterior Pituitary Hormones	Target	Action
	thyroid	
	adrenal cortex	
	breasts	
	ovaries, testes	
	mature ovarian follicles testis interstitial cells	
	general body long bones at puberty	
	melanocytes	

Posterior Pituitary Hormones	Target	Action
	breasts, uterus	
	kidney	

2. In this chapter there are many hormones from many sources. To practice keeping them straight, match the hormones with their sources by placing the correct letter by each entry.

<u>Hormone Sources</u>

a. adenohypophysis
b. adrenal cortex
c. adrenal medulla
d. hypothalamus
e. islets of Langerhans
f. neurohypophysis
g. nonspecific source
h. ovaries
i. parathyroid
j. pineal body
k. testes
l. thyroid
m. thymus

<u>Hormones</u>

_____ adrenocorticotropic hormone (ACTH).
_____ growth hormone (GH).
_____ oxytocin.
_____ aldosterone.
_____ testosterone.
_____ thyroxin
_____ luteinizing hormone (LH).
_____ calcitonin.
_____ prolactin.
_____ epinephrine.
_____ melatonin.
_____ glucagon.

_____ triiodothyronine.
_____ cortisol.
_____ prostaglandins.
_____ parathyroid hormone (PTH)
_____ estrogen.
_____ insulin.
_____ thyroid-stimulating hormone (TSH).
_____ antidiuretic hormone (ADH).
_____ testosterone.
_____ follicle-stimulating hormone (FSH).
_____ thyroxin
_____ releasing hormones

3. Locate the endocrine structures listed below in the diagram. Check your work with Figure 37.11 in the text.

adrenals
hypothalamus
islets of Langerhans
ovaries
parathyroid
pineal body
pituitary
thymus
thyroid

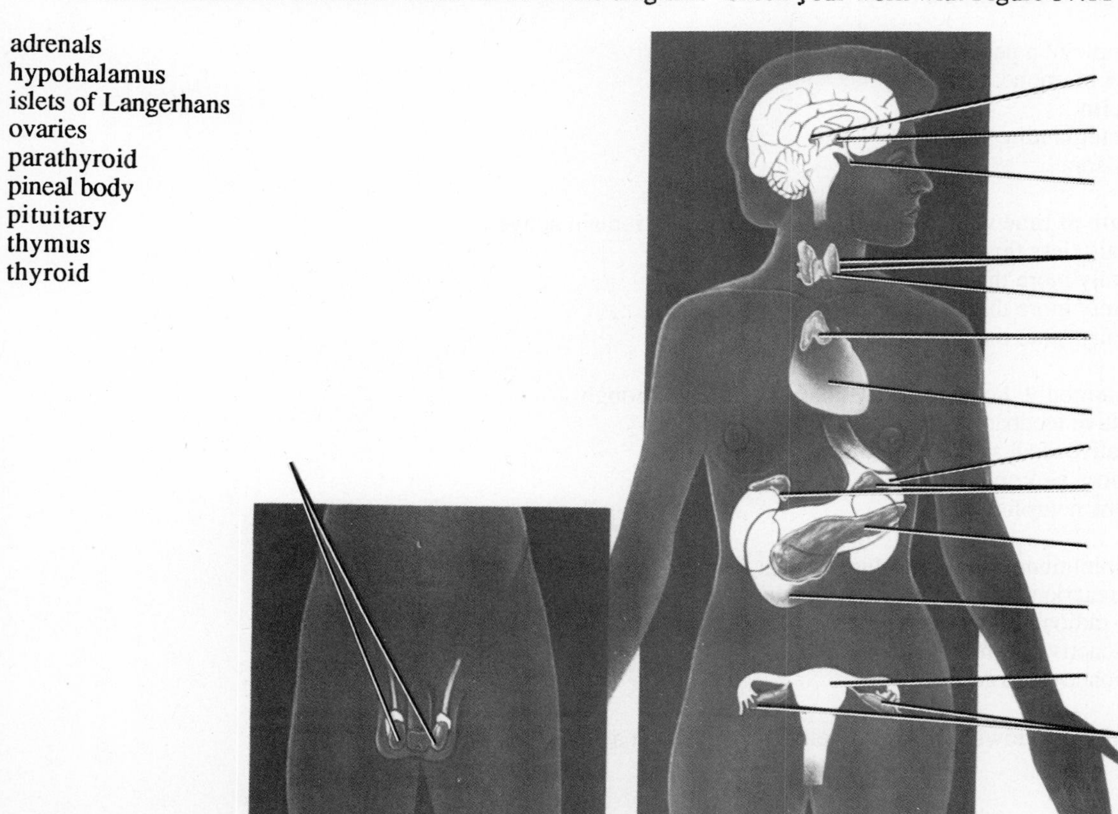

Self-Exam

You should be able to easily answer the following questions after learning the material in Chapter 37. If you have difficulty with any question, study the appropriate section in the text and try again.

A. Multiple Choice Questions

Circle one alternative that best completes the statement or answers the question.

1. Hormones that act in the cells in which they are produced are
   a. endocrine hormones.
   b. paracrine hormones.
   c. autocrine hormones.
   d. exocrine hormones.

2. Chemical messengers that act in areas of the body other than where they are produced are
   a. endocrine hormones.
   b. paracrine hormones.
   c. autocrine hormones.
   d. exocrine hormones.

3. Secretions such as sweat and mucus are produced by
   a. endocrine glands.
   b. paracrine glands.
   c. autocrine glands.
   d. exocrine glands.

4. In general, we can say that neurohormones
   a. are never released from the cells where they are produced.
   b. travel through neural axons to the terminals, where they are released.
   c. circulate via the bloodstream.
   d. are excreted from the body.

5. An example of a peptide hormone is
   a. a sex hormone.
   b. insulin.
   c. prostaglandin.
   d. cortisone.

6. The length of time which many hormones typically remain active is
   a. usually less than one hour.
   b. usually more than one hour.
   c. usually more than an hour but less than one day.
   d. usually more than one day.

7. Most commonly, hormone secretion is regulated through
   a. positive feedback.
   b. negative feedback.
   c. sensory neurons.
   d. motor neurons.

8. The radioimmunoassay procedure detects
   a. the reaction between antibody and antigen.
   b. any radioactivity.
   c. radioactive hormone receptor sites.
   d. radioactive target cells.

9. Which of the following is most likely to function via a second messenger?

a. a steroid hormone.
b. a peptide hormone.
c. an exocrine secretion.
d. the immune system.

10. Which of the following is an example of a second messenger?
    a. ATP.
    b. cyclic AMP.
    c. ADH.
    d. insulin.

11. The correct sequence describing how epinephrine works in the liver is
    a. epinephrine ⇒ cAMP ⇒ protein kinase ⇒ phosphorylase kinase ⇒ glycogen phosphorylase *a*.
    b. epinephrine ⇒ protein kinase ⇒ cAMP ⇒ phosphorylase kinase ⇒ glycogen phosphorylase *a*.
    c. epinephrine ⇒ cAMP ⇒ phosphorylase kinase ⇒ protein kinase ⇒ glycogen phosphorylase *a*.
    d. epinephrine ⇒ phosphorylase kinase ⇒ protein kinase ⇒ cAMP ⇒ glycogen phosphorylase *a*.

12. Which statement is accurate concerning hormonal control of gut and blood vessel smooth muscle?
    a. cAMP brings about muscle relaxation, while cGMP causes smooth muscle contraction.
    b. cAMP leads to muscle contraction, while cGMP brings smooth muscle relaxation.
    c. A single hormone brings about both contraction and relaxation.
    d. Second messengers are not involved in smooth muscle contraction and relaxation.

13. It is known that environmental cues are involved in hormonal responses in invertebrates. Which is correct?
    a. Both crabs and crayfish will molt if kept in constant light.
    b. Neither crabs nor crayfish will molt unless kept in the dark.
    c. Crabs will not molt if kept in constant darkness, nor will crayfish molt if kept in constant light.
    d. Crabs will not molt if kept in constant light, nor will crayfish molt if kept in constant darkness.

14. Metamorphosis in insects is controlled by
    a. ecdysone alone.
    b. juvenile hormone alone.
    c. molting hormone alone.
    d. ecdysone and juvenile hormone.

15. Another name for the pituitary gland is
    a. the hypophysis.
    b. the hypothalamus.
    c. the pineal body.
    d. the third eye.

16. The hypothalamus is anatomically and hormonally linked to the anterior pituitary by
    a. axons from hypothalamus neurons.
    b. the third ventricle.
    c. a portal vein.
    d. plasmodesmata.

17. Which group of hormones includes those controlled by negative feedback control?
    a. GH, prolactin, and MSH.
    b. ACTH, thyrotropin, and gonadotropin.
    c. GH, prolactin, MSH, ACTH, thyrotropin, and gonadotropin are all controlled by negative feedback.
    d. None of these; negative feedback control is not involved with regulation of pituitary hormones.

18. The connection between the hypothalamus and the posterior pituitary is via
    a. axons from hypothalamus neurons.
    b. the third ventricle.
    c. a portal vein.
    d. plasmodesmata.

19. ACTH, TSH, FSH, and LH are all secreted by
    a. the hypothalamus.
    b. the anterior pituitary.
    c. the posterior pituitary.
    d. the thyroid.

20. Pituitary dwarfism results from
    a. a deficiency of GH.
    b. a deficiency of FSH.
    c. a deficiency of gonadotropins.
    d. a deficiency of ACTH.

21. In some birds such as the pigeon, prolactin stimulates production of
    a. milk.
    b. crop milk.
    c. eggs.
    d. butter.

22. The neurosecretions of the posterior pituitary are
    a. steroids.
    b. sent to the hypothalamus.
    c. short neuropeptides of 9 amino acids.
    d. long polypeptides of dozens of amino acids.

23. A deficiency of iodine in the diet of individuals with a normal thyroid can lead to
    a. formation of a goiter.
    b. Grave's disease.
    c. myxedema.
    d. cretinism.

24. The thyroid hormone involved in regulating Ca^{++} levels in the body is
    a. TSH.
    b. triiodothyronine (T_3).
    c. thyroxin (T_4).
    d. calcitonin.

25. Which statement below is accurate with respect to calcium ion regulation?
    a. PTH decreases the blood calcium level, while calcitonin increases it.
    b. PTH increases the blood calcium level, while calcitonin decreases it.
    c. Both PTH and calcitonin act to increase the blood calcium level.
    d. Both PTH and calcitonin act to decrease the blood calcium level.

26. In bone, what is the action of calcitonin?
    a. Calcitonin stimulates osteoclasts.
    b. Calcitonin inhibits osteoclasts.
    c. Calcitonin stimulates osteoblasts.
    d. Calcitonin inhibits osteoblasts.

27. In the islets of Langerhans of the pancreas,
    a. alpha cells secrete glucagon.
    b. alpha cells secrete insulin.
    c. alpha cells secrete both glucagon and insulin.
    d. beta cells secrete both glucagon and insulin.

28. The most striking aspects of the "fight or flight" response are brought about by the release of
    a. aldosterone from the adrenal cortex.
    b. epinephrine from the adrenal medulla.

c. thyroid-stimulating hormone.
d. calcitonin.

29. Which hormone is produced principally by the testis?
   a. progesterone.
   b. testosterone.
   c. estrogen.
   d. estradiol.

30. Which structure is sometimes called the "third eye"
   a. pituitary.
   b. hypophysis.
   c. pineal body.
   d. thymus.

B. True or False Questions

Mark the following statements either T (True) or F (False).

_____ 31. Chemical control is generally much slower than neural control, and its effects are longer lasting.

_____ 32. The "newcomer" group of hormones, the prostaglandins, are classified as neuropeptides.

_____ 33. In dogs, surgical removal of the pancreas results in diabetes.

_____ 34. Steroid hormones seem to act at the level of gene regulation.

_____ 35. Peptide hormones can themselves be considered first messengers.

_____ 36. A significant feature of second messengers is their long persistence; they do not degrade rapidly.

_____ 37. In the autonomic nervous system, acetylcholine activates the second messenger cAMP.

_____ 38. As molting begins in crustaceans, the hard exoskeleton is weakened by the absorption of calcium.

_____ 39. In lobsters, MIH (molt-inhibiting hormone) is produced by the Y-organ.

_____ 40. High levels of juvenile hormone can prevent metamorphosis in insects.

_____ 41. The adenohypophysis is a true gland, while the neurohypophysis is a neural structure.

_____ 42. A portal circuit refers to an opening in the wall between two structures.

_____ 43. The pituitary can be regarded as subservient to the hypothalamus.

_____ 44. Another term for FSH and LH is somatotropin.

_____ 45. The actual release of milk is under the influence of oxytocin.

_____ 46. Thyroid hormone is essential in bringing about metamorphosis in amphibians.

_____ 47. Thyroxin (T_4) is an inactive form of triiodothyronine (T_3), which is the active hormone.

_____ 48. The parathyroid glands are pea-sized bodies embedded in the tissue of the thymus.

_____ 49. Glucagon increases blood glucose levels, while insulin decreases blood glucose levels.

_____ 50. Aldosterone is on of the familiar mineralocorticoids.

C. Fill in the Blanks

Answer the question or complete the statement by filling in the blanks with the correct word or words.

51. Hormones circulate widely, but they elicit a response only in specific _________________.

52. Peptide hormones trigger cytoplasmic events through intermediaries called _________________.

53. When it enters the cytoplasm of a target cell, a steroid hormone becomes tightly bound to a _________________.

54. The periodic shedding of the exoskeleton among crustaceans and insects is molting or _________________.

55. The anterior pituitary and the posterior pituitary are the _________________ and _________________.

56. GHRH and GHRIH are secreted by the _________________.

57. FSH and LH are also known as _________________.

58. _________________ promotes milk production in mammals, including humans.

59. Another term for human vasopressin is _________________.

60. A lack of thyroid hormone from birth can lead to _________________ in humans.

61. Insulin and glucagon are produced in _________________.

62. Adrenal cortex hormones include mineralocorticoids, glucocorticoids, and _________________.

63. The hormones estradiol, estriol, and estrone are all considered _________________.

64. Behavioral or physiological patterns that occur on a daily cycle are termed _________________.

65. A structure that seems to be essential to daily cycles in some birds is the _________________.

66. The New Zealand reptile _________________ has a well developed third eye.

67. The synthesis of _________________ is apparently inhibited by aspirin.

68. Deficiencies in insulin produce _________________, which characterizes diabetes mellitus.

Questions for Discussion

1. There are many similarities between neural control of animal functions and hormonal control. Discuss the similarities, citing examples of parallel or identical features. What are the evolutionary implications of the similarities.

2. Discuss how the hormone controlling a particular process can be identified. Give an example of how such a search might work. How are the identifications of a hormone and its source confirmed?

3. What are second messengers? What types of hormones act through second messengers? Why is understanding the activity of second messengers so important?

4. Outline the hormonal regulation of larval development and metamorphosis in an insect with complete metamorphosis. Can you think of ways that such hormonal regulation might be modified by humans as an approach toward control of insect pests?

5. What are the actions of calcitonin and PTH on calcium ions in the human body? Include in your discussion the effects of these hormones on blood, kidney, and bone.

Answers to Self-Exam

Multiple Choice Questions

1. c
2. a
3. d
4. b
5. b
6. a
7. b
8. a
9. b
10. b
11. a
12. a
13. d
14. d
15. a
16. c
17. b
18. a
19. a
20. b
21. b
22. c
23. a
24. d
25. b
26. c
27. a
28. b
29. b
30. c

True or False Questions

31. T
32. F
33. T
34. T
35. T
36. F
37. F
38. T
39. F
40. T
41. T
42. F
43. T
44. F
45. T
46. T
47. F
48. F
49. T
50. T

Fill in the Blank Questions

51. target cells
52. second messengers
53. cytoplasmic binding protein
54. ecdysis
55. adenohypophysis, neurohypophysis
56. hypothalamus
57. gonadotropins
58. prolactin
59. antidiuretic hormone (ADH)
60. cretinism
61. islets of Langerhans of the pancreas
62. sex steroids
63. estrogens
64. circadian rhythms
65. pineal body
66. *Sphenodon punctatus* or tuatara
67. prostaglandins
68. hyperglycemia

CHAPTER 38 Digestion and Nutrition

Learning Objectives

After mastering the material covered in Chapter 38, you should be able to confidently do the following tasks:

- Describe the basic anatomy of the digestive systems for representative groups of invertebrates.
- Explain the patterns of feeding and digestion observed in the major vertebrate groups.
- Identify and describe the principal structures and features of the human digestive system.
- Outline the chemical processes involved in human digestion of carbohydrates, fats, proteins, and nucleic acids.
- List the basic nutritional requirements that must be met by the human diet and show how these requirements are fulfilled.

Chapter Outline

I. Digestive Systems.

A. Saclike systems in invertebrates.
B. The tube-within-a-tube plan.
C. Feeding and digestive structures in vertebrates.
1. sharks and bony fishes.
2. amphibians, reptiles, and birds.
3. foraging and digestive structures in mammals.

II. The Digestive System of Humans.

A. The oral cavity and the esophagus.
B. The stomach.
C. The small intestine.
D. The liver and pancreas.
E. The large intestine.

III. The Chemistry of Digestion.

A. Carbohydrate digestion.
B. Fat digestion.
C. Protein digestion.
D. Nucleic acid digestion.
E. Integration and control of the digestive process.

IV. Essentials of Nutrition.

A. Carbohydrates.
B. Fats.
C. Protein.
D. Vitamins.
E. Mineral requirements.

Key Words

digestion
nutrition
extracellular digestion
intracellular digestion
digestive vacuoles
collar cells
choanocytes
amebocytes
gastrovascular cavity
tube-within-a-tube
mouth
pharynx
esophagus
crop
gizzard
intestine
typhlosole
chloragen cells
deamination
palps
mandibles
maxillae
labrum
labium
foregut
midgut
hindgut
gastric cecum
spiral valve
fangs
Jacobson's organ
pit viper
raptor
crop
stomach
proventriculus
ventriculus
incisors
canines
premolars
molars
ruminant
cellulose
beta linkages
rumen
ulum
omasum
abomasum
cud chewing
taste buds
saliva
parotids
sublinguals
submandibulars
amylase
larynx
Adam's apple
laryngopharynx
soft palate
epiglottis
glottis
serosa
mucosa
submucosa
muscularis
peristalsis
bolus
cardiac sphincter
heartburn
chief cells
pepsinogen
]parietal cells
hydrochloric acid (HCl)
pepsin
gastric lipase
mucin
chyme
pyloric sphincter
small intestine
duodenum
jejunum
ileum
villi
microvilli
capillary bed
lacteal
glycocalyx
liver
bile
gall bladder
cholecystekinin-pancreozymin
bile duct
pancreatic duct
pancreas
sodium bicarbonate
colon
bowel
cecum
ascending colon
transverse colon
descending colon
sigmoid colon
rectum
anal canal
anus
appendix
ileocecal valve
Escherichia coli
rectal valve
anal sphincter
hydrolysis
salivary amylase
maltase
amylopectin
1-4 linkages
1-6 linkages
alpha glycosidase
sucrose
sucrase
lactose
lactase
lactose intolerance
bile salts
lipase
chylomicrons
protease
endopeptidase
exopeptidase
trypsin
chymotrypsin
carboxypeptidase
aminopeptidase
dipeptidase
trypsinogen
chymotrypsinogen
procarboxypeptidase
enterokinase
pancreatitis
nuclease
endonuclease
exonuclease

vagus nerve
gastrin
enterogastrone
secretin
essential fatty acids
essential amino acids
retinol (vitamin A)
thiamine (vitamin B_1)
riboflavin (vitamin B_2)
niacin
nicotinic acid
folic acid
vitamin B_6
pantothenic acid
vitamin B_{12}
biotin
choline
ascorbic acid (vitamin C)
vitamin D
vitamin E
vitamin K
antioxidants
minerals
trace elements
goiter
cholesterol
plaques
free radicals

Exercises

1. The various digestive enzymes produced by the human digestive system act on many different dietary foodstuffs. Complete the table below to give the correct enzyme for each substrate listed and to show the source of each enzyme. Check your work with Table 38.1 in the text.

Substrate	Enzyme	Enzyme Source
starch		
protein		stomach lining
casein		
triglycerides		stomach lining
peptides		pancreas
RNA and DNA		pancreas
starch		pancreas
triglycerides		pancreas
C-terminal bond of peptides		
N-terminal bond of peptides		
dipeptides		
nucleotides		intestinal lining
maltose		
sucrose		
lactose		

2. Humans need a dietary source of a number of vitamins, and they all have important functions. Match the vitamins listed below with their main functions. Check your work with Table 38.3 in the text.

Vitamin

a. vitamin A, retinol
b. vitamin B_1, thiamine
c. vitamin B_2, riboflavin
d. vitamin B_6
e. vitamin B_{12}
f. vitamin C, ascorbic acid
g. vitamin D
h. vitamin E
i. vitamin K
j. biotin
k. choline
l. folic acid
m. niacin
n. pantothenic acid

Functions of Vitamins

_____ respiratory coenzyme.

_____ oxidative chains in cell respiration.

_____ synthesis of blood cells.

_____ part of coenzyme A of cell respiration.

_____ in coenzymes.

_____ connective tissues and matrix antioxidant.

_____ antioxidant.

_____ synthesis of visual pigments.

_____ part of NAD and FAD in cell respiration.

_____ active transport.

_____ red blood cell production.

_____ fat, carbohydrate, protein metabolism.

_____ absorption of calcium.

_____ blood-clotting factors.

3. Please provide labels for this diagram showing the parts of the human digestive system. Use terms in the list below. Check your work with Figure 38.11 in the text.

anus
appendix
ascending colon
cecum
descending colon
esophagus
laryngopharynx
larynx
liver
parotid gland
pharynx
rectum
sigmoid colon
small intestine
spleen
stomach
sublingual gland
submandibular gland
transverse colon

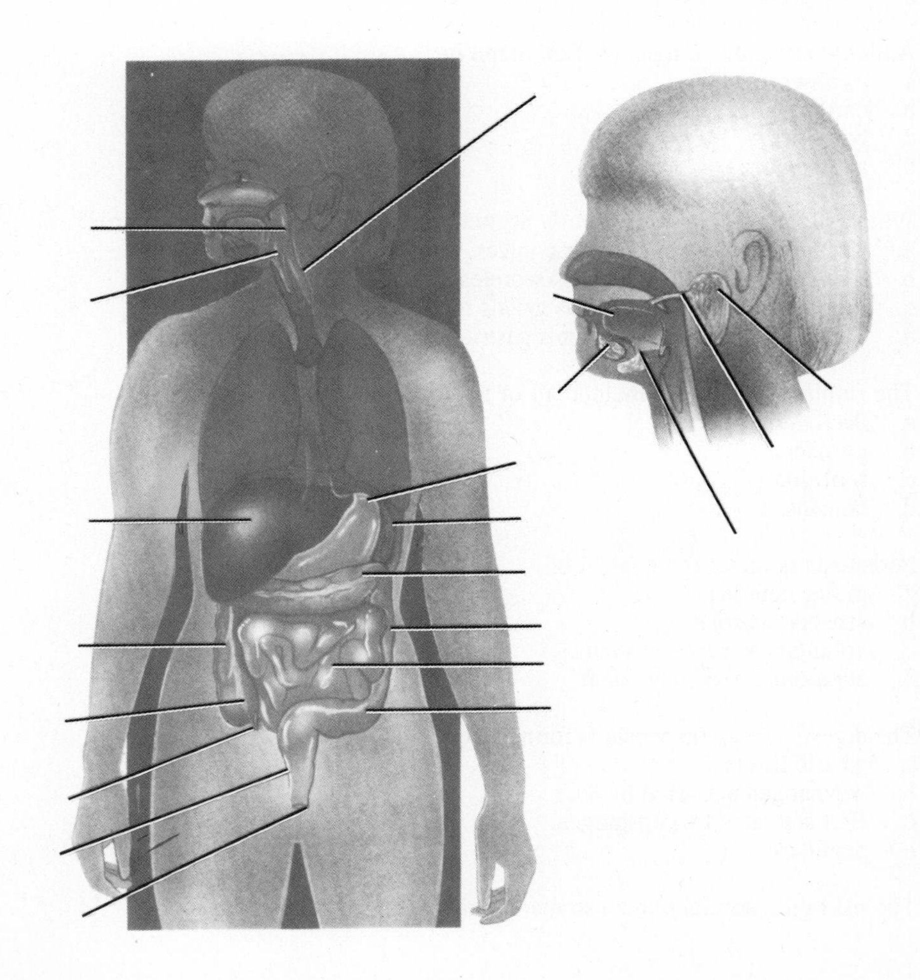

Self-Exam

You should be able to easily answer the following questions after learning the material in Chapter 38. If you have difficulty with any question, study the appropriate section in the text and try again.

A. Multiple Choice Questions

Circle one alternative that best completes the statement or answers the question.

1. The flatworm digestive system can be characterized as
   a. collar cells.
   b. comprising choanocytes.
   c. a gastrovascular cavity.
   d. a tube-within-a-tube system.

2. The sequence of structures encountered in the earthworm digestive system is
   a. mouth ⇒ pharynx ⇒ esophagus ⇒ crop ⇒ gizzard ⇒ intestine.
   b. mouth ⇒ esophagus ⇒ pharynx ⇒ crop ⇒ gizzard ⇒ intestine.
   c. mouth ⇒ pharynx ⇒ esophagus ⇒ gizzard ⇒ crop ⇒ intestine.
   d. mouth ⇒ pharynx ⇒ crop ⇒ esophagus ⇒ gizzard ⇒ intestine.

3. Insects which feed predominantly on plant phloem sap have
   a. mainly biting mouthparts.
   b. mainly chewing mouthparts.
   c. mainly piercing and sucking mouthparts.
   d. unspecialized mouthparts.

4. A highly moveable tongue is often found in
   a. earthworms.
   b. cartilaginous fish.
   c. bony fish.
   d. amphibians.

5. Which statement is most generally accurate about digestive systems in birds?
   a. The large crop secretes gastric juices.
   b. The glandular proventriculus secretes gastric juices.
   c. The muscular proventriculus grinds the food.
   d. The glandular gizzard secretes gastric juices.

6. The ruminant mammals include all of the following except
   a. deer.
   b. giraffes.
   c. buffalo.
   d. humans.

7. Peristalsis is under the control of
   a. motor neurons.
   b. sensory neurons.
   c. voluntary nervous system.
   d. autonomic nervous system.

8. The digestive enzyme pepsin is formed from
   a. gastric lipase.
   b. pepsinogen activated by HCl.
   c. HCl activated by pepsinogen.
   d. peptidase.

9. The pH in the normal human stomach fluids during digestion is about

a. pH 1.6 to 2.4.
b. pH 2.4 to 4.6.
c. pH 4.6 to 7.0.
d. pH 7.0 to 8.6.

10. The length of the small intestine in humans is about
a. 6 feet.
b. 6 meters.
c. 20 meters.
d. 60 feet.

11. What is the location of the intestinal digestive enzymes in humans?
a. the lumen of the intestine.
b. the cytoplasm of villi cells.
c. bound to the plasma membrane of columnar epithelium.
d. the lymphatic vessels.

12. The 1-6 linkages of amylopectin are sheared by the enzyme
a. salivary amylase.
b. pancreatic amylase.
c. alpha glucosidase.
d. lactase.

13. In general, fat digestion in humans takes place
a. in the stomach.
b. in the liver.
c. in the small intestine.
d. in the large intestine.

14. As peptides are digested, carboxypeptidase cleaves peptide bonds
a. at the N-terminus.
b. at the C-terminus.
c. in the middle.
d. at either end.

15. What is the source of the hormone gastrin in humans?
a. cells of the stomach wall.
b. cells of the pituitary.
c. cells of the tongue.
d. cells of the vagus nerve.

16. The coenzymes NAD and FAD are derived from
a. niacin and riboflavin.
b. vitamins A and C.
c. thiamin and folic acid.
d. vitamins D and K.

17. A mineral needed as a cross-linking agent in the elastic walls if major arteries is
a. arsenic.
b. silicon.
c. selenium.
d. zinc.

B. True or False Questions

Mark the following statements either T (True) or F (False).

_____ 18. Some digestion of food for the housefly takes place outside the gut.

_____ 19. The jaws of sharks and of bony fishes can move in three different planes.

_____ 20. Herbivorous fishes lack tearing teeth.

_____ 21. Birds such as the domestic chicken have very tiny teeth that are hard to find.

_____ 22. The dentition of humans is considered by evolutionary biologists to be very advanced and specialized.

_____ 23. In the human digestive system, most digestion and all absorption of foods occurs in the small intestine.

_____ 24. Considerable digestion occurs in the human large intestine, aided by *E. coli*.

_____ 25. Pepsin is a generalized peptidase, attacking virtually any peptide bond.

_____ 26. Exonucleases are DNAases which attack only the exon portion of DNA molecules.

_____ 27. The hormone gastrin stimulates the release of gastric juice.

_____ 28. Polyunsaturated fats are not needed at all in the human diet.

_____ 29. Essential fatty acids must be provided in the human diet.

_____ 30. Green vegetables provide a good dietary source of the trace element magnesium.

C. Fill in the Blanks

Answer the question or complete the statement by filling in the blanks with the correct word or words.

31. In sponges, the task of obtaining food is done by _________________ cells.

32. There are epithelial cells in earthworms that perform some functions similar to the vertebrate liver. They are _________________ cells.

33. In the shark intestine, food passage is slowed and absorptive surface increased by the _________________.

34. The roof of a snake's mouth has a specialized olfactory organ called _________________.

35. The three pairs of salivary glands in humans are the _________________, _________________, and _________________.

36. The _________________ cells secret pepsinogen and the _________________ cell secrete HCl.

37. The fat-splitting enzyme secreted in the stomach is _________________.

38. The term for the liquified food that reaches the small intestine is _________________.

39. Many digestive enzymes are secreted into the small intestine from the _________________.

40. Deficiencies of vitamin K and folic acid are prevented in humans by the presence of _________________ in the large intestine.

41. The inactive form of pancreatic carboxypeptidase is _________________.

42. The condition _________________ occurs when enzymes attack and destroy pancreatic tissue itself.

43. The eight amino acids that must be supplied in the human diet are the _______________.

44. Small amounts of _______________ in the diet seems to prevent tooth decay in humans.

Questions for Discussion

1. Compare the teeth of sharks, herbivorous fish, grazing mammals, and humans. What types of specialization are seen in teeth from these different groups? What might be told concerning the diet of prehistoric human ancestors by examining the teeth which have been discovered?

2. Some vertebrates have relatively short digestive systems, while others have very long, highly folded digestive systems. What differences in diet might account for digestive systems of different lengths?

3. When you learn the essentials of nutrition, you learn that some fairly complex molecules must be provided in your diet, such as the essential fatty acids and the essential amino acids. Other complex molecules that the body needs are not required in the diet. Why must some be in the food we eat? Why are others not essential in our diet?

4. Take a typical (for some) meal: hamburger with lettuce and tomato, french fried potatoes cooked in beef tallow, and milk shake. Trace this completely through the process of digestion. State what digestive processes occur and where they occur. Be as detailed as you can.

Answers to Self-Exam

Multiple Choice Questions

1. c
2. a
3. c
4. d
5. b
6. d
7. d
8. b
9. a
10. b
11. c
12. c
13. c
14. b
15. a
16. a
17. b

True or False Questions

18. T
19. F
20. T
21. F
22. F
23. T
24. F
25. F
26. F
27. T
28. F
29. T
30. T

Fill in the Blank Questions

31. collar cells or choanocytes
32. chloragen
33. spiral valve
34. Jacobson's organ
35. parotids, submandibulars, and sublinguals
36. chief cells, parietal cells
37. gastric lipase
38. chyme
39. pancreas
40. bacteria
41. procarboxypeptidase
42. pancreatitis
43. essential amino acids
44. fluoride

CHAPTER 39 Circulation

Learning Objectives

After mastering the material covered in Chapter 39, you should be able to confidently do the following tasks:

- Show how some animals exchange materials with their environments without a circulatory system.
- Explain the advantage of circulatory systems, and distinguish open and closed circulatory systems.
- Compare the circulatory systems in the major groups of vertebrates, and trace the development of the two-, three-, and four-chambered hearts.
- Describe the structure, function, and control of the human heart.
- Identify the major human circulatory system circuits.
- Explain how exchange of nutrients, gases, and wastes takes place in capillary beds.
- List the principal cellular and plasma components of blood.
- Describe the human lymphatic system, and explain its function.

Chapter Outline

I. Animals Without Circulatory Systems.

II. Circulatory Systems in Invertebrates.

A. Open and closed circulatory systems.

III. Transport in Vertebrates.

A. Fishes and the two-chambered heart.
B. Amphibians, reptiles, and the three-chambered heart.
C. Birds, crocodiles, and mammals and the four-chambered heart.

IV. The Human Circulatory System.

A. Circulation through the heart.
B. Control of the heart.
C. The working heart.
D. Blood pressure.

E. Circuits in the human circulatory system.
    1. hepatic portal circuit.
    2. renal circuit.
    3. cardiac circuit.
    4. systemic circuit.
F. Capillaries.
G. Veins.
H. Blood.
    1. red blood cells.
    2. white blood cells.
    3. platelets.
    4. clotting.

V. The Lymphatic System

Key Words

intracellular fluids
interstitial fluids
closed circulatory system
open circulatory system
hemocoels
ostium (-ia)
aortic arches
two-chambered heart
atrium
auricle
ventricle
sinus venosus
conus arteriosus
ventral aorta
hepatic portal vein
three-chambered heart
pulmonary circuit
systemic circuit
right atrium
left atrium
septum
four-chambered heart
right ventricle
left ventricle
pulmonary arteries
arterioles
capillaries
venules
pulmonary veins
superior vena cava
inferior vena cava
tricuspid valve
chordae tendineae
pulmonary semilunar valve
bicuspid valve
mitral valve
aortic semilunar valve
aorta
iliac arteries
extrinsic control
intrinsic control
sympathetic nerves
parasympathetic nerves
epinephrine
sinoatrial node (SA)
pacemaker
atrioventricular node (AV)
bundle of His
Purkinje fibers
systole
diastole
stroke volume
cardiac output
blood pressure
systolic pressure
diastolic pressure
sphyngmomanometer
arteriosclerosis
atherosclerosis
vasodilation
vasoconstriction
sinusoids
renal arteries
renal veins
coronary arteries
anastomoses
coronary thrombosis
coronary veins
coronary sinus
precapillary sphincters
plasma
erythrocytes
albumins
globulins
antibodies
immunoglobins
red bone morrow
leukocytes
neutrophils
basophils
eosinophils
lymphocytes
monocytes
macrophages
natural killer cells
platelets
thrombocytes
platelet mother cells
megakaryocytes
hemocytoblast
stem cell
prothrombin
fibrinogen
thromboplastins
thrombin
fibrin
hemophilia
lymph vessels
lymph nodes
lymph capillaries
lymphatic collecting ducts
lymphatics

Exercises

1. A diagrammatic section of the human heart is shown below. Label the parts of the heart, using the terms listed below. Then use a blue pencil to draw arrows indicating the circulation of blood through the heart prior to flowing to the lungs. Then use a red pencil to draw arrows showing circulation of blood through the heart after it had returned from pulmonary circuit. Check your work with Figure 39.9 in the text.

aorta
aortic semilunar valve
bicuspid valve
inferior vena cava
left atrium
left ventricle
pulmonary artery
pulmonary semilunar valve
right atrium
right ventricle
sinoatrial node
superior vena cava
tricuspid valve

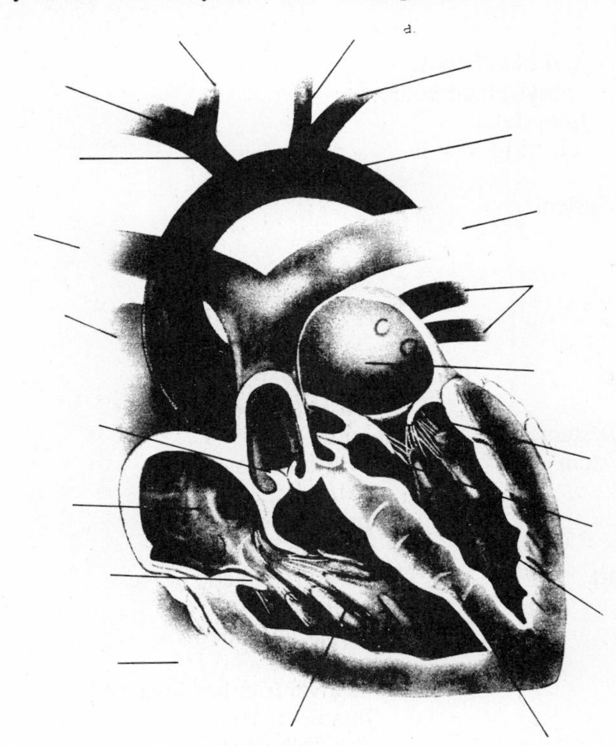

2. The blood pressure in the brachial artery varies through the cardiac cycle as the chambers of the heart contract and relax. Assume an individual has a heart rate of 72 beats per minute. How many beats would there be in a 5-second period? Sketch a tracing of the variations in blood pressure that might be seen in a normal young adult over a period of 5 seconds. Show the systolic and diastolic pressures. Check your work with Figure 39.12(b) in the text.

Self-Exam

You should be able to easily answer the following questions after learning the material in Chapter 39. If you have difficulty with any question, study the appropriate section in the text and try again.

A. Multiple Choice Questions

Circle one alternative that best completes the statement or answers the question.

1. An example of a multicellular animal that relies only on its cell surfaces for exchanging materials is
   a. the earthworm.
   b. a sponge.
   c. an insect.
   d. a crayfish.

2. Which of the following typically have organized circulatory systems?
   a. cnidarians.
   b. hydroids.
   c. jellyfish.
   d. arthropods.

3. In general, arthropods have
   a. no specialized circulatory system.
   b. an open circulatory system.
   c. a closed circulatory system.
   d. a portal circulatory system.

4. The sequence of blood flow through a typical fish heart would be
   a. sinus venosus ⇒ atrium ⇒ ventricle ⇒ conus arteriosus.
   b. sinus venosus ⇒ ventricle⇒ atrium ⇒ conus arteriosus.
   c. conus arteriosus ⇒ atrium ⇒ ventricle ⇒ sinus venosus.
   d. conus arteriosus ⇒ ventricle⇒ atrium ⇒ sinus venosus.

5. Which of the following groups of vertebrates typically has a two-chambered heart?
   a. fish.
   b. amphibians.
   c. reptiles.
   d. birds.

6. Which of the following groups of vertebrates typically has a four-chambered heart?
   a. fish.
   b. frogs.
   c. snakes.
   d. birds.

7. Which accurately describes the way blood arrives at the human heart?
   a. Blood arrives at the right atrium via the superior and inferior vena cava.
   b. Blood arrives at the left atrium via the superior and inferior vena cava.
   c. Blood arrives at the right ventricle via the superior and inferior vena cava.
   d. Blood arrives at the left ventricle via the superior and inferior vena cava.

8. The atrioventricular node (AV)
   a. transmits impulses across the atrial walls.
   b. initiates contraction of the two ventricles.
   c. causes the atria and ventricles to contract simultaneously.
   d. allows the ventricular contraction to precede atrial contraction.

9. During a human heartbeat, systole is

a. the period during which the chambers of the heart fill with blood.
b. the period of ventricular relaxation.
c. the period of ventricular contraction.
d. the period of atrial contraction.

10. For a healthy young human adult, typical systolic and diastolic blood pressures are, respectively, about
    a. 72 mmHg and 160 mmHg.
    b. 80 mmHg and 120 mmHg.
    c. 120 mmHg and 80 mmHg.
    d. 72 mmHg and 40 mmHg.

11. The disorder "hardening of the arteries" is technically called
    a. arteriosclerosis.
    b. atherosclerosis.
    c. a coronary.
    d. a stroke.

12. Relaxing the smooth muscle of arterial walls should lead to
    a. vasodilation and increased blood pressure.
    b. vasoconstriction and decreased blood pressure.
    c. vasodilation and decreased blood pressure.
    d. vasoconstriction and increased blood pressure.

13. The hepatic portal vein connects
    a. the heart and the liver.
    b. the heart and the small intestine.
    c. the small intestine and the liver.
    d. the small intestine and the kidney.

14. Kidneys are supplied with blood by the
    a. hepatic portal vein.
    b. renal artery.
    c. pulmonary vein.
    d. coronary artery.

15. Which sequence below correctly describes blood flow in the coronary circuit?
    a. right ventricle ⇒ coronary arteries ⇒ coronary veins ⇒ coronary sinus ⇒ left atrium.
    b. left ventricle ⇒ coronary arteries ⇒ coronary veins ⇒ coronary sinus ⇒ right atrium.
    c. right atrium ⇒ coronary arteries ⇒ coronary veins ⇒ coronary sinus ⇒ left ventricle.
    d. left atrium ⇒ coronary arteries ⇒ coronary veins ⇒ coronary sinus ⇒ right ventricle.

16. Which of the following is a major component of plasma?
    a. leukocytes.
    b. platelets.
    c. erythrocytes.
    d. water.

17. Numerically, the majority of leukocytes are
    a. neutrophils.
    b. basophils.
    c. eosinophils.
    d. macrophages.

18. Which of the following is *not* generally considered to be a role of the lymphatic system?
    a. assist in the work of the immune system.
    b. maintain fluid and electrolyte balance in the body.
    c. transport certain fatty acids from the intestinal villi to the blood.
    d. transport glucose to cells for respiratory demands.

B. True or False Questions

Mark the following statements either T (True) or F (False).

_____ 19. The larger an animal is, the larger its surface area to volume ratio, and the more easily it accommodates exchange processes via its surface alone.

_____ 20. Some arthropods often have several hearts.

_____ 21. In an arthropod with an open circulatory system, blood is drawn into the heart through openings called ostia.

_____ 22. Annelid worms typically have a open circulatory system with five pairs of aortic arches.

_____ 23. The hepatic portal vein carries blood to the kidney.

_____ 24. In the three-chambered reptile heart, deoxygenated blood returning from the body enters the left atrium.

_____ 25. The pulmonary arteries of humans carry deoxygenated blood.

_____ 26. In the human heart, the bicuspid or mitral valve is between the left atrium and left ventricle.

_____ 27. Human hearts will not beat unless connected to at least one major nerve.

_____ 28. In periods of great activity, the human heart can pump as much as 30 to 35 liters per minute.

_____ 29. In general, vasodilation increases blood pressure.

_____ 30. As blood flows through capillaries, most water returns to the bloodstream through osmosis.

_____ 31. Because it is a fluid, blood is not considered a tissue in the body.

_____ 32. Like red blood cells, white blood cells lack nuclei at maturity.

_____ 33. Both blood cells and platelets originate from hemocytoblasts.

C. Fill in the Blanks

Answer the question or complete the statement by filling in the blanks with the correct word or words.

34. The fluid in the body that is not found within cells is termed the ________________ fluid.

35. When blood remains within vessels throughout its entire circulation, an animal has a ________________ circulatory system.

36. In arthropods, beyond the ends of arteries, blood enters ________________, or blood cavities.

37. The blood leaving a fish's two-chambered heart passes through the ________________.

38. The tricuspid valve of the human heart is prevented from collapsing backward by the ________________.

39. The only vein which carries oxygenated blood is the ________________ vein.

40. Impulses from the AV node pass through a specialized muscle called the ________________.

41. The amount of blood passing through the heart with each heartbeat is the ________________.

42. Blood pressure is usually measured with a ________________.

43. Arterial lesions or plaques, deposits of lipids such as cholesterol, are termed ________________.

44. Minute cavities in the liver through which blood passes are called ________________.

45. A blood clot occurring n a vessel of the coronary circuit is a ________________.

46. Opening and closing selected capillary beds can be accomplished by ________________ of smooth muscle.

47. In the entire circulatory system, blood pressure is usually lowest in the ________________.

48. Red and white blood cells are produced continuously in the ________________.

49. Blood clots are made up of damaged platelets and fibers of the protein ________________.

50. Vitamin ________________ is important for blood clotting, because it is required by the liver for the synthesis of prothrombin.

Questions for Discussion

1. Some evolutionary biologists believe that specialized circulatory systems exist because there is an upper limit to the size of an organism for which diffusion alone is adequate for transport of nutrients, gases and waste. This has been examined with respect to surface area-to-volume ratios. Explain this reasoning. What are some of the ways that larger organisms have effectively increases the surface area available for exchange.

2. Among vertebrates there seems to be a progression from a two-chambered heart to a three-chambered heart to a four chambered heart. Presumably there is an evolutionary path that connects the various heart types. Which type of heart do you think is the ancestral type? Why? Explain why the more "advanced" type of heart might have developed through evolutionary change.

3. Assume you are in a science-fiction movie, and you have been reduced to a tiny scientist assigned to ride in a tiny "submarine" through a human's circulatory system. Assume you enter through the lungs: trace a complete journey through the circulatory system, visiting as many of the viscera as you can. Your instruments allow you to measure blood chemistry as you travel. What do you observe with respect to oxygen, carbon dioxide, food, and waste molecules?

4. Why is the cuff of a sphygmomanometer usually placed around a person's upper arm rather than somewhere else? Would the blood pressure reading be the same in a person's fingers or toes? Is it more informative to measure arterial or venous pressure? Why?

Answers to Self-Exam

Multiple Choice Questions

1. b
2. d
3. b
4. a
5. a
6. d
7. a
8. b
9. c
10. c
11. a
12. c
13. c
14. b
15. b
16. d
17. a
18. d

True or False Questions

19. F
20. T
21. T
22. F
23. F
24. F
25. T
26. T
27. F
28. T
29. T
30. T
31. F
32. F
33. F

Fill in the Blank Questions

34. interstitial
35. closed
36. hemocoels
37. ventral aorta
38. chordae tendineae
39. pulmonary
40. bundle of His
41. stroke volume
42. sphyngmomanometer
43. atherosclerosis
44. sinusoids
45. coronary thrombosis
46. precapillary sphincters
47. veins
48. red bone marrow
49. fibrin
50. vitamin K

CHAPTER 40 Respiration

Learning Objectives

After mastering the material covered in Chapter 40, you should be able to confidently do the following tasks:

- Identify the types of gas exchange surfaces seen in animals, from simple skin surface exchange to tracheae, gills, and lungs.
- Explain the specialized features of gills that make them efficient exchange surfaces.
- Show the adaptations of the lungs and explain its importance for terrestrial animals.
- Describe the anatomy of the human respiratory system.
- Explain how gas exchange takes place between air and the bloodstream in the human lungs, and describe how oxygen and carbon dioxide are transported between the lungs and other tissue.
- Give an explanation of the control mechanisms involved in regulation of the human respiratory system.

Chapter Outline

I. Gas Exchange Surfaces.

- A. The simple body interface.
- B. Tracheae.
- C. Gills.
  1. gills in mollusks and arthropods.
  2. gills in fishes.
- D. Lungs.
  1. evolution of the vertebrate lung.
  2. the vertebrate lung.

II. The Human Respiratory System.

- A. The breathing movements.
- B. The exchange of gases.
  1. partial pressures.
  2. exchange in the alveoli.
- C. Oxygen transport.
- D. Carbon dioxide transport.
- E. The control of respiration.
  1. neural control.
  2. chemical control.

Key Words

respiratory system
respiratory interface
dermal branchae
tracheal system
tracheae
tracheoles
spiracles
air sacs
tracheal gills
gill
hemocyanin
carapace
bailer
open circulatory system
gill arches
gill filaments
lamellae
afferent vessels
efferent vessels
countercurrent exchange
operculum (-a)
preadaptation
internal nares
palate
swim bladder
crosscurrent flow
pharynx
larynx
trachea
bronchi
lungs
diaphragm
soft palate
goblet cells
voice box
vocal cords
primary bronchi
bronchioles
respiratory tree
alveolus (-i)
pleura (-ae)
pleural cavity
ventilation
inspiration
expiration
intercostal muscles
vital capacity
residual air
partial pressure
standard conditions
heme group
deoxyhemoglobin
oxyhemoglobin
erythropoietin
Bohr effect
carbaminohemoglobin
carbonic acid
carbonic anhydrase
sodium bicarbonate
acid-base buffering system
inspiratory centers
expiratory centers
aortic bodies
carotid bodies

Exercises

1. Provide labels for the following diagram of the human respiratory system. Use terms from the list given below. Check your work with Figure 40.9 in the text.

alveoli
bronchi
bronchiole
diaphragm
epiglottis
larynx
lung
nasal cavity
nostril
pharynx
trachea

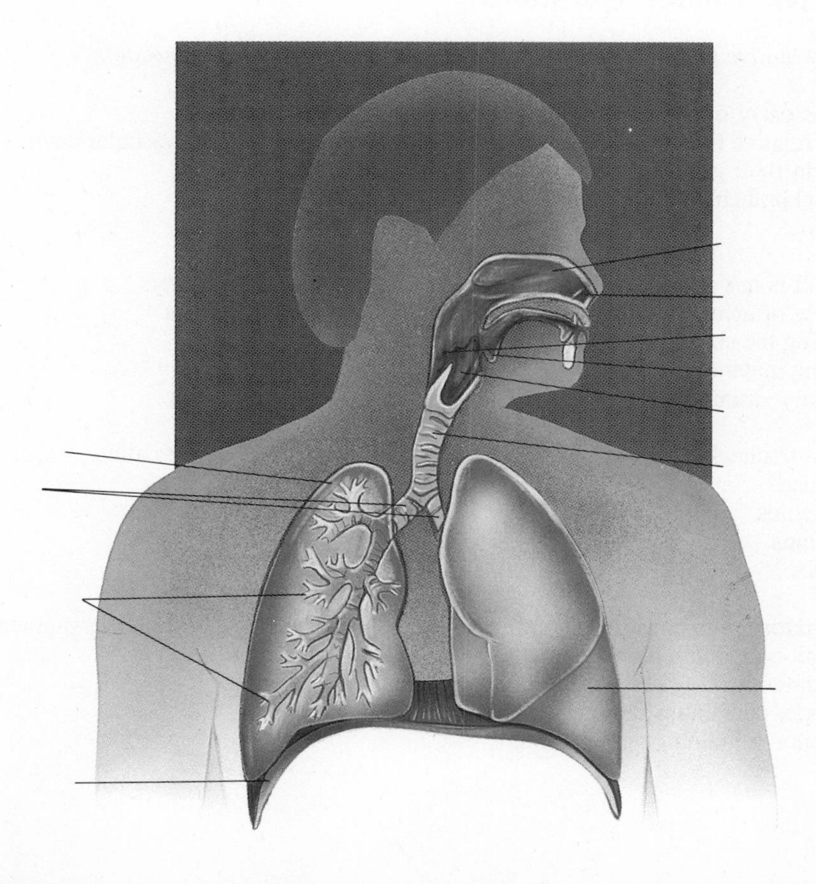

2. The atmospheric gases oxygen and carbon dioxide are exchanges and transported to tissue by the human respiratory system. The relative amounts of these two gases vary. The partial pressures of O2 and CO_2 may be measured at different points. On the axes below, plot these partial pressures for both gases.

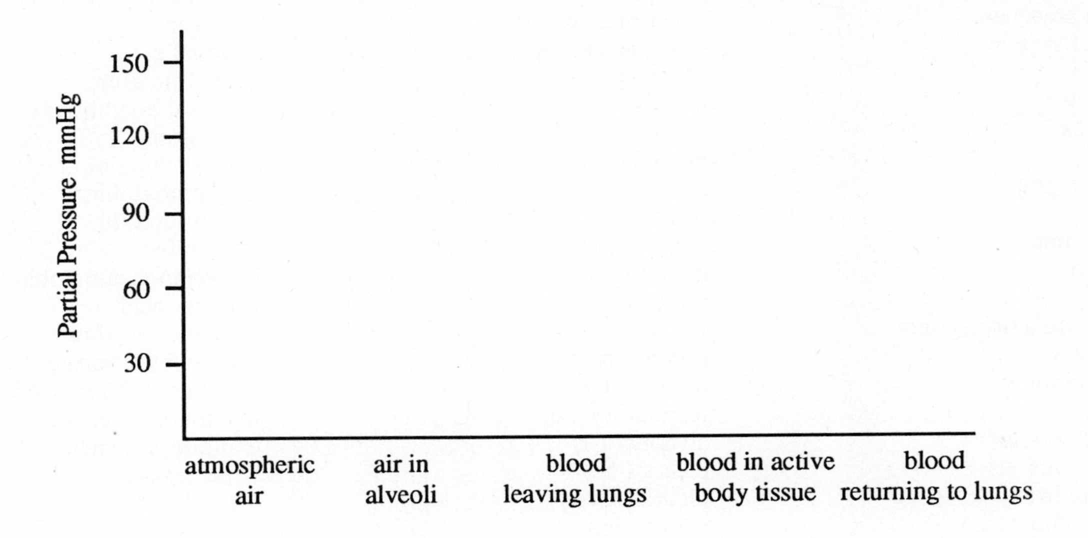

Self-Exam

You should be able to easily answer the following questions after learning the material in chapter 40. If you have difficulty with any question, study the appropriate section in the text and try again.

A. Multiple Choice Questions

Circle one alternative that best completes the statement or answers the question.

1. Flatworms carry out respiratory exchange via
   a. their relative large surface area and their highly branched gastrovascular cavity.
   b. gills in their gastrovascular cavity.
   c. dermal branchae.
   d. lungs.

2. Which trait is *not* shared by both earthworms and lungless salamanders?
   a. the use of hemoglobin as an oxygen-carrying respiratory pigment.
   b. keeping the skin moist with a slimy layer of mucus.
   c. staying in damp places.
   d. a highly vascularized pharynx for gas exchange.

3. The insect tracheal system has tiny valve-like openings to the outside called
   a. tracheae.
   b. tracheoles.
   c. spiracles.
   d. air sacs.

4. The respiratory pigment in some mollusks turns blue when oxygenated; this pigment is
   a. an iron-containing hemoglobin.
   b. an iron-containing hemocyanin.
   c. a copper-containing hemocyanin.
   d. a copper-containing hemoglobin.

5. Which is generally true concerning the amount of oxygen dissolved in water?
   a. There is a greater percentage of oxygen in water than in air.
   b. There is more oxygen in warm water than in cold water.
   c. There is more oxygen in cold water than in hot water.
   d. Oxygen is insoluble in water.

6. The correct sequence of structures encountered as air flows into the human lung is
   a. mouth or nasal passages ⇒ pharynx ⇒ larynx ⇒ trachea ⇒ bronchi ⇒ lung.
   b. mouth or.nasal passages ⇒ larynx ⇒ pharynx ⇒ trachea ⇒ bronchi ⇒ lung.
   c. mouth or nasal passages ⇒ pharynx ⇒ larynx ⇒ bronchi ⇒ trachea ⇒ lung.
   d. mouth or nasal passages ⇒ larynx ⇒ pharynx ⇒ bronchi ⇒ trachea ⇒ lung.

7. Inhaled dust is move out of the respiratory passages mainly by
   a. exhaling forcefully.
   b. mucus and cilia.
   c. coughing.
   d. hiccuping.

8. The residual volume of the lungs is about
   a. 1.5 gallons.
   b. 1.5 liters.
   c. 0.15 gallons.
   d. 0.15 liters.

9. At standard conditions, the partial pressure of oxygen in the atmosphere at sea level is about
   a. 760mm Hg.
   b. 0.04% of 760mm or about 0.3mm Hg.
   c. 21% of 760mm or about 160mm Hg.
   d. 100mm Hg.

10. On a mountaintop the partial pressure of oxygen in the atmosphere
    a. is the same as at sea level.
    b. is less than at sea level.
    c. is more than at sea level.
    d. may be either leas than or more than at sea level.

11. Which statement about the oxygen-carrying capacity of respiratory pigments is most accurate?
    a. Each hemoglobin can carry 8 oxygen atoms.
    b. Each heme can carry 8 oxygen atoms.
    c. Each hemoglobin can carry two oxygen atoms.
    d. Each heme can carry two O_2 molecules.

12. Which statement best describes the Bohr effect?
    a. When the CO_2 level is low, hemoglobin holds its O_2 more tightly.
    b. When the CO_2 level is low, hemoglobin holds its O_2 less tightly.
    c. When the partial pressure of CO_2 is high, hemoglobin has a higher affinity for O_2.
    d. When the partial pressure of CO_2 is high, hemoglobin has a higher affinity for CO_2.

13. Which of the following is likely to happen in the blood of the alveolar capillaries?
    a. Hemoglobin and carbon dioxide form carbaminohemoglobin.
    b. Carbon dioxide and water form carbonate ions.
    c. Sodium ions and bicarbonate ions form sodium bicarbonate.
    d. Carbaminohemoglobin releases its carbon dioxide.

14. Where are the neural inspiratory centers and expiratory centers located?
    a. pituitary.
    b. cerebellum.

c. cortex.
d. medulla.

15. The breathing rate increases in response to chemoreceptors when
    a. the carbon dioxide level increases and blood acidity decreases.
    b. the carbon dioxide level increases and blood acidity increases.
    c. the carbon dioxide level decreases and blood acidity increases.
    d. the carbon dioxide level decreases and blood acidity decreases.

B. True or False Questions

Mark the following statements either T (True) or F (False).

_____ 16. Much of the oxygen present in the atmosphere today may have arisen as a toxic product of early photosynthetic life.

_____ 17. Although many respiratory surfaces are moist, gases normally cross dry membranes just as easily.

_____ 18. Some salamanders use their skins as respiratory gas exchange surfaces.

_____ 19. If groups of animals that have similarly constructed gills, we can always conclude that they share common ancestry.

_____ 20. Very cold water, when saturated with oxygen, will have 21% oxygen.

_____ 21. An early ancestor of bony fishes probably obtained oxygen by gulping air through its mouth.

_____ 22. In many of today's bony fish, the primitive swim bladder has become the modern lung.

_____ 23. Some amphibians breathe through their skins when they are dormant in cold or dry seasons.

_____ 24. The active bird lung inflates and deflates rapidly and frequently during breathing.

_____ 25. Most humans can easily breathe and swallow at the same time.

_____ 26. When a human breathes, the weight of the atmosphere forces air down the trachea to inflate the lungs.

_____ 27. A typical partial pressure for carbon dioxide in blood entering the alveoli might be about 45mm Hg.

_____ 28. Equal weights of water and hemoglobin can carry about the same amount of oxygen.

_____ 29. The affinity of hemoglobin for oxygen is unaffected by such factors as temperature or pH of the blood.

_____ 30. In terms of regulation, respiration is strictly under neural control.

C. Fill in the Blanks

Answer the question or complete the statement by filling in the blanks with the correct word or words.

31. In echinoderms, the respiratory interface is increased by _________________ outpocketing of the coelomic wall that protrude through pores in the endoskeleton.

32. _________________ are thin-walled, feathery structures with extensive capillary beds that function as exchange surfaces for respiratory gases.

33. The gill serves in both food filtering and respiration in mollusks such as the _________________.

34. Carbon dioxide-laden blood enters the gill filaments from _________________ vessels, and upon leaving the gills, the oxygen-rich blood passes through _________________ vessels.

35. In gills, the opposing flow of water and blood sets up an efficient _________________ exchange system.

36. In mammals, the nasal cavity is separated from the mouth by the _________________.

37. In the bird lung, air passages and capillaries are at right angles, setting up a _________________ flow.

38. In humans, the nasal cavity is lined with mucus glands, cells bearing cilia, and mucus-secreting _________________ cells.

39. In humans, the airtight, bag-like membranes that enclose the lungs are the _________________.

40. The maximum volume of air that can be moved into or out of the lungs in a single breath is termed the _________________.

41. Most hemoglobin entering the lungs is called _________________ while most hemoglobin leaving the lungs is termed _________________.

42. If the oxygen level in blood passing through the kidney is below a certain level, the hormone _________________ is released.

43. The combined form of carbon dioxide and hemoglobin is called _________________.

44. The enzyme _________________ speeds formation of carbonic acid in the bloodstream.

45. The arterial chemoreceptors that detect elevated carbon dioxide levels in the blood are the _________________ and the _________________.

Questions for Discussion

1. As we examine respiratory exchange systems from very simple small animals to larger more complex animals, we see a trend of adaptations to increase the surface area available for respiratory exchange. Outline the ways that exchange surface area is increased from the simplest invertebrates, through the insects and aquatic vertebrates, to the air-breathing terrestrial vertebrates.

2. How do fish "breathe"? Do all fish do it the same way? Give examples of specific mechanisms. What would happen to a shark if held stationary?

3. Some birds are able to fly at altitudes so great that a human would have inadequate oxygen. What is there about these birds' respiratory systems that enables them to remain active at such oxygen-poor elevations?

4. In the human respiratory system, exchanges of oxygen and carbon dioxide occur between the atmosphere and the bloodstream on one hand, and between the bloodstream and metabolically active tissue on the other. The actual exchanges involve diffusion of gases into and out of the blood cells. Discuss the role of the partial pressure of each gas on the exchange process. Be sure to consider all the movements of oxygen and carbon dioxide into and out of the blood.

5. What adaptations would be necessary for humans to become able to swim freely underwater as many marine and freshwater vertebrates do? Can you think of different ways this might be accomplished? Which do you think is most reasonable, based on your knowledge of marine mammals?

Answers to Self-Exam

Multiple Choice Questions

1. a
2. d
3. c
4. c
5. c
6. a
7. b
8. b
9. c
10. b
11. a
12. a
13. d
14. d
15. b

True or False Questions

16. T
17. F
18. T
19. F
20. F
21. T
22. F
23. T
24. F
25. F
26. T
27. T
28. F
29. F
30. F

Fill in the Blank Questions

31. dermal branchae
32. gills
33. bivalves or clams
34. afferent; efferent.
35. countercurrent
36. palate
37. crosscurrent
38. goblet
39. pleurae
40. vital capacity
41. deoxyhemoglobin; oxyhemoglobin
42. erythropoietin
43. carbaminohemoglobin
44. carbonic anhydrase
45. carotid bodies; aortic bodies

CHAPTER 41 The Immune System

Learning Objectives

After mastering the material covered in Chapter 41, you should be able to confidently do the following tasks:

- Identify the components of the human immune system.

- List the elements of the first line of defense against infection.

- Describe the non-specific cellular and chemical defenses of the immune system.

- Distinguish the classes of cells comprising the immune system and explain their roles in the immune response.

- Outline the sequence of events that occur in the primary and secondary responses of the immune system.

- Describe antibody structure in general and give the basis for specific antibody diversity.

- Explain the basis for immunization..

- Describe how immune system function is affected by AIDS, and explain why the immune response is ineffective against HIV.

Chapter Outline

I. The First Line of Defense.

II Secondary Defenses: Non-specific and Specific.

  A. Non-specific chemical defenses.
    1. histamine.
    2. kinins.
    3. complement.
    4. interferon.
  B. Non-specific cellular defenses.
    1. phagocytes.
    2. natural killer cells.

III. Specific Chemical and Cellular Defenses.

  A. Lymphatic tissues and lymphocytes.
  B. B and T lymphocytes.
  C. Antibodies and the humoral response.
  D. Lymphocytes and cell recognition.

IV. Clonal Selection and the Primary Immune Response.

- A. Antigen-presenting cells.
- B. Aroused T-cells.
- C. B-cells.
- D. Suppressor T-cells.
- E. Memory cells and the quick response.

V. The Secondary Immune Response.

VI. The Road to Lymphocyte Diversity.

VII. Monoclonal Antibodies.

VIII. When the Immune System Goes Wrong.

- A. Autoimmune disease.
- B. AIDS.

IX. Immunity in Other Animals.

Key Words

integument
keratin
lysozymes
non-specific defenses
histamine
kinins
complement
opsonization
C3b receptor
interferon
leukocyte
eosinophil
neutrophil
monocyte
basophil
lymphocyte
phagocyte
pus
macrophage
natural killer (NK) cell
specific defenses
immune responses
lymphoid tissue
primary lymphoid tissue
thymus
red bone marrow
secondary lymphoid tissue
lymph node
tonsils
adenoids
spleen
B-cells
T-cells
thymus
bursa of Fabricus
cell-mediated response
humoral response
antibody
immunoglobin
antigen
immunogen
antigenic determinant
heavy chain
light chain
variable region
constant region
IgA
IgM
IgD
IgG
IgE
mast cells
Fc region
Fc receptor
major histocompatibility complex (MHC)
human leukocyte associated antigen gene cluster (HLA)
cytotoxic T-cells
helper T-cells
virgin T-cells
virgin B-cells
primary immune response
clonal selection
antigen-presenting macrophages
interleukin I
metastasis
plasma cells
memory B-cells
suppressor T-cells
secondary immune response
artificial active immunity
antiserum
passive immunization
gene rearrangement
monoclonal antibodies
myeloma cells
hybridoma
autoimmunity
autoimmune disease
arthritis
nephritis
rheumatic fever
systemic lupus erythematosus
AIDS
Kaposi's sarcoma
pneumocystic pneumonia
retrovirus
human immunodeficiency virus (HIV)
human T-cell lymphotrophic virus (HTLV-III)
CD4
azidothymidine (AZT)
reverse transcriptase

Exercises

1. In the space below, complete a table to summarize the different types of cells involved in the immune system and the functions they perform in defending the body against infection. A couple of cell types are given to help start the list, but you should make it as complete and well-organized as possible.

Cell Type	Source	Function
I. Non-specific cellular defenses:		
Leucocytes	Red bone marrow	
eosinophils		
neutrophils		
II. Specific cellular defenses:		
T-cells		
cytotoxic T-cells		
B-cells		
virgin B-cells		

2. In the space below, draw a schematic diagram of an antibody such as immunoglobin G. Show all four polypeptides. Label the light chains and heavy chains, and show where disulfide bonding occurs. Indicate the constant regions and the variable regions. Show the antigen-binding sites. Check your work with Figure 41.9 in the text.

Self-Exam

You should be able to easily answer the following questions after learning the material in Chapter 41. If you have difficulty with any question, study the appropriate section in the text and try again.

A. Multiple Choice Questions

Circle one alternative that best completes the statement or answers the question.

1. Which of the following is *not* considered part of the body's first line of defense against invasion?
   a. The integument of the body.
   b. Secretions from sweat and oil glands.
   c. Lysozymes in tears and other body secretions.
   d. Specialized leucocytes.

2. Which of these elements of the immune system causes the redness associated with inflammation?
   a. histamine.
   b. interferon.
   c. B-cells.
   d. T-cells.

3. Which cells have C3b receptors and can bind to complement?
   a. NK cells.
   b. phagocytes.
   c. lymphocytes.
   d. basophils.

4. Which of the following cell types is *not* a type of leucocyte?
   a. eosinophil.
   b. monocyte.
   c. lymphocyte.
   d. erythrocyte.

5. Which of the following cell types is *not* a type of phagocyte?
   a. eosinophil.
   b. basophil.
   c. neutrophil.
   d. monocyte.

6. The most numerous class of phagocytes in the human includes the
   a. eosinophils.
   b. basophils.
   c. neutrophils.
   d. monocytes.

7. The primary lymphoid tissue exists mainly in
   a. the thymus and red bone marrow.
   b. the spleen and pancreas.
   c. the lymph nodes.
   d. the liver.

8. Lymphocytes which are T-cells differentiate
   a. in the thymus.
   b. in red bone marrow and the fetal liver.
   c. in the Bursa of Fabricus.
   d. in the spleen.

9. Immunologically speaking, an antigen is
   a. produced by B-cells.
   b. produced by T-cells.
   c. a foreign substance to which the immune system responds.
   d. the same as an antibody.

10. An antibody is made up of how many polypeptides?
    a. one.
    b. two.
    c. three.
    d. four.

11. The part of an antibody which binds to a phagocyte during opsonization is
    a. the variable region.
    b. the constant region.
    c. the antigen region.
    d. the interaction region.

12. The least common antibody type associates with mast cells in the allergic response; it is
    a. IgA.
    b. IgD.
    c. IgE.
    d. IgG.

13. Which cell type is destroyed preferentially by the AIDS virus?
    a. certain B-cells.
    b. certain T-cells.
    c. certain macrophages.
    d. certain monocytes.

14. The human major histocompatibility complex (MHC) involves
    a. a single gene.
    b. six genes.

c. hundreds of genes on the sixth chromosome.
d. no genes at all.

15. Complement is coded by which MHC class(es)?
a. MHC Class I.
b. MHC Class II.
c. MHC Class III.
d. MHC Classes I and II.

16. T-cells with receptors that match Class II MHC proteins are
a. cytotoxic T-cells.
b. helper T-cells.
c. dual T-cells.
d. Class II T-cells.

17. In the early stages of an infection, often one of the first immune cells to respond is
a. a giant macrophage.
b. a virgin T-cell.
c. a virgin B-cell.
d. a basophil.

18. Metastatic cancer cells seem to escape cytotoxic T-cells by
a. changing the T-cell antibody.
b. changing the cancer cell's MHC cell surface protein.
c. changing the B-cell recognition site.
d. inactivating T-cells.

19. Which statement best describes an antiserum?
a. An antiserum contains specific antibodies against an agent.
b. An antiserum contains T-cells.
c. An antiserum contains B-cells.
d. An antiserum contains specific antigens against an agent.

20. Which process leads to the high diversity of antibodies a human can produce?
a. Production of new genes.
b. Rearrangement of existing genes.
c. Opsonization.
d. Vaccination.

21. Systemic lupus erythematosus is an example of
a. an infectious disease.
b. an autoimmune disease.
c. a type of vaccination.
d. a pathogenic organism.

22. The frequency of AIDS in the United States is lowest among which of the following groups?
a. Homosexual males.
b. Intravenous drug users.
c. Blood recipients.
d. Married women.

23. In the United States, AIDS is transmitted most frequently via
a. sexual contact.
b. kissing.
c. hospital contact.
d. sharing tooth brushes and other personal items.

24. The greatest specialization in T-cell and B-cell function is found among
    a. annelid worms.
    b. fish.
    c. amphibians
    d. mammals.

B. True or False Questions

Mark the following statements either T (True) or F (False).

_____ 25. Complement may destroy an invader directly by forming holes in a bacterium's plasma membrane.

_____ 26. Interferon is expensive, since it is obtained only from natural sources, and cannot be produce through laboratory techniques.

_____ 27. Some phagocytes are residents of the brain.

_____ 28. Eosinophils can attact and kill some parasitic worms.

_____ 29. The non-specific defenses of the human body provide what are termed the immune responses.

_____ 30. B-cells specialize in cell-mediated responses of the immune system.

_____ 31. Free-floating IgM antibodies occur in large clusters of five "Y-shapes."

_____ 32. The process of opsonization makes it possible for phagocytes to bind to an invading cell.

_____ 33. Phagocytes are very specific in their recognition of invaders.

_____ 34. Identical twins are identical in the chemical nature of their MHC proteins.

_____ 35. Prevailing theory suggests that B-cells mature in the thymus.

_____ 36. Macrophages can present virgin T-cells with a sample of antigen from an invader.

_____ 37. Masses of antigen, clumped by antibodies, can be engulfed and destroyed by phagocytes.

_____ 38. Until recently, the rabies virus was treated with passive immunization.

_____ 39. Since B-cells grow well in tissue culture, it is easy to obtain antibodies from colonies of pure B-cells.

_____ 40. Kaposi's sarcoma is one of the causes of AIDS.

_____ 41. When attacking a helper T-cell, the AIDS virus first binds to the CD4 cell surface protein.

_____ 42. The AIDS-causing HIV is well known to invade brain tissue.

_____ 43. Condoms made of animal membranes are impervious to both sperms and viruses (like the HIV).

_____ 44. The immune response is known only from the vertebrate groups of animals.

_____ 45. Sponges and other invertebrates are known to reject tissue grafts from other individuals.

C. Fill in the Blanks

Answer the question or complete the statement by filling in the blanks with the correct word or words.

46. Injured cells release _________________, polypeptides that increase circulation and capillary permeability.

47. In the process of _________________, complement may come to coat the surface of a bacterial cell.

48. _________________ is the chemical produced by some cells invaded by double-stranded RNA viruses.

49. The _________________ are white blood cells that are mobilized in non-specific cellular defenses.

50. The remains of phagocytes which accumulate at the site of an infection contribute to the formation of _________________.

51. A type of lymphocyte that roams the body and kills cancer cells or cells containing viruses is termed a _________________ cell.

52. Regions where lymphocytes congregate are known as _________________.

53. Antigens contain localized regions called _________________ that trigger a response by a matching antibody.

54. The part of an antibody that binds to its matching antigen is found in the _________________ regions.

55. New T-cells that have completed development of dual receptors are termed _________________.

56. The initial response by B-cells and T-cells to an invader is called the _________________.

57. A macrophage can release _________________ which stimulates an attached T-cell to divide repeatedly.

58. The production and secretion of copious amounts of antibody by plasma cells is termed the _________________ response.

59. The _________________ T-cell modulates the immune response, eventually bringing it to a halt.

60. A large amount of antibody produced under lab conditions from a single B-cell line is _________________.

61. Monoclonal antibodies are produced by _________________ cells, the result of fusing a B-cell and a myeloma cell.

62. The virus responsible for AIDS is termed _________________.

Questions for Discussion

1. When a non-immunized person is exposed to a pathogenic virus such as the one causing measles, the disease may follow. If the victim survives, however, he is immune to subsequent infections by the same pathogenic virus. Outline in sequence the events in humoral and cellular immune response which defeat the initial infection and lead to subsequent immunity.

2. Many "colds" are caused by viruses, yet we do not seem to become immune to "colds." Why is this?

3. Describe the elements of a human's first line of defense against infection. How effective are these first line defenses?

4. How can B-cells make antibodies against thousands and thousands of different antigens?

5. What is the difference between active and passive immunization?

6. A person with Rh negative blood has no Rh antigen, while an Rh positive person does have an Rh antigen circulating in the bloodstream. An Rh negative individual will produce antibodies against an Rh antigen if exposed to Rh positive blood. There is risk involved when an Rh negative woman conceives Rh positive babies, because some fetal Rh positive blood cells may enter the mother's circulation. Often the first such pregnancy causes no serious health effects on mother or baby, but subsequent pregnancies are at great risk. Why is this? Explain the problem in terms of immune system response. Today, mothers with an Rh incompatible baby are often treated with "rhogam" at the time of delivery to prevent problems in futures pregnancies. What do you think "rhogam" is in terms of elements of the immune system?

Answers to Self-Exam

Multiple Choice Questions

1. d
2. a
3. b
4. d
5. b
6. c
7. a
8. a
9. c
10. d
11. b
12. c
13. b
14. c
15. c
16. b
17. a
18. b
19. a
20. b
21. b
22. d
23. a
24. d

True or False Questions

25. T
26. F
27. T
28. T
29. F
30. F
31. T
32. T
33. F
34. T
35. F
36. T
37. T
38. T
39. F
40. F
41. T
42. T
43. F
44. F
45. T

Fill in the Blank Questions

46. kinins
47. opsonization
48. Interferon
49. leukocytes
50. pus
51. natural killer (NK)
52. lymphoid tissues
53. antigenic determinants
54. variable
55. virgin T-cells
56. primary immune response
57. interleukin I
58. humoral
59. suppressor
60. monoclonal antibody
61. hybridoma
62. HIV or HTLV-III

CHAPTER 42 Reproduction

Learning Objectives

After mastering the material covered in Chapter 42, you should be able to confidently do the following tasks:

- Outline the different patterns of reproduction found among animals.
- Give examples of reproduction among the invertebrates.
- Compare the fertilization process among the major groups, tracing the trend from external to internal fertilization.
- Describe the structure and function of the human reproductive system in detail
- Compare human male and female reproductive anatomy.
- Explain how hormones regulate the human reproductive cycle.
- List the various approaches used for human contraception and evaluate the effectiveness of each method.

Chapter Outline

I. Modes of Reproduction.

A. Asexual reproduction.
B. Sexual reproduction.
1. one sex versus two sexes.
2. external fertilization.
3. internal fertilization.

II. Invertebrate Reproductive Patterns.

III. Sexual Reproduction in Vertebrates.

A. External fertilization in vertebrates.
B. Internal fertilization in vertebrates.

IV. Human Reproduction.

A. The male reproductive system.
1. anatomy.
2. spermatogenesis.

B. The female reproductive system.
   1. anatomy.
   2. oogenesis.

C. Hormonal control of human reproduction.
   1. male hormonal action.
   2. female hormonal action.
   3. the menstrual cycle.

D. Contraception
   1. natural methods.
   2. mechanical devices.
   3. chemical methods.
   4. surgical intervention.

Key Words

fission
budding
fragmentation
gemmules
parthenogenesis
Daphnia
monoecious
hermaphroditic
dioecious
external fertilization
internal fertilization
copulation
penis
Helix
testis (-es)
spermatophores
vagina
spermatheca
ovary
oviduct
salmon
stickleback
cloaca
glans
hemipenes
placenta
monotreme
spiny anteater
echidna
duck-billed platypus
marsupial
scrotum
seminiferous tubule
germinal epithelium
interstitial cells
rete testis
epididymis
vas deferens
spermatic cord
seminal vesicles
semen
urethra
prostate gland
bulbourethral gland
Cowper's gland
erectile tissue
corpora cavernosa
corpus spongiosum
spermatogenesis
spermatogonia
primary spermatocyte
secondary spermatocyte
spermatid
spermatozoa
Sertoli cells
sperm head
acrosome
vulva
mons pubis
mons veneris
labia majora
labia minora
clitoris
glans clitoridis
urethral meatus
introitus
hymen
vagina
vestibular glands
Bartholin's glands
uterus
oviducts
fallopian tubes
cervix
mucosa
endometrium
fimbriae
oocyte
ovum (-a)
oogenesis
gonadotropic releasing hormones
gonadotropins
follicle-stimulating hormone (FSH)
luteinizing hormone (LH)
interstitial cell-stimulating hormone (ICSH)
testosterone
inhibin
menstrual cycle
estrous cycle
ovulation
menstruation
period
primary follicle
Graafian follicle
estrogens
estradiol
estrone
corona radiata
corpus luteum
progesterone
human chorionic gonadotropin (HCG)
contraception
coitus interruptus
orgasm
rhythm method
condom
diaphragm
intrauterine device (IUD)
birth control pill
9-nonoxynol
contraceptive sponge
abortion
dilation and curettage (D&C)
vasectomy
tubal ligation

Exercises

1. The human menstrual cycle is under hormonal control. On the axes below, plot the relative increases and decreases in the pituitary hormones LH and FSH and the ovarian estrogen and progesterone levels over a complete menstrual cycle. Use different colors of pencil for each hormone, and label the lines clearly. Would these be the same if a pregnancy occurred? Check your work with Figure 42.22 in the text.

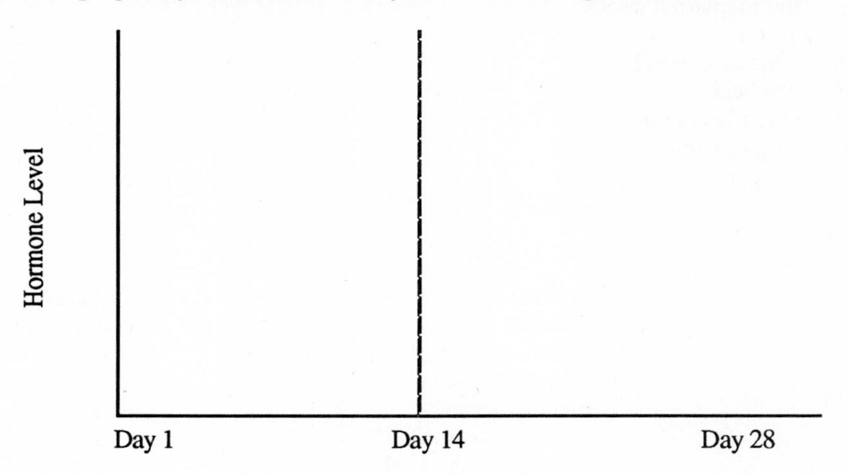

2. Provide labels for the following diagrams of human male and female reproductive anatomy, utilizing terms from the list below. Check your work with Figures 42.15 and 42.19 in the text.

bulbourethral gland
cervix
clitoris
epididymis
fimbriae
glans penis
labia majora
labia minora
mons pubis
ovary
oviduct
penis
prostate gland
scrotum
seminal vesicles
seminiferous tubule
testis
urethra
uterus
vagina
vas deferens
vestibular gland

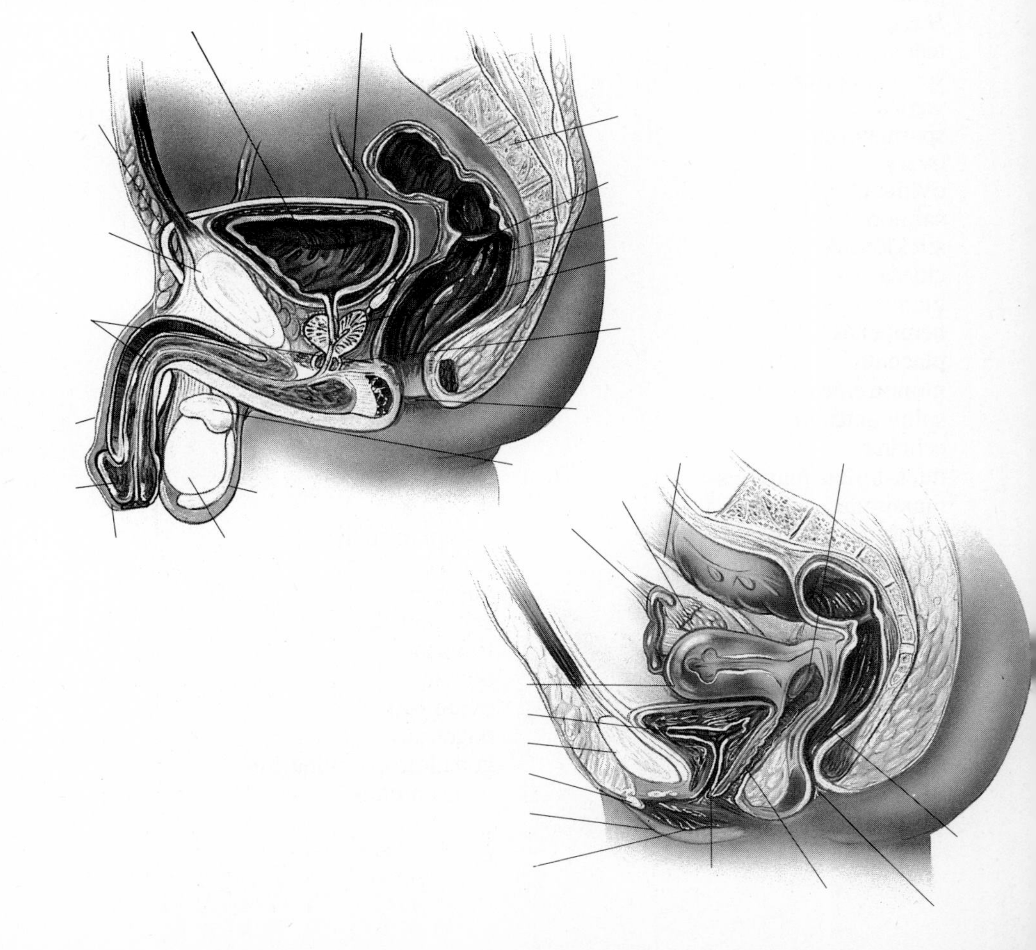

Self-Exam

You should be able to easily answer the following questions after learning the material in Chapter 42. If you have difficulty with any question, study the appropriate section in the text and try again.

A. Multiple Choice Questions

Circle one alternative that best completes the statement or answers the question.

1. Which mode of reproduction involves eggs?
   a. fission.
   b. budding.
   c. gemmules.
   d. parthenogenesis.

2. An animal species such as the dog or cat might be termed
   a. monoecious.
   b. hermaphroditic.
   c. dioecious.
   d. asexual.

3. In the pattern of reproduction, salmon
   a. mate and have internal fertilization.
   b. have external fertilization which occurs on the floor of the ocean.
   c. have external fertilization which occurs in nests in freshwater streams.
   d. are parthenogenic.

4. Which statement is accurate concerning stickleback reproduction?
   a. The females are parthenogenic.
   b. The fish are hermaphroditic.
   c. The females lay eggs and guard them.
   d. The females lay eggs but males guard them.

5. Most frogs exhibit which reproductive behavior?
   a. There is no contact between males and females, and fertilization is external.
   b. The male clasps the female, but fertilization is external.
   c. The male clasps the female, and fertilization is internal.
   d. Most frogs are parthenogenic.

6. The male shark copulatory organ is derived from
   a. modified pelvic fins
   b. modified tail fins.
   c. a structure above its eye-stalks.
   d. its modified dorsal fin.

7. In humans, the sperm storage structure in the male scrotum is
   a. the testis.
   b. the seminiferous tubule.
   c. the rete testis.
   d. the epididymis.

8. Most of the volume of human semen is produced
   a. in the testes.
   b. in the vas deferens.
   c. in the seminal vesicles.
   d. in the bulbourethral gland.

9. Which sequence correctly describes events of spermatogenesis?

a. primary spermatocyte ⇒ meiosis I ⇒ secondary spermatocyte ⇒ meiosis II ⇒ spermatid ⇒ spermatozoa.
b. meiosis I ⇒ primary spermatocyte ⇒ meiosis II ⇒ secondary spermatocyte ⇒ spermatid ⇒ spermatozoa.
c. spermatid ⇒ primary spermatocyte ⇒ meiosis I ⇒ secondary spermatocyte ⇒ meiosis II ⇒ spermatozoa.
d.. primary spermatocyte ⇒ secondary spermatocyte ⇒ meiosis I ⇒ meiosis II ⇒ spermatid ⇒ spermatozoa.

10. Which statement concerning human oogenesis is most accurate?
a. It is completed during fetal life, before birth.
b. It begins during fetal development, but stops during prophase I.
c. It begins during fetal development, but stops after meiosis II.
d. It begins during fetal development, but stops during prophase II.

11. Which statement is correct concerning the hormones LH and FSH?
a. Both are steroid hormones.
b. LH is a steroid hormone and FSH is a polypeptide.
c. LH is a polypeptide and FSH is a steroid hormone.
d. Both are polypeptides.

12. The male hormone testosterone is produce mainly by
a. the hypothalamus.
b. the testes.
c. the anterior pituitary.
d. the posterior pituitary.

13. High levels of testosterone
a. stimulate additional production of testosterone.
b. stimulate release of LH.
c. inhibits release of LH.
d. affects neither testosterone production nor LH release.

14. The ovaries respond to FSH by producing
a. estrogens.
b. testosterone.
c. LH.
d. gonadotropins.

15. In human females, androgens are secreted
a. by the ovaries.
b. by the pituitary.
c. by the adrenal glands.
d. not at all.

16. In human females, a mature ovum is released from one of the ovaries
a. once each calendar month.
b. about every 28 days.
c. about every 35 days.
d. upon copulation.

17. During the proliferative phase of the menstrual cycle, the hypothalamus stimulates the pituitary to
a. increase its output of both FSH and LH.
b. increase its output of FSH and decrease its output of LH.
c. decrease its output of FSH and increase its output of LH.
d. decrease its output of both FSH and LH.

18. Ovulation occurs at which point during the human menstrual cycle?
a. About 1 week after the menstrual flow begins.
b. About 14 days after menstrual flow begins.
c. At the midpoint of the menstrual flow period.
d. Immediately before the menstrual flow begins.

19. The corpus luteum secretes
    a. LH and estrogen.
    b. estrogen and testosterone.
    c. progesterone and estrogen.
    d. progesterone and testosterone.

20. The hormone detected by some pregnancy tests, human chorionic gonadotropin (HCG), is secreted by
    a. the embryo.
    b. extraembryonic tissue.
    c. the ovary.
    d. the corpus luteum.

21. Which contraceptive device is generally the least likely to result in pregnancy when used correctly?
    a. condom.
    b. diaphragm.
    c. intrauterine device (IUD).
    d. all are equally ineffective.

B. True or False Questions

Mark the following statements either T (True) or F (False).

_____ 22. In parthenogenesis, an embryo develops from a fertilized egg.

_____ 23. Some species routinely alternate between sexual and asexual reproduction.

_____ 24. The monoecious condition is considered evolutionarily more advanced than the dioecious condition.

_____ 25. Although they are quite simple, internal fertilization can occur in the sponges.

_____ 26. Many female insects mate only once, although they may produce eggs over a lengthy lifespan.

_____ 27. In terms of the small numbers of gametes they must produce, the fertilization process for ocean fish such as cod is very efficient.

_____ 28. The tailed frog uses its tail in copulation.

_____ 29. Most male birds have a penis to facilitate internal fertilization.

_____ 30. A male opossum's penis is forked, corresponding to a female opossum's divided uterus.

_____ 31. The seminal vesicles in human males function mainly to store sperm.

_____ 32. Mature sperm cells are tiny structures that lack mitochondria.

_____ 33. Unlike most mammals, humans have a single uterus; most mammals have paired uteri.

_____ 34. Among humans, both males and females produce LH.

_____ 35. Testosterone is essential in humans for the development of characteristics of "maleness."

_____ 36. In human females, a body temperature drop of about 1°F indicates the "fertile period" of the menstrual cycle.

_____ 37. The rhythm method of birth control has a success rate of about 98%.

_____ 38. The birth control pill is considered to be a virtually risk-free method of contraception.

_____ 39. The common surgical sterilization technique for men is called a tubal ligation.

C. Fill in the Blanks

Answer the question or complete the statement by filling in the blanks with the correct word or words.

40. Reproduction by _________________ involves breaking into to or more parts, each part regenerating a new individual.

41. Monoecious animals having both sexes on the same individual are termed _________________.

42. Male insects often package sperm cells into packets called _________________.

43. In many female insects, sperm cells are stored in a _________________ after copulation.

44. The structure in many vertebrates that carries digestive waste, excretory waste, and gametes is the _________________.

45. Male snakes are known to have two penises or _________________.

46. The egg-laying mammals are termed the _________________.

47. The part of the testis with cells which undergo meiosis is the _________________.

48. Spermatids are apparently nourished by"nurse" cells called _________________.

49. The human female structure that is homologous with the male penis is the _________________.

50. Under the influence if FSH and LH, the testis produces the hormone _________________.

51. The reproductive cycle in many female mammals other than humans is termed an _________________ cycle.

52. The lining of the human uterus which is shed during the menstrual cycle is termed an _________________.

53. Under the influence of FSH and LH, an ovarian follicle becomes a _________________.

54. The contraceptive device that is a mechanical barrier to sperm used by women is a _________________.

55. The active substances in birth control pills are _________________ and _________________.

56. The surgical male sterilization technique is a _________________, and involves cutting the _________________.

Questions for Discussion

1. What are the disadvantages of external fertilization and what are the advantages of internal fertilization? If internal fertilization is so advantageous, why are so many organisms able to successfully reproduce with external fertilization?

2. Outline the trends in reproductive patterns in vertebrates, from fish through amphibians, reptiles, and birds to mammals. Is there any pattern in the degree of development or maturity in offspring at birth?

3. Describe in detail the hormonal control of reproduction in humans, both male and female. Begin with the hormonal control of male and female development., the control of sexual maturation, and the control of the reproductive cycle. Indicate feedback regulation mechanisms when they exist.

4. A couple wishes to maximize their chances of conceiving a child. How could they determine the times that intercourse would be most likely to result in conception?

5. Why is the rhythm method of birth control unsuccessful in many cases?

Answers to Self-Exam

Multiple Choice Questions

1. d
2. c
3. c
4. d
5. b
6. a
7. d
8. c
9. a
10. b
11. d
12. b
13. c
14. a
15. c
16. b
17. a
18. b
19. c
20. b
21. c

True or False Questions

22. F
23. T
24. F
25. T
26. T
27. F
28. T
29. F
30. T
31. F
32. F
33. T
34. T
35. T
36. F
37. F
38. F
39. F

Fill in the Blank Questions

40. fragmentation
41. hermaphroditic
42. spermatophores
43. spermatheca
44. cloaca
45. hemipenes
46. monotremes
47. germinal epithelium
48. Sertoli cells
49. clitoris
50. testosterone
51. estrous
52. endometrium
53. corpus luteum
54. diaphragm
55. estrogen, progesterone
56. vasectomy, vas deferens

CHAPTER 43 Development

Learning Objectives

After mastering the material covered in Chapter 43, you should be able to confidently do the following tasks:

- Describe the structure and biology of the sperm and egg cells, and give details of the fertilization process.
- Outline the early processes in development: cleavage, blastula formation, and gastrulation.
- Contrast early embryogenesis in representative major groups of animals.
- Describe the extraembryonic supporting structures in the developing eggs of reptiles and birds, and the special extraembryonic structures that support developing mammals.
- Give a chronological account of human development through three trimesters of gestation and birth.
- Provide an analysis of our understanding of how development is directed by determination in zygotes, by tissue interactions, and by developmental regulatory mechanisms at the molecular level.

Chapter Outline

I. Gametes.

A. The sperm.
B. The egg.

II. Fertilization.

III. Early Developmental Events.

A. Cleavage.
B. The blastula.
C. Gastrulation.

IV. Organogenesis.

A. Establishing the body axis and neurulation.
B. Further vertebrate development.

V. Vertebrate Supporting Structures.

A. Support system of reptiles and birds.
B. Support system of mammals.

VI. Human Organogenesis.

A. The first trimester.
B. The second and third trimesters.
C. Birth.

VII. Analysis of Development.

A. Determination in eggs and zygotes.
B. Tissue interaction in development.
C. The determining role of migrating cells.
D. Imaginal disks and homeotic mutants in *Drosophila*.
E. The homeobox: molecular biology of development.

Key Words

embryogenesis
gametogenesis
fertilization
cleavage
gastrulation
neurulation
organogenesis
determination
differentiation
morphogenesis
sperm
acrosome
egg
yolk
blastodisc
polarity
animal hemisphere
vegetal hemisphere
placenta
vitelline membrane
chorion
zona pellucida
corona radiata
vitelline layer
jelly coat
acrosomal reaction
acrosomal process
fertilization cone
pronuclei
cortical reaction
polyspermy
fertilization membrane
cortical granules
second polar body
blastomeres
morula
micromeres
macromeres
egg white
blastoderm
yolk sac
protostome
deuterostome
pole cells
determinate cleavage
indeterminate cleavage
blastocoel
blastula
blastocyst
blastocyst cavity
trophoblast
inner cell mass
gastrulation
ectoderm
endoderm
mesoderm
gastrula
invagination
archenteron
blastopore
mesenchyme
pluteus larva
dorsal lip
yolk plug
epiblast
hypoblast
primitive streak
primitive groove
neural plate
neural groove
neural tube
neural crest
notochord
somites
gill apparatus
gill arches
gill slits
yolk sac
chorion
allantois
amnion
chorioallantois
human chorionic gonadotropin (HCG)
amniotic cavity
chorionic villi
umbilicus
umbilical cord
trimester
genital tubercle
urogenital groove
urogenital folds
genital swellings
Wolffian ducts
Mullerian ducts
meconuim
ossification
oxytocin
dilation
labor
crown
fetal expulsion
placental separation
foramen ovale
ductus arteriosus
colostrum
gray crescent
Hans Spemann
totipotent
Briggs and King
regeneration
blastema
fate maps
embryonic induction
tissue interaction
epidermal placode
dermal papilla
feather bud
scale bud
lens placode

optic cup
morphogen
nerve-growth factor
hyaluronic acid
mosaic development
imaginal disks
imago
homeotic mutations
homeotic genes
antennapedia
bithorax
homeobox

Exercises

1. Please provide labels for the below diagram of a human sperm, using the terms listed. Check your work with Figure 43.1 in the text.

acrosome
centriole
cytoplasm
flagellum
head
midpiece
mitochondrion
nucleus
tail

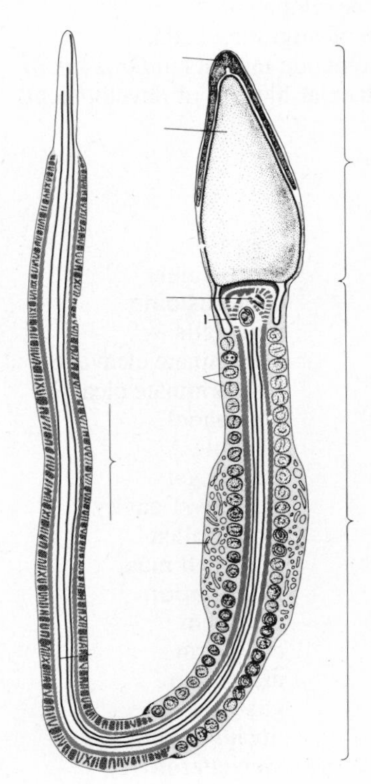

2. Use the space below to draw simple pencil diagrams of the following: A. Section of a frog blastula, labeled to indicate the animal pole, vegetal pole, blastocoel, and regions which will give rise to ectoderm, mesoderm, and endoderm. B. Section of a frog gastrula, labeled to indicate the animal and vegetal poles, yolk cells, archenteron, dorsal and ventral lips of the blastopore, the yolk plug, and the distribution of ectoderm, mesoderm, and endoderm. Check your work with Figures 43.8 and 43.9 in the text.

A. Frog Blastula

B. Frog Gastrula

3. Indicate on the following diagram of the human embryonic support system the structures listed below. Check your work with Figure 43.15 in the text.

amnion
amniotic fluid
chorion
chorionic villi
embryo
microvilli
placenta
umbilicus
uterus

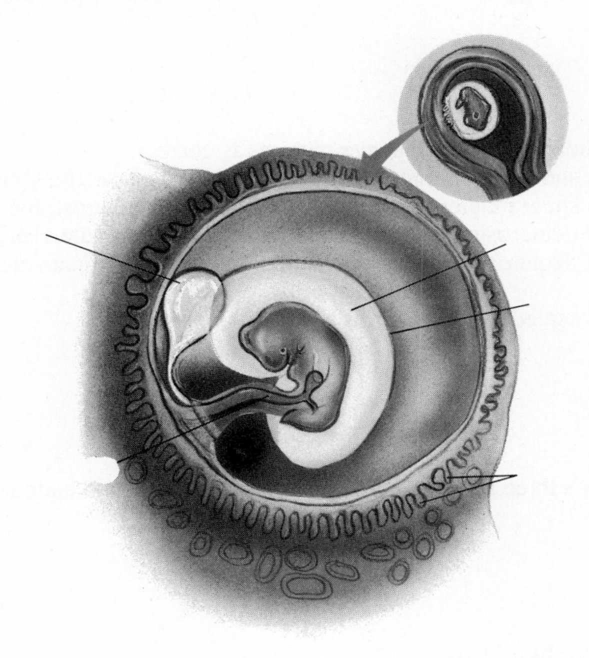

Self-Exam

You should be able to easily answer the following questions after learning the material in Chapter 43. If you have difficulty with any question, study the appropriate section in the text and try again.

A. Multiple Choice Questions

Circle one alternative that best completes the statement or answers the question.

1. The three major parts or subdivisions of a sperm cell are
   a. the head, the acrosome, and the tail.
   b. the head, the midpiece, and the tail.
   c. the mitochondrion, the midpiece, and the tail.
   d. the acrosome, the mitochondrion, and the centriole.

2. The yolky part of an egg is generally found
   a. in the animal hemisphere.
   b. in the vegetable hemisphere.
   c. in the vegetal hemisphere.
   d. in the cerebral hemisphere.

3. In mammals, there is a clear area around the egg periphery termed
   a. the vitelline membrane.
   b. the chorion.
   c. the zona pellucida.
   d. the jelly coat.

4. The simple ball of cells stage in mammalian embryogenesis is
   a. the morula stage.
   b. the blastula stage.
   c. the gastrula stage.
   d. the macromere stage.

5. In bird eggs, the nutrient-rich part is
   a. the egg white.
   b. the yolk.
   c. the blastodisc.
   d. the chorion.

6. Which following generalization about cleavage patterns is correct?
   a. Protostome embryos show determinate cleavage, while in deuterostome embryos it is indeterminate.
   b. Protostome embryos show indeterminate cleavage, while in deuterostome embryos it is determinate.
   c. Both protostome and deuterostome embryos tend to have determinate cleavage.
   d. Both protostome and deuterostome embryos tend to have indeterminate cleavage.

7. The blastula stage is represented in mammals as a
   a. morula.
   b. blastocyst.
   c. trophoblast.
   d. gastrula.

8. The formation of the body's three germ layers, ectoderm, mesoderm, and endoderm, occurs during
   a. cleavage.
   b. blastula formation.
   c. gastrula formation.
   d. oogenesis.

9. The amphibian blastopore is closed by
   a. the yolk plug.
   b. the neural groove.
   c. the blastocyst.
   d. the archenteron.

10. The neural cells that are known to migrate far and wide are called
    a. the neural plate.
    b. the neural fold cells.
    c. the neural tube cells.
    d. the neural crest cells.

11. In vertebrates, the entire gill apparatus is derived from
    a. neural crest.
    b. the rudimentary gut.
    c. ectoderm.
    d. the notochord.

12. In bird eggs, which extraembryonic membrane forms a fluid-filled protective sac that acts as a shock absorber?
    a. the yolk sac.
    b. the chorion.
    c. the allantois.
    d. the amnion.

13. In a mature bird egg, the network of blood vessels providing gas exchange with the outside permeates the
    a. yolk sac.
    b. chorion.
    c. amnion.
    d. chorioallantois.

14. How many extraembryonic membranes are produced by the typical ,mammalian embryo?
    a. 1.
    b. 2.

c. 3.
d. 4.

15. In a pregnant human, the hormone HCG is secreted by
    a. the uterus.
    b. the trophoblast.
    c. the endometrium.
    d. the corpus luteum.

16. So-called "pregnancy test kits" designed for home use detect pregnancies by indicating presence of
    a. LH in the urine.
    b. progesterone in the urine.
    c. estrogen in the urine.
    d. human chorionic gonadotropin in the urine.

17. The embryonic human heart starts contractions by about
    a. day 2.
    b. day 22.
    c. week 22.
    d. the 2nd trimester.

18. Limb buds of human embryos first appear about
    a. the 4th day.
    b. the 4th week.
    c. the 4th month.
    d. the 2nd trimester.

19. Which of the following statements about development of male and female reproductive anatomy is correct?
    a. The Wolffian ducts form male epididymis and vas deferens, while Mullerian ducts become female oviducts.
    b. The Mullerian ducts form male epididymis and vas deferens, while Wolffian ducts become female oviducts.
    c. The Wolffian ducts form male epididymis and vas deferens, and they also form the female oviducts.
    d. The Mullerian ducts form male epididymis and vas deferens, and they also form the female oviducts.

20. A mother often first begins to feel a fetus move and "kick" by about
    a. week 6.
    b. week 16.
    c. week 26.
    d. 3rd trimester.

21. The first breath taken by a newborn infant is a response to
    a. increased oxytocin.
    b. increased HCG levels.
    c. increased carbon dioxide level.
    d. increased oxygen level.

22. In their work with leopard frogs, Briggs and King found totipotent cells from
    a. blastulas.
    b. embryo tail buds.
    c. tadpoles.
    d. frogs.

23. In development of the vertebrate eye, the retina is derived from
    a. lens placode.
    b. cornea.
    c. optic cup.
    d. epidermis.

24. In *Drosophila*, the developmental abnormality resulting in legs on the head is termed
    a. antennapedia.
    b. bithorax.
    c. aristapedia.
    d. setapedia.

B. True or False Questions

Mark the following statements either T (True) or F (False).

_____ 25. Only half of a zygote's chromosomes come from the sperm nucleus.

_____ 26. In general, bird eggs have little yolk while mammal eggs have considerable yolk.

_____ 27. Typically a number of sperms penetrate the sea urchin egg when fertilization occurs.

_____ 28. Since only one sperm is necessary to produce a human zygote, men with low sperm counts are typically fully fertile.

_____ 29. The first cleavage in the frog zygote passes through both the animal and vegetal hemispheres.

_____ 30. Protostome cleavage follows a radial pattern, but deuterostomes show a spiral cleavage pattern.

_____ 31. In the frog blastula, most of the yolk is at the vegetal pole.

_____ 32. Gastrulation in mammals is almost identical to that of birds and reptiles.

_____ 33. The vertebrate brain is principally mesodermal in origin.

_____ 34. The notochord which takes form at the start of neurulation is present in all chordates at some point.

_____ 35. Structures in the human inner ear are derived from embryonic gill arches.

_____ 36. Human milk is sufficient in all vitamins excepts vitamin D.

_____ 37. Regeneration is common in invertebrates, but it is not known to occur in any vertebrate.

_____ 38. The primary organizer is presumed to be molecular in nature, but no specific molecule is known.

_____ 39. There are cells in insect larvae that make little or no contribution to adult structure.

_____ 40. The homeobox has been characterized from invertebrates, but it has not been found in any vertebrate.

C. Fill in the Blanks

Answer the question or complete the statement by filling in the blanks with the correct word or words.

41. In amphibians and reptiles, the _________________ forms just outside the plasma membrane of the egg cell.

42. As the jelly coat of the egg is penetrated by a sea urchin sperm, the enzyme-laden cap over the sperm head shows changes termed the _________________ reaction.

43. After fertilization, the egg and sperm nuclei are termed _________________.

44. The earliest series of cell divisions in an embryo following fertilization is termed _________________.

45. The cavity created by invagination of the blastula surface is termed the ________________ or primitive gut.

46. Mesoderm is formed from ________________, cells that arise from the blastopore region and move into the interior of the old blastocoel.

47. Gastrulation in birds begins with the formation of a crease, the ________________.

48. Formation of the brain and spinal cord occur through ________________.

49. On both sides of the notochord, mesoderm is organized into blocks called ________________.

50. During human development, the chorion fuses with the mother's endometrium to form the ________________.

51. In human development, the ________________ ducts degenerate in males but form oviducts in females.

52. The fluids and cellular debris that accumulates in a human fetal gut is called ________________.

53. The clear yellowish fluid secreted by a mother's breast prior to lactation is ________________.

54. In the process of regeneration, masses of simple cells form an undifferentiated tissue called ________________.

55. Diagrams showing the tissues that will arise from different parts of an early embryo are termed ________________.

56. The substance in a regenerating limb stump that stimulates neurons to grow is ________________.

57. In an insect like *Drosophila*, clusters of cells that early in larval development have their adult fate fully determined are called ________________.

58. In *Drosophila*, ________________ genes regulate whole blocks of genes, each of which is responsible for some part of the structure arising from certain cells of the imaginal disk.

59. Many homeotic genes have a highly conserved region of about 180 nucleotides termed the ________________.

Questions for Discussion

1. In earlier days, some biologist would say "ontogeny recapitulates phylogeny," which was to say that a relatively "advanced" vertebrate, such as a mammal, would during its development, pass through embryonic stages that appeared to be very similar to embryos of "lower" vertebrates. In fact, there is much less visible difference in early vertebrate embryos than there is in the animals into which they develop. Give examples of "primitive" structures present in embryos of humans. Do such structures tell us about human ancestors directly? How might comparative studies of embryo development help us understand the relationships between various animal groups and their ancestry?

2. In humans, development is not complete by the time of birth. In fact, human infants are quite incompletely developed and helpless. What further developmental changes must occur in humans following birth? What are some mammals that are more developed and less helpless following birth? Why might some animals be more self reliant so much earlier? Or, conversely, why might humans be so helpless for so long?

3. Outline the steps in development of a frog, beginning with the egg. List each stage in the developmental pathway, and describe the events that occur as development proceeds.

4. What is meant by a primary organizer? What types of evidence support the existence of primary organizers? What types of molecules are the likely candidates for role of organizer?

5. What are homeotic genes and homeotic mutants? Give examples from *Drosophila*. Why are they important in our understanding of development? What is the role of the homeobox?

Answers to Self-Exam

Multiple Choice Questions

1. b
2. c
3. c
4. a
5. b
6. a
7. b
8. c
9. a
10. d
11. b
12. d
13. d
14. d
15. b
16. d
17. b
18. b
19. a
20. b
21. c
22. a
23. c
24. a

True or False Questions

25. T
26. F
27. F
28. F
29. T
30. F
31. T
32. T
33. F
34. T
35. T
36. T
37. F
38. T
39. T
40. F

Fill in the Blank Questions

41. vitelline membrane
42. acrosomal
43. pronuclei
44. cleavage
45. archenteron
46. mesenchyme
47. primitive groove
48. neurulation
49. somites
50. placenta
51. Mullerian
52. meconium
53. colostrum
54. a blastema
55. fate maps
56. nerve growth factor
57. imaginal disks
58. homeotic
59. homeobox

CHAPTER 44 The Development and Control of Animal Behavior

Learning Objectives

After mastering the material covered in Chapter 44, you should be able to confidently do the following tasks:

- Explain what is known about the genetic basis of behavior, and cite examples of supporting experimental evidence.
- Describe the ways that hormones are known to influence behavior.
- Distinguish instinct from learning as a basis for behavior.
- List and distinguish the types of learning that have been described.
- Contrast short-term memory and long-term memory.
- Present current theories on the biological basis of long-term and short-term memory.

Chapter Outline

I. The Development of Behavior.

A. Genes and behavior.
1. inbreeding experiments.
2. artificial selection experiments.
3. hybridization experiments.

II. Hormones and Behavior.

A. Gland removal and hormone replacement.
B. How hormones can influence behavior.

III. Instinct.

A. Sign stimulus.
B. Fixed action patterns.

IV. Learning.

A. Kinds of learning.
B. Latent learning.
C. Insight learning.
D. Imprinting.

V. How Instinct and Learning Can Interact.

VI. Memory and Learning.

A. Theories on information storage.
B. Physiological differences in long- and short-term memory.

Key Words

avoidance learning
shuttle box
artificial selection
honey bee (*Apis mellifera*)
Bacillus larvae
American foulbrood disease
Peach-faced lovebird (*Agapornis roseicollis*)
Fisher's lovebird (*Agapornis personata fischeri*)
retrieval behavior
territorial behavior
three-spined stickleback (*Gasterosterus aculeatus*)
instinct
Konrad Lorenz
Nika Tinbergen
Karl von Frisch
ethologist
sign stimulus
innate releasing mechanism (IRM)
fixed action pattern
orienting
learning
reinforcement
reward
habituation
classical conditioning
Ivan Pavlov
generalization
discrimination
operant conditioning
B.F. Skinner
Skinner box
latent learning
insight learning
imprinting
memory
engram
long-term memory
short-term memory
consolidation
consolidation hypothesis
calpain
fodrin

Self-Exam

You should be able to easily answer the following questions after learning the material in Chapter 44. If you have difficulty with any question, study the appropriate section in the text and try again.

A. Multiple Choice Questions

Circle one alternative that best completes the statement or answers the question.

1. If a particular behavior is strongly genetically based, which of the below statements concerning two different inbred lines with very different adaptive behaviors is most likely to be supported?
   a. The two lines will behave identically in the same environment.
   b. The two lines will behave differently in the same environment.
   c. The two lines will behave identically in different environments.
   d. The behavior of the two lines will be unpredictable.

2. Rothenbuler's work with hygienic behavior of bees in response to foulbrood disease suggested the hygienic behavior is controlled by
   a. one gene.
   b. two genes.
   c. four genes.
   d. *Bacillus larvae.*

3. The hybrid offspring of crosses between Peach-faced and Fisher's lovebirds display
   a. The nesting behavior of Peach-faced lovebirds.
   b. The nesting behavior of Fisher's lovebirds.
   c. The nesting behavior of neither lovebird species.
   d. a compromise between the two parental forms of nesting behavior.

4. In testing the role of hormones in behavior, in some mammals, castration is followed by
   a. decreased copulation behavior.
   b. increased copulation behavior.
   c. no change in copulation behavior.
   d. an increase in testosterone level.

5. As an example of a sign stimulus, fighting behavior in territorial male European robins may be released by
   a. sight of a female robin.
   b. the proper birdsong.
   c. sight of a tuft of red feathers.
   d. a complex set of events.

6. The type of learning involved in learning *not* to respond to a stimulus is termed
   a. habituation.
   b. classical conditioning.
   c. operant conditioning.
   d. imprinting.

7. The type of learning involved in solving problems without the benefit of experience is termed
   a. habituation.
   b. imprinting.
   c. latent learning.
   d. insight learning.

8. A group of goslings followed Konrad Lorenz around the farm because they saw him first after hatching. Their behavior was due to
   a. habituation.
   b. imprinting.
   c. latent learning.
   d. insight learning.

9. Among birds, flight is largely
   a. an innate pattern.
   b. learned by observation.
   c. learned through trial-and-error.
   d. learned by habituation.

10. Which is apparently essential for learning to occur?
    a. instinct.
    b. short-term memory alone.
    c. short-term storage followed by long-term storage.
    d. long-term memory alone.

B. True or False Questions

Mark the following statements either T (True) or F (False).

_____ 11. Genes code for proteins, not for behavior directly.

_____ 12. Different strains of mice all have the same ability in learning to avoid a shock.

_____ 13. In honeybees there is a genetic basis for hygienic and unhygienic behavior.

_____ 14. A mother seal's retrieval behavior (to protect young pups) is related to presence of certain hormones.

_____ 15. A reward is defined as a result of an action that increases the probability of that action being repeated.

_____ 16. Habituation involves learning to respond to a stimulus.

_____ 17. Operant conditioning is the same as classical conditioning.

_____ 18. Latent learning occurs in the absence of an immediate reward.

_____ 19. In memory, short-term storage is apparently necessary before long-term storage can occur.

_____ 20. Both long-term and short-term memory form engrams.

C. Fill in the Blanks

Answer the question or complete the statement by filling in the blanks with the correct word or words.

21. A shuttle box, a two-compartment device with an electrified floor, can be used to study _________________ learning in mice.

22. A honeybee hive infected with *Bacillus larvae* can develop _________________ disease.

23. Studies of male song sparrows have shown a correlation between maximum levels of testosterone and male _________________ behavior.

24. Environmental factors that release instinctive behavior are called _________________.

25. Innate behavior patterns which are independent of the environment, such as nest building behavior techniques of birds, are termed _________________ patterns.

26. _________________ is the process of developing a behavioral response based on experience.

27. The process where a substitute stimulus can replace the normal stimulus and release a behavior is termed _________________.

28. The apparatus developed in the 1930's that was used to demonstrate operant conditioning is called a _________________.

29. The presumed physical change in the brain associated with learning is an _________________ or memory trace.

30. The process of transferring memory from short-term storage to long-term storage is called _________________.

31. In one model of the chemical basis for the brain changes associated with memory, calcium activates a dormant enzyme called _________________, which can break down proteins such as _________________, which is involved in structural form of dendrites.

Questions for Discussion

1. Most animal species probably have a number of fixed action patterns. For example, human infants exhibit a set of characteristic head and mouth movements in order to suckle; stimulate a hungry infant's face anywhere near its mouth with an object, and the infant will turn toward the object and attempt to nurse. Can you think of other fixed action patterns exhibited by humans? Give as many examples as you can.

2. How can you distinguish between instinctive behavior and learned behavior? Using a typical pet, such as a dog or cat, give several examples of instinctive behaviors and learned behaviors. What fraction of human behavior do you think is learned? Why?

3. Some accidents or diseases can cause brain lesions which interfere with memory, both short-term and long term. In some rare cases, viral infections of the brain have apparently completely destroyed the ability to convert short-term storage to long term storage. Such individuals recognize their families, can read and write, remember how to perform their jobs, and have normal memories of events that happened before their brain injuries occurred. However, after the damage has occurred, they remember nothing for more than ten or fifteen minutes: What do you think a typical day might be like for such a person, from the time he wakes up in the morning until he finally returns to sleep? Can you think of any way for him to "remember" the day's events?

4. In terms of human behavior, there has been considerable controversy regarding the relative roles of "nature versus nurture," that is, how much behavior is inherited and how much is a consequence of the environment where a person is raised. Identical twins are genetically identical. How could twin studies provide insights in the "nature versus nurture" controversy?

Answers to Self-Exam

Multiple Choice Questions

1. b
2. b
3. d
4. a
5. c
6. a
7. d
8. b
9. a
10. c

True or False Questions

11. T
12. F
13. T
14. T
15. F
16. F
17. F
18. T
19. T
20. F

Fill in the Blank Questions

21. avoidance
22. American foulbrood
23. territorial
24. sign stimuli
25. fixed action
26. learning
27. classical conditioning
28. Skinner box
29. engram
30. consolidation
31. calpain, fodrin

CHAPTER 45 Adaptiveness of Behavior

Learning Objectives

After mastering the material covered in Chapter 45 you should be able to confidently do the following tasks:

- Give examples of how habitat selection and foraging behavior can be adaptive for animals.
- Describe the types of biological clocks that are known from animals.
- Explain why migratory behavior may be important for some animals, and explain what is known about how animals can navigate over long distances.
- Categorize the ways animals can communicate with one another, give examples of various means of communication, and show how communication behavior is adaptive.
- Explain the role of aggression in animal behavior, and give examples of animal behavior where aggression is stylized.
- Distinguish such interactions as cooperation, mutualism, and commensalism, and give examples.
- Explain altruism and how it may be involved with kin selection.

Chapter Outline

I. Proximate and Ultimate Causation.

II. Behavioral Ecology.

A. Habitat selection.
B. Foraging behavior.

III. Biological Clocks.

IV. Navigation.

V. Communication.

A. Visual communication.
B. Sound communication.
C. Chemical communication.

VI. Communication and Recognition.

A. Species recognition.
B. Individual recognition.
C. Kin recognition.

VII. Agonistic Behavior.

A. Fighting, a form of aggression.
B. Is aggression instinctive?

VIII. Social Behavior.

A. Cooperation.
B. Symbiosis.

IX. Altruism.

X. Sociobiology.

Key Words

proximate causation
ultimate causation
behavioral ecology
habitat selection
foraging behavior
generalist
specialist
Everglade kite
great tit
foraging efficiency
Bluegill sunfish
Daphnia
striped bass
shiners
crayfish
biological clock
annual rhythm
circadian cycle
clock shifting
Acetabularia
compass sense
time-direction mechanism
map sense
orientation
navigation
Gustav Kramer
sun-compass orientation
homing pigeon
migration
Lincoln Brower
monarch butterfly
graded display
mobbing call
pheromone
bombykol
Bombyx
species recognition
golden-fronted woodpecker
red-bellied woodpecker
dominance hierarchy
kin recognition
agonistic behavior
aggression
threat
submission
fighting
stylized fighting
evolutionary stable strategy (ESS)
retaliator strategy
bourgeois strategy
social behavior
Jane Goodall
chimpanzee
cooperation
interspecific cooperation
rhinoceros
tickbird
intraspecific cooperation
porpoise
symbiosis
mutualism
commensalism
altruism
kin selection
J.B.S. Haldane
W.D. Hamilton
eusocial insects
Robert Trivers
reciprocal altruism
sociobiology
group foraging
group protection
"selfish herd" effect

Exercises

1. For each of the following concepts, give an example by naming an animal which shows the type of behavior listed, and describe the specific behavior involved.

A. Generalized foraging.

B. Specialized foraging.

C. Sun-compass orientation.

D. Graded display.

E. Mobbing call.

F. Pheromones.

G. Dominance hierarchy.

H. Stylized fighting.

I. Interspecific cooperation.

J. Intraspecific cooperation

K. Mutualism.

L. Commensalism.

Self-Exam

You should be able to easily answer the following questions after learning the material in Chapter 45. If you have difficulty with any question, study the appropriate section in the text and try again.

A. Multiple Choice Questions

Circle one alternative that best completes the statement or answers the question.

1. In examining proximate and ultimate causation of animal behavior, we are usually
   a. asking *why* questions concerning both proximate causation and ultimate causation.
   b. asking *how* questions concerning both proximate causation and ultimate causation.
   c. asking *why* questions concerning proximate causation and *how* questions concerning ultimate causation.
   d. asking *how* questions concerning proximate causation and *why* questions concerning ultimate causation.

2. In Wecker's experiments with habitat preference in prairie deer mice, he found
   a. "forest"-reared laboratory mice preferred grassland habitat.
   b. "forest"-reared laboratory mice preferred "forest" habitat.
   c. "forest"-reared laboratory mice preferred laboratory habitat.
   d. "forest"-reared laboratory mice showed no preference whatsoever.

3. In studying animal foraging behavior, how is energy expenditure measured?
   a. By the food value of the item.
   b. By the energy the animal expends searching for food.
   c. By the energy the animal expends handling food.
   d. By the energy expended searching and handling the food.

4. When a striped bass is foraging, why is it better off chasing a shiner than catching crayfish?
   a. It is easier to handle a crayfish.
   b. It is easier to catch a shiner.
   c. It is easier to handle a shiner once it is caught.
   d. It is easier to handle a crayfish once it is caught.

5. With respect to an animal's circadian rhythm, how is clock shifting done?
   a. By altering the timing of cues associated with some phase of the day.
   b. By waking up earlier.
   c. By crossbreeding.
   d. By adjusting the clock mechanism.

6. An animal which seems to know its precise longitude and latitude no matter where is it has
   a. a compass sense.
   b. a map sense.
   c. an internal clock.
   d. a time-direction mechanism.

7. How do we think homing pigeons navigate?
   a. They use sun compass clues alone.
   b. They use magnetic force field clues alone.
   c. They use visual details for clues.
   d. The use all of the above clues.

8. As a generalization about birds, the farther north from the tropics a species breeds,
   a. the larger is its brood.
   b. the smaller is its brood.
   c. the less likely it is to raise its young.
   d. the more slowly it raises its brood.

9. Which of the following is *not* one of the ways communication is generally adaptive for animals?
   a. It is involved with reproductive success as a component of mating behavior.
   b. It helps offspring avoid danger when parents give warning cries.
   c. It helps animals carry out their reproductive imperative.
   d. It entertains other species and makes them less aggressive.

10. Fireflies are attracted to one another on the basis of
    a. their flash intervals.
    b. the way they fly.
    c. the sounds they emit.
    d. assuming a head-up posture while fluttering drooping wings.

11. Sound communication is generally most important for which groups?
    a. echinoderms and vertebrates.
    b. echinoderms and arthropods.
    c. arthropods and vertebrates.
    d. annelids and vertebrates.

12. Bombykol is an example of
    a. a method of visual communication by birds.
    b. a method of sound communication by crickets.
    c. a molecule of chemical communication by moths.
    d. a type of behavioral communication by vertebrates.

13. For a dominance hierarchy to be maintained depends on
    a. repeated fighting.
    b. individuals recognizing other individuals.
    c. individuals recognizing the size of larger animals.
    d. recognizing individuals of other species.

14. Agonistic behavior does *not* include
    a. aggression.
    b. submission.
    c. fighting.
    d. predation.

15. In general, fighting as a form of aggression is most common
    a. between members of the same sex of the same species.
    b. between members of opposite sexes of the same species.
    c. between members of the same sex of different species.
    d. between members of opposite sexes of different species.

16. In stylized fighting, once dominance is established,
    a. the loser is usually killed.
    b. the loser is usually pursued.
    c. the loser is usually permitted to retreat.
    d. there is no loser.

17. Hyenas often hunt in packs and sometimes cooperate in bringing down their prey. This is an example of
    a. intraspecific cooperation.
    b. interspecific cooperation.
    c. altruism.
    d. sociobiology.

18. Cattle harbor cellulase-producing bacteria in their digestive systems to help them derive energy from cellulose. This is an example of
    a. commensalism.
    b. mutualism.

c. parasitism.
d. altruism.

19. In which of the following groups would we find the most extreme example of altruism?
    a. fish.
    b. eusocial insects.
    c. antisocial humans.
    d. dogs.

B. True or False Questions

Mark the following statements either T (True) or F (False).

_____ 20. Behavioral ecology is the study of how behavior affects the environment.

_____ 21. An example of a feeding generalist is the Everglades kite, which feeds on snails.

_____ 22. In Smith and Sweatman's experiment with great tits, the birds continued to sample other areas in their aviary even though they were able to feed extensively where food density was greatest.

_____ 23. In foraging, most species tend to maximize the benefit while minimizing the cost to themselves.

_____ 24. Scientists have found evidence for sun-compass orientation only in homing pigeons.

_____ 25. All homing pigeons return faultlessly to their home lofts; pigeons are almost never lost.

_____ 26. North America's familiar monarch butterflies overwinter in the Mexican Highlands.

_____ 27. Smelling the urine of a strange male can cause pregnant rats to abort their fetuses and become sexually receptive.

_____ 28. Golden-fronted woodpeckers and red-belied woodpeckers are known to form hybrids in areas where they coexist in Texas.

_____ 29. Sea gulls are able to recognize their mates calls as distinct from the calls of other gulls.

_____ 30. John Maynard Smith identified the "retaliator" strategy of aggression to be an evolutionary stable strategy (ESS).

_____ 31. Behavioral biologists have found that with the exception of humans, animals never fight to the death with others of their own species.

_____ 32. In examples of commensalism between two animals, one species benefits while the other is harmed.

_____ 33. Kin selection provides a simple explanation for birds issuing warning calls to alert others.

_____ 34. For birds to group closely together may offer protection by confusing a predator with sheer numbers.

C. Fill in the Blanks

Answer the question or complete the statement by filling in the blanks with the correct word or words.

35. One of the longest and most pronounced rhythms in nature is the earth's ________________ due to the earth's inclined axis and its orbit of the sun.

36. A behavior pattern that has a daily rhythm follows a ________________ cycle.

37. __________________ is the ability to face the right direction, while __________________ is the ability to successfully get between two specific points is __________________.

38. Periodic movement between two habitats is __________________.

39. Behavioral displays that can be given at different levels of intensity are termed __________________.

40. Chemicals produced by one animal that influence the behavior of another individual of the same species are termed __________________.

41. For some vertebrates, __________________ enables individuals to give preferential treatment to relatives, to avoid competing with relatives, and to perhaps avoid inbreeding.

42. Behavior which helps to resolve conflict among members of a species is termed __________________.

43. Maynard Smith's retaliator strategy and bourgeois strategy are examples of __________________.

44. When members of different species live together in close association. the relationship is termed __________________.

45. An act that benefits one individual at the expense of the first individual is called __________________.

46. When several animals are more efficient at finding food than is a single individual, they may live together and display __________________ behavior.

Questions for Discussion

1. Why do animals migrate? What are the major theories concerning how migratory behavior evolved? Do humans exhibit migratory behavior? Why?

2. We know a man who says he awakens each morning at 7 am (plus or minus 2 or 3 minutes) without the aid of an alarm clock. He say he has a "biological clock" that awakens him. We think he is merely responding to a set of external events that occur regularly: the newspaper arrives at 5 am, his neighbor the grocer starts his car and leaves for work at 6:15 am, the widow across the street lets her dog out at 6:30 am when she arises, and the dog barks to go back inside at 6:45 am. Interestingly, our friend sleeps late on Sunday (he credits his sophisticated biological clock). How could we determine if his punctual waking is genuinely due to an internal clock? What do you think happens when there is a shift from standard time to daylight savings time?

3. Choose a domestic animal which you are familiar with, such as a dog, cat, or horse. Tell the ways this animal communicates such important messages as friendly greeting, uncertainty, caution, anger or aggression, submissiveness, playfulness, territorial possession, hunger. Is such communication directed at other animals or at humans? Do you think such communication is effective?

4. Explain what Robert Trivers means by "reciprocal altruism." Can you give examples of reciprocal altruism in action for humans?

5. In many societies, orphans are adopted and raised by aunts or uncles. Is there any biological reason that this should occur? Is kin selection involved?

Answers to Self-Exam

Multiple Choice Questions

1. d
2. d
3. d
4. c
5. a
6. b
7. d
8. a
9. d
10. a
11. c
12. c
13. b
14. d
15. a
16. c
17. a
18. b
19. b

True or False Questions

20. F
21. F
22. T
23. T
24. F
25. F
26. T
27. T
28. F
29. T
30. T
31. F
32. F
33. T
34. T

Fill in the Blank Questions

35. annual cycle or rhythm
36. circadian
37. orientation, navigation
38. migration
39. graded displays
40. pheromones
41. kin recognition
42. agonistic
43. evolutionary stable strategies
44. symbiosis
45. altruism
46. group foraging

CHAPTER 46 Biosphere and Biomes

Learning Objectives

After mastering the material covered in Chapter 46, you should be able to confidently do the following tasks:

- Describe the physical elements that shape the biosphere.

- Show how sunlight interacts with a tilted, rotating, revolving earth to produce climate.

- Identify the major biomes of earth and show where they occur.

- For each of earth's biomes, give the characteristic temperature and rainfall distributions, and identify the dominant or characteristic plant and animal species.

- Identify the major divisions of the marine environment and the types of organisms supported in each.

Chapter Outline

I. The Biosphere.

A. Water.
B. The atmosphere.
C. Solar energy and climate.

II. Terrestrial Biomes.

A. Deserts.
B. Grasslands.
C. Tropical savannas.
D. Tropical rain forests.
E. Chaparral.
F. Temperate deciduous forest.
G. Taiga.
H. Tundra.

III. Water Communities.

A. The freshwater province.
1. rivers and streams.
2. lakes and ponds.
B. The marine environment.
1. The oceanic province.
2. The neritic province.

Key Words

environment
ecology
biosphere
hydrologic cycle
Coriolis effect
rain shadow
Great American Desert
gyres
Gulf Stream
biomes
communities
transition zones
Ruwenzori mountain range
deserts
xerophytes
cactus
ocotillo
joshua tree
creosote bush
palo verde
kangaroo rat (*Dipodomys deserti*)
estivation
grasslands
steppes
prairies
pampas
veldts
savannas
diffuse root system
rhizomes
stolons
"sodbusters"
bison
pronghorn antelope
wildebeest
zebra
jackrabbits
prairie dogs
kea (*Nestor notabilis*)
tropical savanna
boabab tree
giraffe
cheetah
lion
tropical rain forest
canopy
subcanopy
epiphyte
jungle
wet jungle
arboreal
chaparral
Mediterranean scrub forest
fire subclimax community
temperate deciduous forest
oak
maple
beech
birch
hickory
chestnut
bark fungus (*Endothia parasitica*)
deer
wolves
bear
foxes
squirrels
raccoons
opossums
taiga
boreal forest
muskegs
spruces
firs
conifers
poplar
alder
moose
elk
grizzly bear
wolverines
clear-cutting
tundra
mosses
lichens
permafrost
caribou
musk oxen
reindeer
ptarmigan
lemmings
Laplanders
Eskimos
Estuaries
Lake Baikal
neritic province
oceanic province
euphotic zone
aphotic zone
pelagic
abyssal region
Marianas trench
bathyscaph
benthic
continental shelf
plankton
phytoplankton
nannoplankton
zooplankton
upwellings
littoral zone
intertidal zone
tide pools
salt marshes
mud flats
coral reefs
calcium carbonate
Great Barrier Reef
slash and burn agriculture
laterite

Exercises

1. To the right is a diagram of a mountain tall enough to support a number of different biomes, from tropical rain forest to ice at the summit. Provide labels to show the relative locations of alpine tundra, coniferous forest, deciduous forest, desert, grasses, snow cover, and rain forest. Check your work with Figure 46.9 in the text.

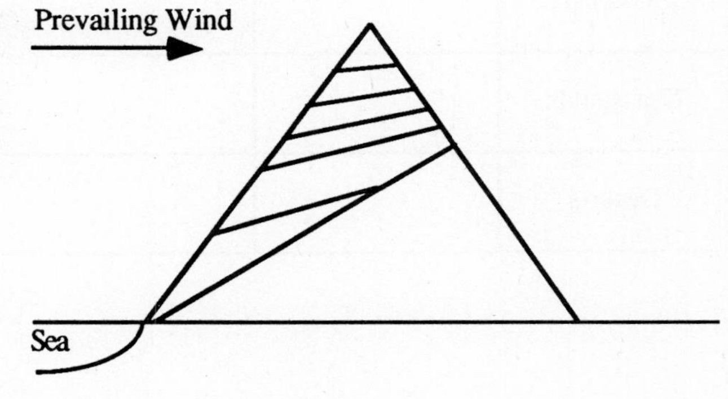

2. Below is a diagram of the marine environment. Please indicate where the major divisions of the marine environment are on this diagram. Show the neritic and oceanic provinces, the intertidal zone, the euphotic and aphotic zones, the continental shelf, the continental slope, and the abyssal region. Indicate where pelagic organisms are most often found and where benthic organisms are found. Check your work with Figure 46.31 in the text.

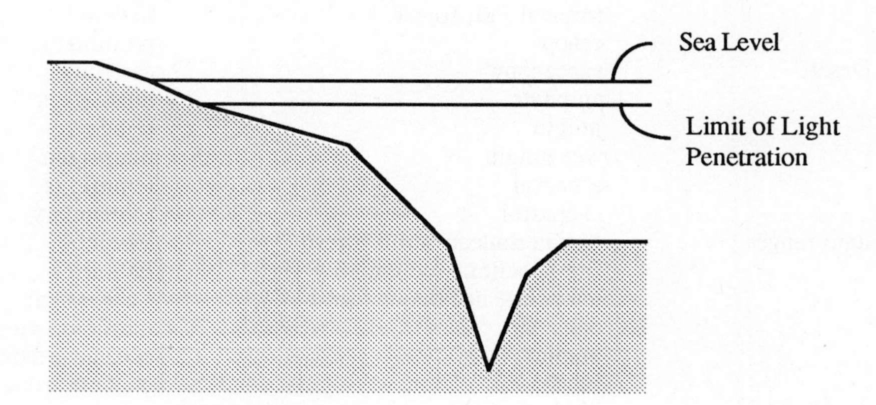

3. Complete the chart below to organize your understanding of the world's major biomes. For each biome, tell where in the world you would expect to find it, the characteristic amounts and distribution of rainfall, the temperature range, and list some of the dominant or characteristic plants and animals.

Biome	Location	Rainfall/Temperature	Characteristic Plants	Characteristic Animals
Tundra				
Taiga				
Temperate Deciduous Forest				
Chaparral				
Tropical Rain Forest				
Tropical Savanna				
Grasslands				
Deserts				

Self-Exam

You should be able to easily answer the following questions after learning the material in Chapter 46. If you have difficulty with any question, study the appropriate section in the text and try again.

A. Multiple Choice Questions

Circle one alternative that best completes the statement or answers the question.

1. The thickness of earth's habitable biosphere is about
   a. 1.4 miles.
   b. 14 miles.
   c. 41 miles.
   d. 140 miles.

2. The gaseous composition of the earth's atmosphere is
   a. almost completely oxygen.
   b. about 78% oxygen and 21% nitrogen.
   c. about 21% oxygen and 78% nitrogen.
   d. about 21% oxygen and 78% carbon dioxide.

3. After accounting for reflection and absorbance in the atmosphere, about what fraction of incoming solar radiation actually reaches the surface of the earth?
   a. about 1/4.
   b. about 1/3.
   c. about 1/2.
   d. about 3/4.

4. Which of the following is *not* a consequence of the 23.5° tilt from vertical of the earth's rotational axis?
   a. Cyclic seasonal changes.
   b. Fluctuating but moderate surface conditions
   c. Tradewinds and rainfall patterns.
   d. An average of half daylight hours and half dark hours over the year.

5. If you were to travel from the North Pole to the equator, which sequence of biomes would you most likely see?
   a. Tundra ⇒ taiga ⇒ deciduous forest ⇒ grassland ⇒ desert ⇒ tropical rain forest.
   b. Tundra ⇒ taiga ⇒ grassland ⇒ deciduous forest ⇒ tropical rain forest ⇒ desert.
   c. Taiga ⇒ tundra ⇒ deciduous forest ⇒ grassland ⇒ desert ⇒ tropical rain forest.
   d. Tundra ⇒ taiga ⇒ deciduous forest ⇒ grassland ⇒ tropical rain forest ⇒ desert.

6. An area which receives less than 25 cm (10 in) of rain each year would most likely be
   a. a grassland.
   b. a Mediterranean scrub forest.
   c. a tropical savanna.
   d. a desert.

7. All of the following plants are xerophytes *except*
   a. mangroves.
   b. cacti.
   c. creosote bushes.
   d. palo verde trees.

8. In many ways grasslands and deserts are similar. The most consistent difference between them is
   a. the average temperature.
   b. the longitudes where they occur.
   c. the average precipitation.
   d. the typical altitude.

9. Bison, pronghorn antelope, and jackrabbits are characteristic of
   a. North American prairies.
   b. Asian steppes.
   c. South American pampas.
   d. the African veldt.

10. Where do the largest savannas occur?
   a. Africa.
   b. South America.
   c. Australia.
   d. Georgia.

11. The world's largest tropical rain forest is found
   a. in the Northern part of Australia.
   b. in the African Congo River basin.
   c. in the South American Amazon River basin.
   d. in remote parts of Central America.

12. The chaparral or Mediterranean scrub forest biome is found in each of the following regions *except*
   a. the Pacific coast of North America.
   b. the coastal hills of Chile.
   c. coastal regions of southernmost Africa.
   d. the east coast of North America.

13. Oak, maple, beech, birch, and hickory might be viewed as characteristic plants of
   a. the tropical rain forest.
   b. the temperate deciduous forest.
   c. the boreal conifer forests.
   d. the Mediterranean scrub forest.

14. Spruce, fir, poplar, alder, and willow are typical of the plants found in
   a. the tropical rain forest.
   b. the temperate deciduous forest.
   c. the boreal conifer forests.
   d. the Mediterranean scrub forest.

15. The northernmost land biome is
   a. taiga.
   b. tundra.
   c. desert.
   d. deciduous forest.

16. Animals that drift in the open sea are termed
   a. pelagic organisms.
   b. benthic organisms.
   c. littoral organisms.
   d. estuarine organisms.

17. The average salinity of the ocean is
   a. about 35 parts per hundred.
   b. about 35 parts per thousand.
   c. about 35 parts per million.
   d. about 0.35%.

18. Deep abyssal regions of the sea support _______________ scavengers.
   a. pelagic.
   b. benthic.

c. neritic.
d. littoral.

19. If we look at the complete hydrologic cycle, where is the greatest "sink" of water?
    a. Water vapor in the atmosphere.
    b. Polar ice caps.
    c. Ground water under the continents.
    d. The world's oceans.

B. True or False Questions

Mark the following statements either T (True) or F (False).

_____ 20. Water vapor in the earth's atmosphere helps keep modern climates on earth relatively moderate.

_____ 21. About half the light reaching the earth's surface enters the photosynthetic process.

_____ 22. The deserts found in eastern North America are among the driest known on earth.

_____ 23. The distribution of biomes on the earth generally follows a regular north-to-south pattern, from arctic tundra to equatorial tropical rain forest.

_____ 24. The world's largest desert is the Sahara.

_____ 25. Although it is a desert dweller, the kangaroo rat like all animals must occasionally drink water.

_____ 26. In its native state, prairie grasslands are maintained by fire and grazing.

_____ 27. In one study of tropical rain forest in Costa Rica, it was found that very few mammals were arboreal.

_____ 28. There is uniform agreement that clear-cutting is the ecologically wisest way to harvest trees.

_____ 29. The tundra receives less solar energy than any other biome, because of its high latitude.

_____ 30. Sediment settles out of rivers primarily in headwaters where they are swift and fast-flowing.

_____ 31. In the ocean, the depths where sunlight never penetrates form the aphotic zone.

_____ 32. The Marianas trench extends deeper below sea level than Mount Everest rises above sea level.

_____ 33. According to best estimates, the majority of earth's photosynthetic activity takes place in the tropical rain forest.

_____ 34. The world's tropical rain forest have only recently come under development, and today they are largely still intact.

C. Fill in the Blanks

Answer the question or complete the statement by filling in the blanks with the correct word or words.

35. The study of interactions between organisms and all those factors, both living and nonliving, that affect them is termed _________________.

36. The part of the earth that supports life is the _________________.

37. The earth's atmosphere acts as a "greenhouse" to trap solar energy; atmospheric ________________ and ________________ are mainly responsible.

38. The east-west movement imparted to moving air masses is a result of the ________________, stemming from the earth's rotation.

39. We think the deepest flow of any major ocean current is found in the ________________.

40. Some desert animals avoid unfavorable circumstances, such as long hot, dry intervals, by ________________ when they become dormant.

41. Grasslands in Africa are called ________________ or ________________.

42. Plants that live on the stems and branches of trees and have no contact with the soil are ________________.

43. Low-growing tangles of plants that arise where sunlight reaches the tropical forest floor are sometimes called ________________.

44. Evergreen scrubby plants with small, tough, waxy leaves are typical of the ________________ biome.

45. Caribou, musk oxen, and reindeer are typical residents of the ________________ biome.

46. The deepest lake in the world is ________________ in the Soviet Union.

47. Photosynthetic algal protists of the oceans are termed ________________.

48. The land exposed between high tide and low tide is referred to as the ________________ zone.

49. Some cnidarians can form huge colonial exoskeletons of calcium carbonate called ________________.

50. The soils of tropical rain forest, when exposed, often turn to very hard, water-resistant ________________.

Questions for Discussion

1. The earth's axis is inclined about 23.5° from perpendicular to its orbital plane. What would be the consequences to earth's climates and seasons if the earth's axis were not inclined at all? What would be the consequences if it were inclined much more steeply, perhaps 75°?

2. If you visit the Grand Canyon in Arizona and hike the Kaibab Trail from the North Rim to Phantom Ranch in the canyon's depths, you pass from a coniferous forest such as might be found in Canada to a desert such as might occur in Mexico. You are passing through a vertical journey equivalent to a much longer journey across latitude. What other biomes might you pass through as you hike down over 5,000 feet? What changes would you expect to see in temperature and rainfall?

3. Your task is to plan an ideal biology field trip which will allow you to visit every North American biome. Try to be efficient in both distance and time. Where would you go? When would you go? (Now, have a good time, and go!)

4. Which biomes are the most productive in terms of total biomass produced? What factors are important in determining productivity?

5. Deserts present very extreme conditions to plants and animals living there. Describe the special adaptations typical desert-dwelling plants and animals use for coping with the scarcity of water and the fluctuation of temperature.

Answers to Self-Exam

Multiple Choice Questions

1. b	8. c	14. c
2. c	9. a	15. b
3. c	10. a	16. a
4. d	11. c	17. b
5. a	12. d	18. b
6. d	13. b	19. d
7. a		

True or False Questions

20. T	25. F	30. F
21. F	26. T	31. F
22. F	27. F	32. T
23. T	28. F	33. F
24. T	29. T	34. F

Fill in the Blank Questions

35. ecology
36. biosphere
37. carbon dioxide, water vapor
38. Coriolis effect
39. Gulf Stream
40. estivation
41. veldt, savanna
42. epiphytes
43. jungles
44. chaparral
45. tundra
46. Lake Baikal
47. phytoplankton
48. intertidal
49. coral reefs
50. laterite

CHAPTER 47 Ecosystems and Communities

Learning Objectives

After mastering the material covered in Chapter 47, you should be able to confidently do the following tasks:

- Identify each of the trophic levels in an ecosystem.
- Show the structure of ecosystems by use of pyramids of numbers, biomass and energy.
- Explain the relationship between energy flow in an ecosystem and the gross and net productivity of the ecosystem.
- Outline the components of the major biogeochemical cycles, and identify the exchange pools and the reservoirs.
- Describe the ecological structure of lakes in terms of life zones and physical characteristics.
- Show the roles of competition and predation in community structure.
- Explain the concept of ecological niche.
- Distinguish primary and secondary succession, and give examples of each process.
- Outline typical pathways of succession for ponds, sand dunes, forest, and old farm fields.

Chapter Outline

I. Energy in Ecosystems.

   A. Trophic levels.
      1. producers.
      2. consumers.
      3. decomposers.
      4. ecological pyramids.
   B. Humans and trophic levels.
   C. Energy and productivity.

II. Nutrient Cycling in Ecosystems.

   A. Nitrogen cycle.
   B. Phosphorus and calcium cycles.
   C. Carbon cycle.

III. The Lake as an Aquatic Community.

A. Lake zones.
B. Thermal overturn and lake productivity.

IV. The Forest as a Terrestrial Community.

V. Community Organization and Dynamics.

A. Competition.
B. Ecological niche.
C. Territoriality and dominance hierarchies.
D. Predation.

VI. Ecological Succession.

A. Primary succession.
B. Secondary succession.
C. Succession in the aquatic community.
D. Forces driving succession.
E. Climax communities.
F. Species richness.

Key Words

community
ecosystem
energy flow
biomass
autotroph
producer
heterotroph
consumer
primary producer
trophic level
photoautotroph
chemoautotroph
primary consumer
herbivore
secondary consumer
carnivore
tertiary consumer
quaternary consumer
food web
decomposer
reducer
scavenger
carrion eater
saprobe
Charles Elton
Eltonian pyramid
pyramid of numbers
pyramid of biomass
pyramid of energy
omnivore
essential amino acids
Kwashiorkor
gross primary productivity (GP)
net primary production (NP)
nutrients
biogeochemical cycle
exchange pool
reservoir
nitrogen cycle
nitrate ions
nitrification
Nitrosomonas
ammonium ions
Nitrobacter
nitrite ions
denitrifiers
Pseudomonas denitrificans
nitrous oxide
nitrogen fixation
alfalfa
legumes
Rhizobium
root nodules
cultural eutrophication
phosphate ions
calcium ions
bicarbonate ions
carbonate rock
limestone
limnologist
littoral zone
limnetic zone
profundal zone
emergent plants
turbidity
Lake Tanganyika
Lake Baikal
thermal overturn
epilimnion
hypolimnion
metalimnion
thermocline
ocean rift community
oak-pine community
open community
closed community
competition
exploitative competition
interference competition
intraspecific competition
ecological niche
fundamental niche
realized niche
narrow niche
wide niche
Gause's law
principle of competitive exclusion
Paramecium aurelia
Paramecium caudatum
Lemna gibba
Lemna polyrhiza
R.H. MacArthur
warblers
territory
hierarchy
pecking order

predation
Lotka-Volterra theory
prey-switching
barnacle
Balanus
Chthamalus
keystone species
intermediate disturbance hypothesis
ecological succession
autogenic succession
allogenic succession
sere
seral stages
primary succession
secondary succession
pioneer organisms
lichens
eutrophication
natural eutrophication
eutrophic lakes
oligotrophic lakes
facilitation model
inhibition model
tolerance model
climax community
species richness

Exercises

1. In the space below, draw ecological pyramids to diagrammatically show the following data:

A. A pyramid of numbers with 2 million producers, 200,000 herbivores, 20,000 secondary consumers, and 200 tertiary consumers.

B. A pyramid of biomass where the producers total 5 g/m^2, the primary consumers 25 g/m^2, and the secondary consumers 1 g/m^2.

C. A pyramid of energy where the producers capture 25,000 kcal/m^2/year, the herbivores harvest 2,500 kcal/m^2/year, the secondary consumers use 250 kcal/m^2/year, and the top predators harvest only 25 kcal/m^2/year.

2. Provide labels for the below diagram of a deep temperate freshwater lake. Show the locations of the littoral zone, the limnetic zone, the profundal zone, and the limit of light penetration. Assuming it is summer, show the epilimnion, the metalimnion, the hypolimnion, and the thermocline. Where is the coldest water?

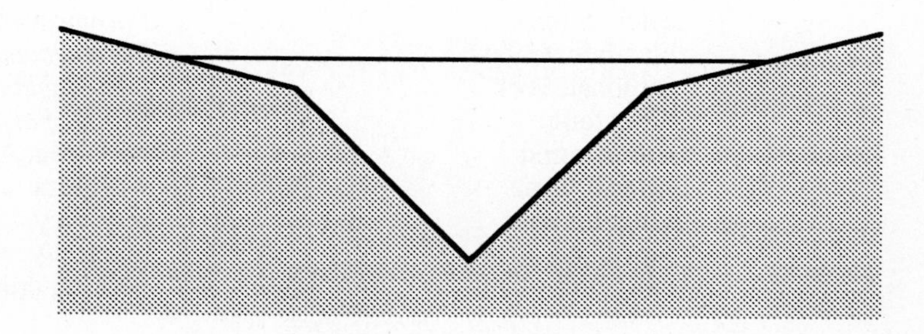

3. Sketch a diagram of the vegetation you might expect to see at different seral stages in development of a mature spruce/fir forest. Show plants such as herbs and shrubs, lichens, spruce, aspen, pine, birch, mosses, fir, etc. Indicate the direction of succession.

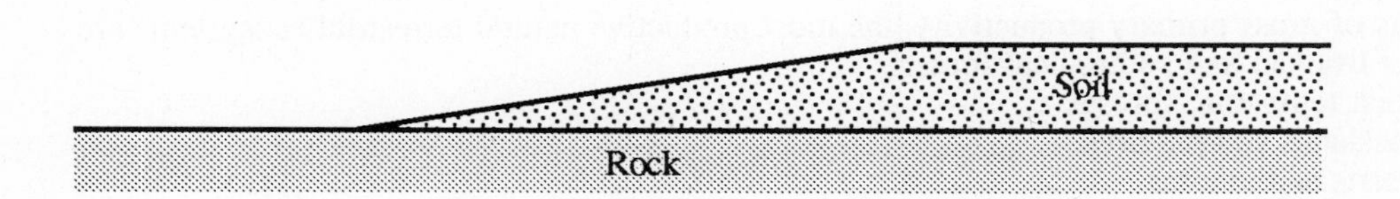

Self-Exam

You should be able to easily answer the following questions after learning the material in Chapter 47. If you have difficulty with any question, study the appropriate section in the text and try again.

A. Multiple Choice Questions

Circle one alternative that best completes the statement or answers the question.

1. The group of organisms called producers includes all of the following *except*
   a. plants.
   b. algal protists.
   c. heterotrophic bacteria.
   d. phototrophic bacteria.

2. The consumer levels of organisms include examples form each of the following groups *except*
   a. animals.
   b. fungi.
   c. bacteria.
   d. autotrophs.

3. Considering trophic levels, humans can best be characterized as
   a. producers.
   b. primary consumers.
   c. secondary consumers.
   d. sometimes primary consumers, sometimes secondary consumers, but never producers.

4. The saprobic fungi are best characterized as
   a. decomposers.
   b. producers.
   c. primary consumers.
   d. secondary consumers.

5. If we look at the pyramid of biomass for a community, we find that typically
   a. the biomass of the consumers is far greater than the biomass of the producers.
   b. the biomass of the consumers is about the same as the biomass of the producers.
   c. the biomass of the producers is far greater than the biomass of the consumers.
   d. we are unable to make generalization about the biomass of producers and consumers.

6. In terms of energy loss as energy is passed from trophic level to trophic level,
   a. only about 1% of the energy in a given level eventually reaches the next trophic level.
   b. only about 10% of the energy in a given level eventually reaches the next trophic level.
   c. about 50% of the energy in a given level eventually reaches the next trophic level.
   d. about 99% of the energy in a given level eventually reaches the next trophic level.

7. In general, the humans are best described nutritionally as
   a. autotrophs.
   b. herbivores.
   c. carnivores.
   d. omnivores.

8. In terms of gross primary productivity, the most productive natural terrestrial ecosystems are
   a. wet tropical and subtropical forests.
   b. moist temperate forests.
   c. grasslands and pastures.
   d. deserts and tundras.

9. In terms of gross primary productivity, the least productive natural terrestrial ecosystems are
   a. wet tropical and subtropical forests.
   b. moist temperate forests.
   c. grasslands and pastures.
   d. deserts and tundras.

10. Plants and nearly all other producers need nitrogen from the environment in the form of
    a. nitrate ions.
    b. atmospheric nitrogen.
    c. amino groups.
    d. urea.

11. Organisms belonging to which of the following groups can convert nitrates and nitrites to nitrous oxide and nitrogen gas?
    a. *Nitrosomonas.*
    b. *Nitrobacter.*
    c. *Pseudomonas denitrificans.*
    d. *Rhizobium.*

12. The bacteria that carry out nitrogen fixation for legumes belong to which group?
    a. *Nitrosomonas.*
    b. *Nitrobacter.*
    c. *Pseudomonas denitrificans.*
    d. *Rhizobium.*

13. Which of the following is probably the largest reservoir of carbon for the carbon cycle?
    a. atmospheric carbon dioxide.
    b. carbonate rock (limestone).
    c. organic matter.
    d. dead plants.

14. What is the source of increased amounts of carbon dioxide released in the atmosphere by industrial development?
    a. decomposition of garbage.
    b. photosynthesis in agricultural lands.
    c. respiration of large numbers of people.
    d. burning fossil fuel.

15. When both photosynthetic production and respiration are considered, which ecosystem has very little or no net community production?

a. an alfalfa field.
b. young pine plantations.
c. medium-aged oak-pine forest.
d. mature rain forest.

16. Many fly larvae feeding on a dead animal provide an example of
   a. interference competition.
   b. exploitative competition.
   c. parasitism.
   d. predation.

17. Of all the possible ways that an organism might interact with its surroundings, the way it actually does so describes its
   a. habitat.
   b. community.
   c. realized niche.
   d. fundamental niche.

18. The Everglade kite is a very specialized feeder; compared to a more generalized feeder, it has
   a. a broader niche.
   b. a narrower niche.
   c. a more restricted habitat.
   d. a more fundamental niche.

19. What was the outcome when Gause tried to grow *Paramecium aurelia* and *P. caudatum* together?
   a. They coexist.
   b. *P. aurelia* excludes *P. caudatum*.
   c. *P. caudatum* excludes *P. aurelia*.
   d. Sometime *P. aurelia* excludes *P. caudatum* , and sometimes *P. caudatum* excludes *P. aurelia*.

20. According to Lotka-Volterra theory, as prey begin to be killed off by predators,
   a. the number of predators should increase.
   b. the number of predators should begin to decrease.
   c. the number of prey should begin to increase.
   d. there should be no change in the predator population number.

21. According to the intermediate disturbance hypothesis concerning species richness, areas with intermediate levels of disturbance should have
   a. more species than areas of higher or lower disturbance.
   b. fewer species than areas of higher or lower disturbance.
   c. the same number of species as areas of higher or lower disturbance.
   d. more species than areas of higher disturbance, few species than areas of lower disturbance.

22. Which of the following is *not* a place where you might expect to see primary succession?
   a. a lava flow.
   b. volcanic islands.
   c. sand dunes.
   d. an old farm field.

23. The time for succession from sand dunes to mature forest on Lake Michigan shores has been estimated to be
   a. about 10 years.
   b. about 100 years.
   c. about 1,000 years.
   d. about 10,000 years.

24. Which two nutrients typically are limiting to plant growth in freshwater lakes?
   a. carbon and nitrates.
   b. nitrates and phosphates.

c. phosphates and carbon.
d. oxygen and nitrates.

25. If we look for trends in species richness, we tend to see species richness increase with
    a. both increased habitat complexity and increased size of the area studied.
    b. increased habitat complexity but not with increased size of the area studied.
    c. increased size of the area studies, but not with increased habitat complexity.
    d. neither increased habitat complexity nor increased size of the area studied.

B. True or False Questions

Mark the following statements either T (True) or F (False).

_____ 26. As producers, deep sea chemotrophs contribute about as much as the phototrophs.

_____ 27. The earth's producers comprise about 99% of the total biomass present in the biosphere.

_____ 28. All of the energy entering ecosystems is eventually released as heat.

_____ 29. Atmospheric nitrogen can be readily used by most nitrogen-requiring plants.

_____ 30. Nitrification involves conversion of ammonium ions to nitrites, and nitrites to nitrates.

_____ 31. In lakes, the littoral zone includes the deepest parts.

_____ 32. Like temperate lakes in the summer, Lake Tanganyika experiences thermal overturn in hot weather.

_____ 33. The realized niche of a species is the part of its fundamental niche that it actually occupies.

_____ 34. MacArthur found that five warbler species could share the same niche in spruce trees.

_____ 35. In territorial birds, a splendid male may often not find a mate if he holds an inferior territory.

_____ 36. Ethologists believe that dominance hierarchies often reduce conflict within groups.

_____ 37. Good predators will typically kill off the prey population they feed on.

_____ 38. Allogenic succession is usually more predictable than autogenic succession.

_____ 39. We think secondary succession may restore a disturbed grassland community more quickly than a disturbed tundra community.

_____ 40. In a climax community, there is an equilibrium between net production and its utilization.

_____ 41. According to our understanding of trends in species richness, the farther an island is from the mainland, the more species it should have.

C. Fill in the Blanks

Answer the question or complete the statement by filling in the blanks with the correct word or words.

42. A ____________________ is a group of organisms that live together and interact, while an ____________________ is this group of organisms together with its abiotic environment.

43. ____________________ are the complex patterns describing who eats whom in a community.

44. An organism that feeds on already dead organisms is termed a ________________.

45. The total rate at which producers accumulate energy is termed ________________.

46. ________________ is gross primary production minus energy used by producer respiration.

47. *Rhizobium* converts atmospheric nitrogen to an organically useful form via ________________.

48. The deepest part of a freshwater lake is termed the ________________ zone.

49. Oxygen is carried to a lake's depths and bottom nutrients are brought to upper reaches during ________________.

50. Communities that blend gradually into others are often called ________________ communities, while those with more definite borders are called ________________ communities,

51. A complete description of the role of a species in nature, including all aspects of its interactions with its environment, is called its ________________.

52. The idea that two species cannot coexist in the same niche is termed Gause's law or the principle of ________________.

53. Some single species in a community, the removal of which would lead to the extinction of other species, is termed a ________________ species.

54. ________________ are often the first pioneers to appear on rocky outcroppings.

55. Lakes that are rich in nutrients and support high productivity are termed ________________, while lakes that have limited nutrients and little productivity are termed ________________.

Questions for Discussion

1. Choose an ecosystem that you are familiar with: your backyard, a golf course, a nearby park, an urban vacant lot, or perhaps some part of your campus. Identify the members of each trophic level that you might find there. Be as specific as you can in identifying the producers, consumers, and decomposers. If you wanted to make Eltonian pyramids for this ecosystems, what would you need to measure? How do you (a human, we hope) fit into this ecosystem?

2. Outline as clearly as you can the workings of the nitrogen cycle. Where is nitrogen found in the biosphere, and what form is it in? What are the reservoirs of nitrogen? Be sure to list the important conversion processes of the nitrogen cycle and identify the agent that carries out each of the processes. Where do humans obtain nitrogen?

3. Succession must occur as an ecosystem changes from supporting a young community to one with a mature community. Discuss the forces believed to be responsible for the changes observed during succession. Compare the facilitation model, the inhibition model, and the tolerance model.

4. We know a biologist studying frog ecology, and he knows a pond where there are three related species of frogs apparently coexisting. All three species seem to have been there for a number of years. They must in some ways occupy different niches. What are some of the ways that similar species could occupy the same habitat and yet avoid competing for limiting resources? If you were this biologist, what types of factors would you study to help understand the frogs' coexistence?

5. Why do some places seem to have more species than others? We have all heard that tropical forests are very species-rich, while the tundra is fairly species-poor. What is there besides the extremes of climate in these two ecosystems that might explain the difference in the numbers of species a given area supports?

Answers to Self-Exam

Multiple Choice Questions

1. c
2. d
3. d
4. a
5. c
6. b
7. d
8. a
9. d
10. a
11. c
12. d
13. b
14. d
15. d
16. b
17. c
18. b
19. b
20. b
21. a
22. d
23. c
24. b
25. a

True or False Questions

26. F
27. T
28. T
29. F
30. T
31. F
32. F
33. T
34. F
35. T
36. T
37. F
38. F
39. T
40. T
41. F

Fill in the Blank Questions

42. community, ecosystem
43. food webs
44. saprobe
45. gross primary productivity
46. net primary productivity
47. nitrogen fixation
48. profundal
49. thermal overturn
50. open, closed
51. ecological niche
52. competitive exclusion
53. keystone
54. lichens
55. eutrophic, oligotrophic

CHAPTER 48 Population Dynamics

Learning Objectives

After mastering the material covered in Chapter 48, you should be able to confidently do the following tasks:

- Show the roles of birth rates and death rates in determining population growth.
- Explain how carrying capacity can limit population growth.
- Describe both exponential and logistic models of population growth.
- Distinguish the characteristics of r-selected and K-selected species, and give examples of each.
- Contrast biotic and abiotic mechanisms of population regulation, and give examples of each.
- Distinguish density-independent and density-dependent population regulation mechanisms.
- Explain how the timing of deaths of individuals may act to increase the fitness of their progeny.

Chapter Outline

I. Population Change: Population Dynamics and Growth Patterns.

A. Exponential increase and the J-shaped curve.
B. Environmental resistance and population crashes.
C. Carrying capacity and the S-shaped curve.
D. Age, life span, and the population.

II. The Evolution of Reproductive Strategies.

A. The theory of K-selection and r-selection.
1. r-selected species.
2. K-selected species.
B. The evolution of human reproductive strategies,

III. Population-Regulating Mechanisms.

A. Abiotic control.
B. Biotic control.
C. Evolution of death as a population control mechanism.

Key Words

population
population rate of change (I)
birth rate (b)
death rate (d)
population number (N)
realized rate of increase (r)
J-shaped curve
S-shaped curve
exponential increase
intrinsic rate of natural increase (r_m)
arithmetic increase
environmental resistance
population crash
carrying capacity (K)
biotic factor
abiotic factor
logistic growth curve
logistic model
life span
age group
survivorship curve
infant mortality
generation length
brood size
multiple broods
cost of reproduction
clutches
K-selection
r-selection
tapeworm
chimpanzee
gull
kittiwake
abiotic population control
drought
density-independent effect
density-dependent effect
biotic population control
programmed death

Exercises

1. A. Assume you are studying a freshwater protozoan that reproduces by fission on a daily basis. That means the population number doubles each day. You wish to study population growth by allowing the population to increase in a large aquarium with plenty of food. Assume the initial number of animals on day 0 (N_0) is 10. Plot the growth curve you would obtain if you made a census each day.

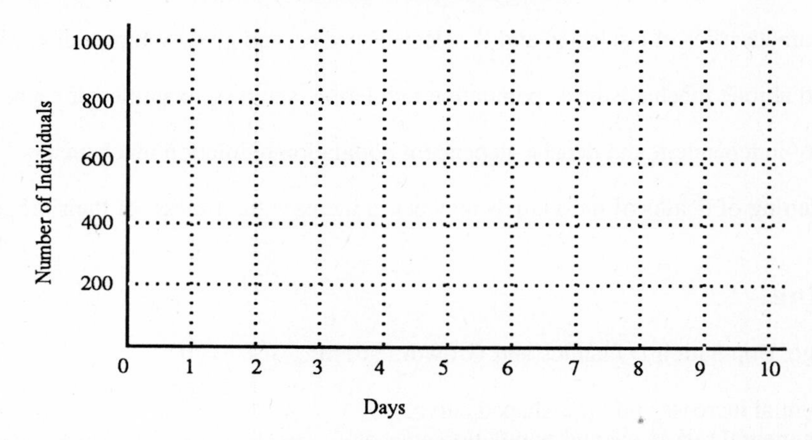

B. Assume your experiment is the same as above, except that because of limited space and food, the aquarium can only support 700 animals. Plot the growth curve you might observe if the population conformed to the logistic growth model. Be sure to indicate K.

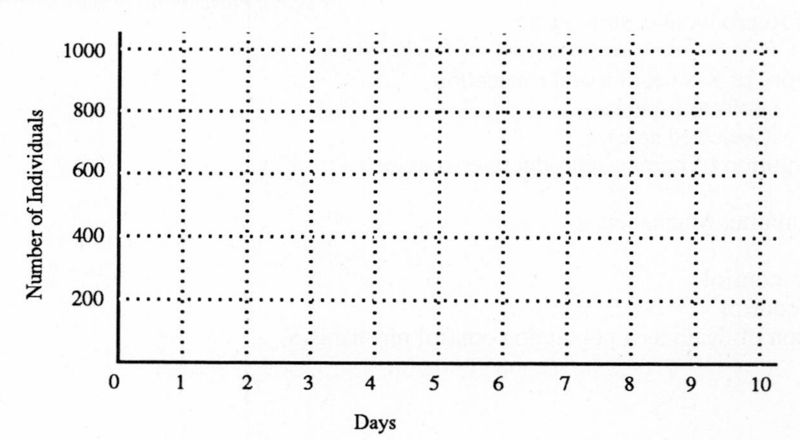

2. Here are census data for three groups of plants that germinated together. They were all dead by the time 100 days had passed. Plot and label their survivorship curves. Which curve is most similar to the human survivorship pattern? To the oyster pattern? Which is most K-selected? Which is most r-selected?

Census Day	Species A N	Species B N	Species C N
0	1000	1000	1000
10	300	40	980
20	120	16	960
30	108	12	950
40	96	8	930
50	84	7	910
60	75	6	880
70	60	4	850
80	36	3	750
90	15	2	200
10	0	0	0

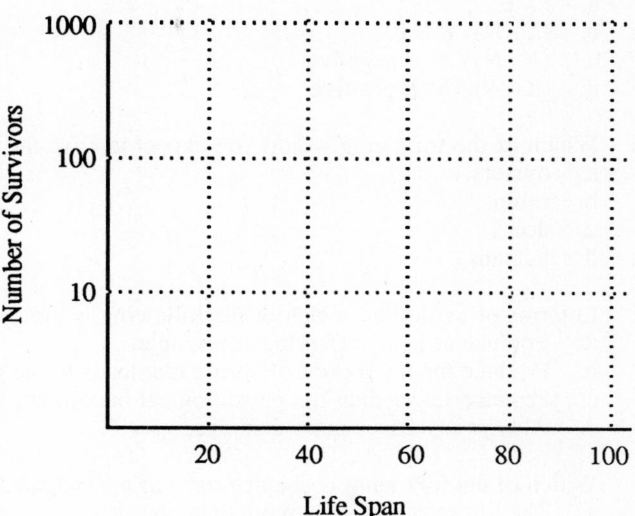

Self-Exam

You should be able to easily answer the following questions after learning the material in Chapter 48. If you have difficulty with any question, study the appropriate section in the text and try again.

A. Multiple Choice Questions

Circle one alternative that best completes the statement or answers the question.

1. In a population, when the realized rate of increase *r* is positive,
   a. the number of individuals *N* will increase.
   b. the number of individuals *N* will not change.
   c. the number of individuals *N* will decrease.
   d. any change in the number of individuals *N* cannot be predicted.

2. In general, the maximum physiological birth rate and the physiologically minimum death rate produce for a population
   a. *r*, the realized rate of increase.
   b. r_m, the intrinsic rate of increase.
   c. R_0, the replacement rate.
   d. *I*, the population rate of change.

3. Which series best illustrates exponential growth?
   a. 2 - 4 - 6 - 8 - 10 - 12.
   b. 1 - 2 - 3 - 5 - 8 - 13 - 21.
   c. 2 - 4 - 8 - 16 - 32.
   d. 8 - 16 - 24 - 32 - 64.

4. If *E. coli* cells in culture divide every 20 minutes, how many bacterial cells would be descended from a single *E. coli* cell after two hours?
   a. 2 x 6 = 12.
   b. 20 x 6 = 120.

c. $2^6 = 64$.
d. $6^2 = 36$.

5. A population should decrease in number when
a. $r = 0$.
b. $(K - N) / K = 0$.
c. $(K - N) / K$ is negative.
d. $(K - N) / K$ is positive.

6. Which of the following would you expect to have the highest infant mortality?
a. oysters.
b. robins.
c. deer.
d. humans.

7. In terms of evolution, which of the following is thought to be the "best" reproductive strategy?
a. Produce as many offspring as possible.
b. Produce the number of offspring that leads to the greatest number surviving to the next generation.
c. Produce fewer than the physiological maximum, so there will be low infant mortality.
d. Always reproduce more than once.

8. Which of the following is characteristic of a K-selected species?
a. The life span is usually more than one year.
b. There is high mortality when young.
c. There is a high rate of population increase.
d. There is a single episode of reproduction.

9. Some bacterial plant diseases do not affect scattered plants, but such a disease might be devastating in a dense garden population. This would be an example of
a. density-dependent abiotic control.
b. density-dependent biotic control.
c. density-independent abiotic control.
d. density-independent biotic control.

10. The population growth curve for a population that approaches the carrying capacity of the environment is
a. a straight line.
b. a J-shaped curve.
c. an S-shaped curve.
d. an L-shaped curve.

11. In the logistic growth model, a populations growth rate is closest to r_m when
a. N is greater than K.
b. N is very close to zero.
c. N is very close to K.
d. $K - N = 0$.

12. A survivorship curve for a species with high levels of infant mortality will be
a. convex.
b. a straight line.
c. concave.
d. S-shaped.

B. True or False Questions

Mark the following statements either T (True) or F (False).

_____ 13. A population will eventually decline in numbers if the death rate exceeds the birth rate.

_____ 14. Most species reproduce near r_m most of the time; K is seldom approached.

_____ 15. If N exceeds K, we predict a population should decline in number.

_____ 16. Oysters have a survivorship pattern very much like that of humans.

_____ 17. If two individuals produce the same number of offspring, but have different ages of first reproduction, the individual that reproduces earlier will leave more descendents in the long run.

_____ 18. Many insects are r-selected species, but some insect species tend in the K-selected direction.

_____ 19. A gull species that nests on a beach accessible to predators will tend to lay fewer eggs than a gull species that nests on steep, inaccessible cliffs.

_____ 20. Human females are incapable of producing more than one offspring per year.

_____ 21. Over the last century, there has been a generally increasing mean life span for humans.

_____ 22. Abiotic controls are usually density-independent, while biotic controls tend to be density-dependent.

_____ 23. The are a number of animals that complete their lives in less than a single year, and then endure some unfavorable period as dormant eggs or larvae.

_____ 24. Humans have the greatest known mean life span for animals. Only trees are known to live longer.

C. Fill in the Blanks

Answer the question or complete the statement by filling in the blanks with the correct word or words.

25. A group of individuals of a species that lives together closely enough to interbreed is a _______________.

26. The tendency of external factors to prevent achieving maximum growth rate is called _______________.

27. The number of individuals an environment will support is the _______________.

28. A graphical plot of the fraction of a population surviving to each age is a _______________.

29. The energy that an organism gains but which is unavailable for growth and maintenance because it is expended in activities like searching for mates or producing eggs is called the _______________.

30. The tapeworm is a good example of an _______________-selected species, while the chimpanzee is an example of a _______________-selected species.

31. When a population is depressed by severe weather, this is usually a density-_______________ effect.

32. Influences on population size brought about by living agents are _______________ population controls.

Questions for Discussion

1. Thomas Malthus suggested that populations tended to increase geometrically while the available food supply often increased only arithmetically. What is the inevitable consequence of these two observations? Do you think Malthus was talking about K, the carrying capacity? Where do you think population numbers should be with respect to K for natural selection to have the greatest impact?

2. Imagine two very similar bird varieties. One variety breeds its first year, produces four successful offspring, and then dies. The following year each offspring does the same: breeds, produces two offspring, and dies. Now, the other variety does not breed until its second season, but then it produces six offspring before dying. If we compare a single pair of each variety, which will leave more descendents after six years? What might be the value of the reproduction pattern that leads to slower growth?

3. What are the population-regulating mechanisms that affect human populations? Are such mechanisms biotic or abiotic? Are they density-dependent or density independent?

Answers to Self-Exam

Multiple Choice Questions

1. a
2. b
3. c
4. c
5. c
6. a
7. b
8. a
9. b
10. c
11. b
12. c

True or False Questions

13. T
14. F
15. T
16. F
17. T
18. T
19. F
20. F
21. T
22. T
23. T
24. F

Fill in the Blank Questions

25. population
26. environmental resistance
27. carrying capacity
28. survivorship curve
29. cost of reproduction
30. *r*, *K*
31. independent
32. biotic

CHAPTER 49 The Human Impact

Learning Objectives

After mastering the material covered in Chapter 49, you should be able to confidently do the following tasks:

- Trace the history of the earth's human population size, from prehistoric times through the present.
- Explain the factors that have either kept the human population in check or have allowed it to expand.
- Describe the basic statistics of today's world population - birth rates, death rates, and growth rates - by region.
- Explain the population changes that accompany a demographic transition.
- Provide reasonable projections for future trends in population size for developed and developing parts of the world.
- Evaluate the prospects for stabilization of the earth's human population.

Chapter Outline

I. The History of the Human Population

A. The first population surge.
B. The second population surge.
C. The third population surge.

II. The Human Population Today.

A. Growth in the developing regions.
B. Demographic transition.

III. The Future of the Human Population.

A. Population structure.
B. Growth predictions and the earth's carrying capacity.

Key Words

crude birth rate
crude death rate
percentage annual growth
population doubling time
agriculture
domestication
disease
famine
war
Black Death
bubonic plague
typhus

influenza
syphilis
germ theory
developed nation
developing nation
demographic transition
baby boom
general fertility rate (GFR)
total fertility rate (TFR)
population structure
age structure histogram
replacement level

Exercises

1. The human population has grown considerably since prehistoric times. Plot the history of the world's total population over the last million years on the axes below. Be sure to notice that both axes have logarithmic scales. Show when the periods of early tool-making, the development of agriculture, and the scientific-industrial revolution occurred. How would this plot look if the axes were arithmetic rather than logarithmic?

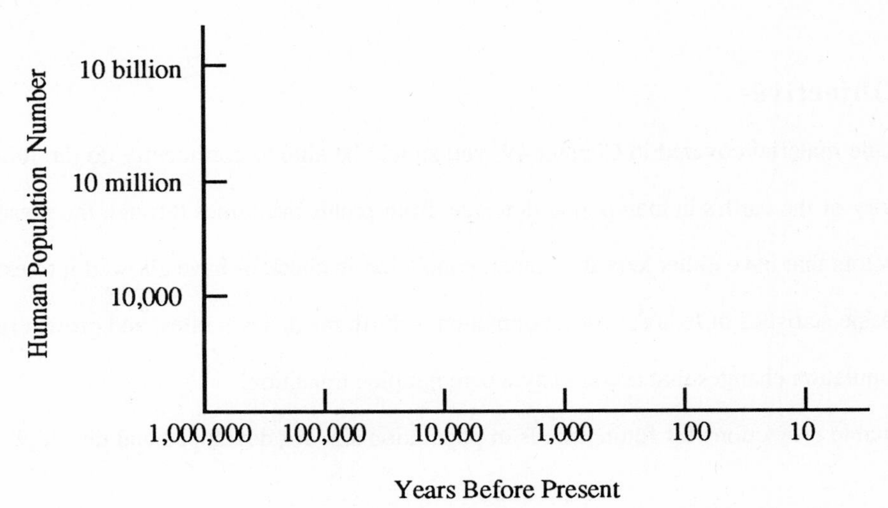

2. Below are given the distributions of ages in populations of nation A and nation B. Assume that there are equal numbers of males and females in each age class. Plot the age structure histograms for each of these populations. Which one might be more typical of a developing country? A developed country?

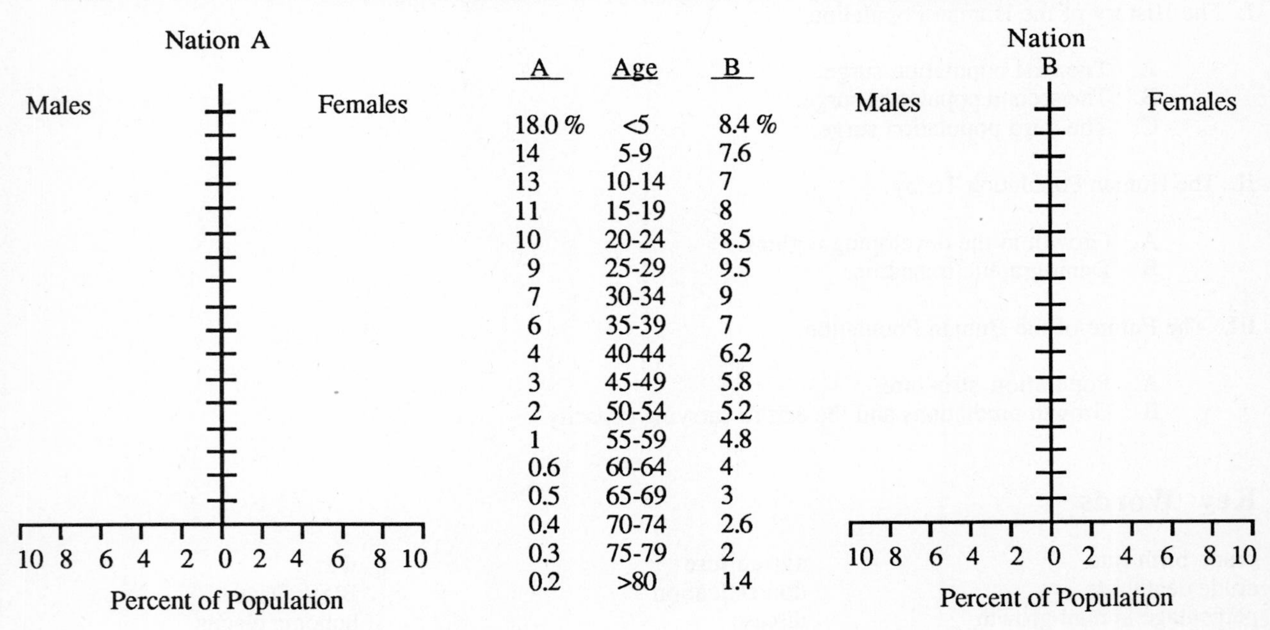

A	Age	B
18.0 %	<5	8.4 %
14	5-9	7.6
13	10-14	7
11	15-19	8
10	20-24	8.5
9	25-29	9.5
7	30-34	9
6	35-39	7
4	40-44	6.2
3	45-49	5.8
2	50-54	5.2
1	55-59	4.8
0.6	60-64	4
0.5	65-69	3
0.4	70-74	2.6
0.3	75-79	2
0.2	>80	1.4

Self-Exam

You should be able to easily answer the following questions after learning the material in Chapter 49. If you have difficulty with any question, study the appropriate section in the text and try again.

A. Multiple Choice Questions

Circle one alternative that best completes the statement or answers the question.

1. Which of these regions has the highest crude birth rate?
   a. Africa.
   b. Asia.
   c. Latin America.
   d. China.

2. Which of these regions has the lowest rate of natural increase?
   a. United States.
   b. Europe.
   c. North America.
   d. Soviet Union.

3. Which of these countries has the greatest percentage of population below 15 years of age?
   a. Soviet Union.
   b. China.
   c. United States.
   d. Mexico.

4. During human history, the second large surge in the growth rate of world's population was probably due to
   a. the adoption of tool use.
   b. the development of agriculture.
   c. the adoption of health care.
   d. urbanization and industrial development.

5. In the mid-19th century, African populations began to grow; this is believed due to
   a. tool use improving agriculture.
   b. agricultural improvement feeding more people.
   c. industrial development stimulating the birth rate.
   d. European medical advances depressing the death rate.

6. Which of the following is most likely to characterize a *developing* nation?
   a. Slow rate of population growth.
   b. Stable industrialized economy.
   c. High percentage of workers employed in agriculture.
   d. High per capita income.

7. Which of the following Asian countries has a doubling time in excess of 70 years?
   a. Philippines.
   b. India.
   c. China.
   d. Japan.

8. In Latin America there has been a decrease in the crude birth rate. Why is its population still growing so fast?
   a. Large numbers of migrants are moving there.
   b. The birth rate is actually higher than reported.
   c. The birth rate is more than three times the death rate.
   d. The population is in fact not growing.

9. Which of the following characterized the final phase of a nation undergoing a demographic transition?
   a. Low birth rate and low death rate.
   b. Low birth rate and high death rate.
   c. High birth rate and low death rate.
   d. High birth rate and high death rate.

10. Which change below is believed to be important in helping a demographic transition proceed?
    a. Increased agricultural activity.
    b. Increased industrial development.
    c. Increased death rate.
    d. Increased population size.

11. During the "baby boom" years, when did the general fertility rate (GFR) reach its maximum in the United States?
    a. 1945.
    b. 1947.
    c. 1957.
    d. 1975.

12. For demographers, reproductive age of women is arbitrarily defined as
    a. more than 15 years of age.
    b. 15-34 years of age.
    c. 15-44 years of age.
    d. 15-54 years of age.

13. If there were 8,000 females of reproductive age in your hometown, and if there were 600 babies born last year, what was the annual GFR for your hometown?
    a. 1.33 births per 1,000 women.
    b. 60 births per 1,000 women.
    c. 75 births per 1,000 women.
    d. 600 births per 1,000 women.

14. Assume the number of students at your school increases 2% per year. According to our rule for approximating doubling times, in how many years will there be twice as many students as today.
    a. 17 years.
    b. 35 years.
    c. 50 years.
    d. 70 years.

15. The formula (number of births in a year / number of women aged 15-44) x 1000 = ______ tells us
    a. the crude birth rate.
    b. the rate of natural increase.
    c. the general fertility rate.
    d. the total fertility rate.

16. What is believed to be responsible for the drop in TFR in the United States during the early 1970's and into the 1980's?
    a. Natural selection.
    b. Changes in attitudes about family size.
    c. Better medical care.
    d. Widespread industrialization.

B. True or False Questions

Mark the following statements either T (True) or F (False).

_____ 17. The population of the world has double since most of today's college students were born.

_____ 18. During most of mankind's 3 to 4 million year history, populations have remained fairly stable.

_____ 19. The development of agriculture and domestication of animals led to increased human population growth about 1,000,000 years ago.

_____ 20. The population of the New World was stable between 1750 and 1850.

_____ 21. Reduction of human population growth by famines no longer occurs on earth.

_____ 22. Although the world population is still growing, its rate of growth slowed near the end of the 1970's.

_____ 23. None of the nations in Europe or North America have population doubling times as short as 25 years or less.

_____ 24. The total fertility rate (TFR) in the United States is at present in excess of 3 births per woman 15-44.

_____ 25. The total fertility rate (TFR) in Mexico declined into the mid-1980's.

_____ 26. Some population biologists believe that the world's human populations is already beyond the earth's carrying capacity.

_____ 27. Any population which has two children per couple will no longer increase in size.

C. Fill in the Blanks

Answer the question or complete the statement by filling in the blanks with the correct word or words.

28. In terms of population numbers, the two largest nations on earth today are India and _________________.

29. The continent with the largest total population today is _________________.

30. Historically, the main factors reducing or limiting human population size have been _________________, _________________, and _________________.

31. The number of years for a population to grow to twice its size is called its _________________.

32. As nations become economically and technologically developed, their population growth rate often declines; this is typical of a _________________ transition.

33. The number of births per thousand women of reproductive age is termed the _________________.

34. The average number of children born to a woman during her reproductive lifetime is called the _________________.

35. The crude birth rate minus the crude death rate yields the rate of _________________.

36. The so-called "Black Death" of 14th century Europe was the disease _________________.

37. The large increase in GFR in the United States following World War II has been called a _________________.

Questions for Discussion

1. Some population historians see three "revolutions" as pivotal in determining the numbers of humans on earth. These are the so-called cultural revolution (inception of tool-making and social existence), the agricultural

revolution, and the industrial revolution. How do these changes relate to the population surges discussed in Chapter 49?

2. Assume that beginning today in Mexico, the average total number of children per couple is 1.9 (which is less than the replacement rate of 2.0). Demographers assure us that the population of Mexico will nevertheless continue to grow for some time. Explain why this is so.

3. We have seen in the 1970's that family size can be greatly influenced by a couple's attitude toward wanting many or few children. It has been suggested that in an agricultural society, children help in production of food so much that additional children are a benefit rather than a liability (in terms of family "income"). But if industrialization occurs and families become urban, extra children may become economic burdens (more mouths to feed). How does this conjecture relate to the frequently observed pattern in demographic transitions, where the birth rate declines in the later phases?

4. What are the main factors that have historically limited human population size? Are such factors in operation today? What is the likelihood that such factors will play an increasing role in the future?

Answers to Self-Exam

Multiple Choice Questions

1. a
2. b
3. d
4. b
5. d
6. c
7. d
8. c
9. a
10. b
11. c
12. c
13. c
14. b
15. c
16. b

True or False Questions

17. F
18. T
19. F
20. F
21. F
22. T
23. T
24. F
25. T
26. T
27. F

Fill in the Blank Questions

28. China
29. Asia
30. disease, famine, war
31. doubling time
32. demographic
33. general fertility rate (GFR)
34. total fertility rate (TFR)
35. natural increase
36. bubonic plague
37. "baby boom"